Biodiversity Conservation Through Access and Benefit Sharing (ABS)

Oommen V. Oommen · K. P. Laladhas ·
Prakash Nelliyat · Balakrishna Pisupati
Editors

Biodiversity Conservation Through Access and Benefit Sharing (ABS)

Himalayas and Indian Sub-Continent

Springer

Editors
Oommen V. Oommen
Computational Biology and Bioinformatics,
Centre for Venom Informatics
University of Kerala
Thiruvananthapuram, Kerala, India

Prakash Nelliyat
Center for Water Resources
Anna University
Chennai, Tamil Nadu, India

K. P. Laladhas
Computational Biology and Bioinformatics
University of Kerala
Thiruvananthapuram, Kerala, India

Balakrishna Pisupati
Head of Biodiversity, Land and Governance
Programme, Environmental Law
and Conventions
United Nations Environment Programme
(UNEP)
Nairobi, Kenya

ISBN 978-3-031-16185-8 ISBN 978-3-031-16186-5 (eBook)
https://doi.org/10.1007/978-3-031-16186-5

This Springer imprint is published by the registered company Springer Nature Switzerland AG
The registered company address is: Gewerbestrasse 11, 6330 Cham, Switzerland

This book is dedicated to Ms. Sugatha Kumarai and Prof. M. K. Prasad, the renowned Environmentalists whose contribution to conservation of Biological Diversity and ABS are remembered with gratitude.

Foreword

A Timely Reminder

This seminal book deals with the multidimensional aspects of a domain of existential significance. Biodiversity conservation through Access and Benefit Sharing (ABS) has been brought to the consciousness of the modern world through Convention on Biological Diversity (CBD). Access and Benefit Sharing of the global public good of biodiversity got its due place in national legislations. The Nagoya Protocol seeks to bring about greater legal certainty and transparency to ensure benefit sharing for both providers and users of genetic resources including Traditional Knowledge (TK). India ratified the Nagoya Protocol in 2014 and enacted the Bio Diversity Act as early as in 2002 and the Biological Diversity Rules in 2004 and issued guidelines on Access to Biological Resources and Associated Knowledge and Benefit Sharing Regulations in 2014. The net effect of these legislative and regulatory exercises is rather limited and sub-optimal, if not disappointing. The contributors to this volume, though restrained and modest, by virtue of their academic discipline, do not fail to point out the socio-cultural and administrative lacunae and hurdles for achieving the objectives of these legislations.

Through the eighteen papers divided into five parts, this volume is able to address the entire gamut of the idea of ABS, underscoring the complexities and interconnectedness. From discussing the essential details of the CBD the papers go on to explore key aspects like the Indian experience of the implementation of the Nagoya Protocol, India's achievements on ABS, statutory obligations, issues of Traditional Knowledge (TK) and technology transfer, Intellectual Property Rights (IPR), ABS and Sustainable Development, ABS and its Economics, and several such specific issues. The volume is distinct for the sweep of topics and their relevance as well as the array of the contributing stalwarts.

Having ratified the CBD, India has discharged its obligation to put in place legal, administrative and institutional measures. India with its rich biowealth should ideally benefit from the CBD and the Nagoya Protocol on ABS. it could be leveraged not only to safeguard our biodiversity but also to leverage economic opportunities for the stakeholders. We have a three tier regulatory structure with the National Biodiversity Authority (NBA), State Biodiversity Boards and Biodiversity Committees at the local level. Statutory Biodiversity Funds are also contemplated for sharing the benefits to the stakeholders. Admittedly, these institutional arrangements have generated considerable interest and concern about biodiversity and a few states have done commendably well. Kerala Biodiversity Board, for instance, has succeeded in preparing People's Biodiversity Registers and positioning itself as a guardian of biodiversity against powerful subversive forces.

Inclusiveness of the stakeholders and involvement of local community are the essential prerequisites for optimizing the ABS process. That would give the local communities a sense of belonging and ownership. The Nagoya Protocol clarifies how ABS can drive sustainable development through market mechanisms. Traditional Knowledge is yet another major area that can contribute to ABS provided the State is able to arrive at a nuanced and sensitive approach and strategy. Enforcement of the legal provisions calls for comprehension, commitment and conviction. Evidently, the Indian experience of ABS process is a mixed bag of achievements and apathy. In a country with so much of biowealth and diverse stakeholders, a suboptimal enforcement deprives opportunities for sustainability and well-being. Communities like the tribal population, fishers, people living in the Himalayas and other groups living in close proximity with nature have a keen awareness about biowealth and an extremely rich Traditional Knowledge that should be safeguarded and leveraged. Medicinal plants and IPR issues have also been brought to sharper focus in the book.

Though India has conformed with the binding provisions of the CBD, the net outcome, after three decades cannot be termed as impressive. Problems and omissions outnumber achievements. Lack of awareness is evidently the primary cause for lackadaisical implementation but the situation merits the question: have sufficient energy and resources been spent at the national or state levels to create awareness, realizing the formidable counter currents and persuading the votaries of 'development at any cost'?

These insightful papers in different tones and intensities convey the essential message that much remains to be done. Communities have to be sensitized and empowered, political opinion has to be influenced, and administrative mechanisms have to be charged with a new sense of urgency to make CBD, Nagoya Protocol and ABS meaningful. 'Business as usual' approach will not take us far. There is an imminent need to contextualize these responsibilities in the larger framework of global warming and climate change. No positive measures to mitigate climate change can ever be useful if biodiversity is ignored and allowed to diminish.

This book is a timely reminder of the unfinished task of conservation of biodiversity, despite having all legal frameworks in place. The papers are not overtly critical but they don't flinch away from telling the truth, albeit with academic objectivity. What should cascade into the administrative arena as executive decisions necessarily begins as scientific and academic concern and conviction. Every reader who appreciates the critical value of biodiversity and ABS in its myriad complexities should be grateful to the editors, Prof. Oommen V. Oommen, Prof. K. P. Laladhas, Dr. Prakash Nelliyat and Dr. Balakrishna Pisupati. Those of us, who value this earth and silently experience gratefulness, will relish this book for its timeliness, objectivity, clarity, structure and the brilliant contributors. Sooner than later this book will find its way to policymakers even as they would be busy warding off the deprivations likely to be caused by the reality of climate change.

K. Jayakumar
Former Chief Secretary
Government of Kerala and
Director, Institute of Management
in Government
Thiruvananthapuram, Kerala, India

Acknowledgments

The editors acknowledge the generous contributions of all authors and the technical support rendered by the department of Computational Biology University of Kerala, accredited with A^{++} Grade by National Assessment and Accreditation Council and Avantika George, Canada for proof reading, during the publication process.

Contents

Editors and Contributors

About the Editors

Dr. Oommen V. Oommen is an eminent biodiversity conservationist with several years of experience as a teacher, scientist and a policy maker. He has more than 40 years of teaching experience. He was Dean, Faculty of Science, Professor and Head, Department of Zoology, University of Kerala, India and CSIR Emeritus scientist. He served as Chairman of Kerala State Biodiversity Board and during this period he had been responsible for bringingabout policy decisions to enable the conservation and sustainable use of biodiversity. As a recognition of his outstanding research contribution in caecilian biology and molecular taxonomy, Scientists from Natural History Museum, London has named a caecilian amphibian, *Uraeotyphlus oommeni*, in his honour. He has published over 170 research papers, 15 book chapters, 20 Gen Bank submissions, edited four international books in the last seven years and published several popular science articles.

Dr. K. P. Laladhas is the former Principal at St Stephen's College, Pathanapuran, Kollam, Kerala is an eminent academician and researcher in various disciplines of life science. He has 26 years of meritorious academic service in higher education teaching. His areas of research are conservation of biodiversity, bioprospecting, molecular characterization and biological assays.

He has completed different major and minor research projects funded by the Department of Science and Technology, the Government of India and the University Grants commission. He has also worked on consultancy projects funded by the UNDP and National Medicinal Plant Board.

Three International Books have been edited; one book was published by Springer International and two by Bentham Publishers. More than sixty-five research publications were published in peer-reviewed International and National Journals. He has also authored seven books in the area of Environment and Biodiversity and been published by National Publishers.

He served Kerala State Biodiversity Board as its Member Secretary for more than five years. His expertise has been registered in the academic and administrative committees of different Universities in Kerala. He has also functioned as an Expert member in the different policy-making committees of the National Biodiversity Authority, Government of India and Government of Kerala. He has also worked as an expert member of Kerala Coastal Zone Management Authority, Ministry of Environment and Forest, Government of India, and Member of the working group on Biodiversity constituted by Kerala State Planning Board for the 13th Five-year plan.

Dr. Prakash Nelliyat is a Key Expert (Economist) in the Monitoring and Evaluation team of the World Bank sponsored Tamil Nadu Irrigated Agricultural Modernization (TNIAM) Project at Centre for Water Resources, Anna University Chennai. Previously, he worked as Subject Expert (Economic Valuation of Tradable Bio-resources in Kerala) in the Rebuild Kerala Inetiative Programme at the Kerala State Biodiversity Board; Consultant (Biodiversity Finance Initiative) at the UNDP; Fellow, Access and Benefit Sharing (ABS), at the Centre for Biodiversity Policy and Law, Chennai, where he involved in research on various issues on ABS. He also worked as Environmental Economist (valuation of biological resources) at the National Biodiversity Authority; Research Coordinator for 'South Asian Integrated Water Resources Management Programme' at the Centre for Water Resources—Anna University and as 'Research Associate' at Madras School of Economics and Madras Institute of Development Studies.

Prakash has more than 25 years research and teaching experience in Water Resources Issues, Environmental Economics, Biodiversity and Natural Resources Management and received IWMI-TATA Water Policy Research Program's Young Scientist Award, 2006. Prakash received the World Bank Aided India Capacity Building project (Environmental Economics) Overseas Fellowship (2002) and carried out research at the Department of Water and Environmental Studies, Linkoping University, Sweden. He presented number of papers on water/biodiversity and other natural resources/environmental issues in national and international conferences and also published articles in reputed journals and edited books.

After the Post-graduation (M.A. Economics), Prakash undergone interdisciplinary training in Environment and Natural Resources Economics and obtained M.Phil. from Centre for Research on New International Economic Order, Chennai—University of Madras—and Ph.D. from Madras School of Economics—University of Madras. His M.Phil. thesis on 'Coconut Husk Retting and its Environmental Problems in Coastal Kerala' and Ph.D. dissertation on 'Industrial Growth and Environmental Degradation: A Case Study of Industrial Pollution in Tiruppur'.

Dr. Balakrishna Pisupati is an internationally renowned conservation and development specialist with close to three decades of experience working at national, regional and international levels, holding positions such as the Chairperson of FLEDGE, an international think tank, Vice-Chancellor, Trans Disciplinary University (TDU, India), Chief of Biodiversity, Land Law and Governance programmes at United Nations Environment Programme (UNEP, Kenya), Senior Policy Fellow at Fridtj of Nansen Institute (FNI, Norway), Chairman, National Biodiversity Authority-Government of India (NBA, India), Coordinator, Biodiplomacy Programme at United Nations University (UNU-IAS, Japan), Head, Regional Biodiversity Programme for Asia at the World Conservation Union (IUCN, Sri Lanka), Head, Biodiversity and Biotechnology programme at the M S Swaminathan Research Foundation (MSSRF, India). He is currently managing International Environmental Governance issues at the United Nations Environment Programme (UNEP) Law Division where he leads issues related to science policy interface. He also served as the Regional Vice Chair for South and Southeast Asia at IUCN Commission on Education and Communication, Global Chair for IUCN Specialist Group on Traditional Knowledge and Community Engagement. He has advised more than 50 countries on conservation and development policy and served as advisor to international bodies such as China Council for International Cooperation on Environment and Development, China, the Global Environment Facility (GEF), World Bank, United Nations Development Programme (UNDP) and others.

Contributors

Anil Kumar N. M. S. Swaminathan Research Foundation (MSSRF), Wayanad, Kerala, India

Ballal Chandish R. Former Director of ICAR-National Bureau of Agricultural Insect Resources, Bengaluru, India

Burton Geoffrey Jean Shannon and Associates Pty Ltd, Canberra, Australia

Dionysius Joseph Mar St. Stephen's College, Pathanapuram, Kerala, India

George V. Amity Institute for Herbal and Biotech Products Development, Thiruvananthapuram, Kerala, India

Haripriya G. Department of Humanities and Social Sciences, Indian Institute of Technology, Bombay, India

Ijinu T. P. Naturæ Scientific, Kerala University Business Innovation & Incubation Centre, Karyavattom Campus, Thiruvananthapuram, Kerala, India;
Amity Institute for Herbal and Biotech Products Development, Thiruvananthapuram, Kerala, India

James T. C. Visiting Fellow, Research and Information System for Developing Countries and President, NIPO, New Delhi, India

Jayachandran K. V. Department of Aquatic Biology and Fisheries, University of Kerala, Thiruvananthapuram, Kerala, India

Lakshmanan Pushpa Kumar Law Centre-I, Faculty of Law, University of Delhi, Delhi, India

Laladhas K. P. Centre for Venom Informatics, Department of Computational Biology and Bioinformatics, University of Kerala, Thiruvananthapuram, Kerala, India;
Department of Computational Biology and Bioinformatics, University of Kerala, Kariyavattom, Thiruvananthapuram, Kerala, India;
CSEIBA: Centre for Sustainable Environment-India Biodiversity Action, Thiruvananthapuram, Kerala, India

Manchikanti Padmavati Rajiv Gandhi School of IP Law, IIT Kharagpur, Kharagpur, India

Nair Achuthsankar S. Department of Computational Biology and Bioinformatics, University of Kerala, Kariyavattom, Thiruvananthapuram, Kerala, India

Nelliyat Prakash Former Fellow (ABS) Centre for Biodiversity Policy and Law, National Biodiversity Authority, Taramani, Chennai, India

Nilayangode Preetha Kerala State Biodiversity Board, Thiruvananthapuram, Kerala, India

Oldham Paul One World Analytics, Lancaster, UK

Oommen Oommen V. Department of Computational Biology and Bioinformatics, University of Kerala, Kariyavattom, Thiruvananthapuram, Kerala, India;
Centre for Venom Informatics, Department of Computational Biology and Bioinformatics, University of Kerala, Thiruvananthapuram, Kerala, India

Pisupati Balakrishna Law Division, United Nations Environment Programme (UNEP), Nairobi, Kenya

Poojar Ishwar UNEP-GEF-MoEF ABS Project. National Biodiversity Authority, Chennai, India

Pradhan Bharat K. Sikkim Biodiversity Board, Forest and Environment Department, Deorali, Gangtok, Sikkim, India

Pradhan Prakash West Bengal Biodiversity Board, Salt Lake City, Kolkata, India

Prajeesh P. M. S. Swaminathan Research Foundation (MSSRF), Wayanad, Kerala, India

Pushpangadan P. Amity Institute for Herbal and Biotech Products Development, Thiruvananthapuram, Kerala, India

Rajasekharan S. Division of Ethnomedicine and Ethnopharmacology, Kerala State Biodiversity Board and Former Director Grade Scientist, Jawaharlal Nehru Tropical Botanic Garden and Research Institute, Thiruvananthapuram, Kerala, India

Rani J. R. Department of Computational Biology and Bioinformatics, University of Kerala, Kariyavattom, Thiruvananthapuram, Kerala, India

Roy Anirban West Bengal Biodiversity Board, Salt Lake City, Kolkata, India

Sankar U. Madras School of Economics, Chennai, India

Sathyarajan Sachin TDU, Bengaluru, India

Sharma Ghanashyam The Mountain Institute India, Tadong Daragaon, Gangtok, Sikkim, India

Sudhakaran P. R. Department of Computational Biology and Bioinformatics, University of Kerala, Kariyavattom, Thiruvananthapuram, Kerala, India

Suneetha M. S. UNU Institute for the Advanced Study of Sustainability, Tokyo, Japan

Vijaya Sree A. S. Department of Zoology, Fathima Matha National College, Kollam, Kerala, India

Abbreviations

ABNJ	Areas beyond National Jurisdiction
ABS	Access and Benefit Sharing
ABS-CH	Access and Benefit Sharing - Clearing-House
ADRIP	American Declaration on the Rights of Indigenous Peoples
AHTEG	Ad-Hoc technical Expertise Group
AICRPE	All India Coordinated Project on Ethnobiology
AISHE	All India Survey of Higher Education (In refernce)
AUUP	Amity University Uttar Pradesh
AVP	Arya Vaidya Pharmacy
BBNJ	Biodiversity in areas Beyond National Jurisdiction
BCANZ	Biological Control Agents introduced to New Zealand
BCAs	Bio control agents
BCDC	Biodiversity Conservation and Development Committee
BCPs	Bio-Cultural Community Protocols
BDCP	Biodiversity Development and Conservation Programme
BDR	Biological Diversity Rules
BFP	Biodiversity Finance Plan
BHS	Biological Diversity Heritage Sites
BIOFIN	Biodiversity Finance Initiative
BIOMIS	Biodiversity Management Information System
BIRAC	Biotechnology Industry Research Assistance Council
BMCs	Biodiversity Management Committees
BMZ	Economic Cooperation and Development
BOBLME	Bay of Bengal Large Marine Ecosystem
BR	Biological Resources
CADC	Comprehensive Area Development Corporation
CAGR	Compound Annual Growth Rate
CAMP	Conservation Assessment & Management Prioritization
CAMPA	Compensatory Afforestation Fund Management and Planning Authority
CBC	Classical biological control

CBD	Convention on Biological Diversity
CBR	Community Biodiversity Register
CBSG	Conservation Breeding Specialist Group
CCAD	Central American Commission on Environment and Development
CCRAS	Central Council for Research in Ayurveda and Siddha
CEBPOL	Centre for Biodiversity Policy and Law
CETAF	Consortium of European Taxonomic Facilitation
CFRM	Computational Finance and Risk Management
CFRs	Community Forest Resources
CII	Confederation of Indian Industry
CITES	Convention on International Trade in Endangered Species of Wild Fauna and Flora
CNA	Competent National Authority
CNPq	National Council for Scientific and Technological Development
CONAGEBIO	Costa Rican national ABS authority
COP	Conferences of the Parties
CSIR	Council of Scientific and Industrial Research
CSMCRI	Central Salt and Marine Chemicals Research Institute
CSR	Corporate Social Responsibility
DBT	Department of Biotechnology
DGFT	Directorate General of Foreign Trade
DSI	Digital Sequence Information
EEZs	Exclusive Economic Zones
EMRIP	Expert Mechanism on the Rights of Indigenous Peoples
ENM	Ecological Niche Modeling
ePBR	Eastern Panhandle Board of Relators
EPO	European Patent Office
ER	Endemic Rent
FAO-ITPGRFA	Food and Agriculture Organization's - International Treaty on Plant Genetic Resources for Food and Agriculture
FEWMD	Forest, Environment and Wildlife Management Department
FOB	Freight on Board
FRLHT	Foundation for Revitalisation of Local Health Traditions
GATT	General Agreement on Tariffs and Trade
GBF	Global Biodiversity Framework
GCC	Girijan Co-operative Corporation
GGBN	Global Genome Biodiversity Network
GIs	Geographical indications
GIZ	Gesellschaftfür Internationale Zusammenarbeit
GMCL	Gram Mooligai Company Limited
GMOs	Genetically Modified Organisms
GR	Genetic Resources
HIMAWANTI	Himalayan Grassroots Women's Natural Resources Management Association
HKH	Hindu Kush Himalayan

HLWG	High-Level working group
HYVs	High Yielding Varieties
IARI	Indian Agricultural Research Institute
IBAMA	Institute of the Environment and Renewable Natural Resources
IBBI	India Business & Biodiversity Initiative
ICAR	Indian Council of Agricultural Research
ICIMOD	International Centre for Integrated Mountain Development
ICIPE	International Centre of Insect Physiology and Ecology
ICMR	Indian Council of Medical Research
IGC	Intergovernmental Committee
IITA	International Institute of Tropical Agriculture
IK	Indigenous Knowledge
ILCs	Indigenous and local communities
ILO	International Labour Organization
ILPs	Indigenous and local peoples
IPBES	Intergovernmental Science-Policy Platform on Biodiversity and Ecosystem Services
IPEN	International Plant Exchange Network
IPHAN	Institute for National Historical and Artistic Heritage
IPLCs	Indigenous Peoples and Local Communities
IPPC	International Plant Protection Convention
IPR	Intellectual Property Rights
IR	Information Rent
IRCC	Internationally Recognized Certificate of Compliance
ISA	International Seabed Authority
ISPM	International Standards for Phytosanitary Measures
ITPGRFA	International Treaty on Plant Genetic Resources for Food and Agriculture
IUCN	International Union for Conservation of Nature
JFMCs	Joint Forest Management Committees
JNTBGRI	Jawaharlal Nehru Tropical Botanic Garden and Research Institute
JPO	Japan Patent Office
KILA	Kerala Institute for Local Administration
KL	Khangchendzonga Landscape
KNP	Khangchendzonga National Park
KSBB	Kerala State Biodiversity Board
KSBC	Kerala State Biodiversity Congress
KSLCDI	Kailash Landscape Project
KVKs	Krishi Vigyan Kendras
LBFs	Local Biodiversity Funds
LK	Local Knowledge
LMMC	Like-minded Megadiverse Countries
MAPs	Medicinal and Aromatic plants
MAT	Mutually Agreed Terms
MDGs	Millennium Development Goals

MFP	Minor Forest Produce
MGNREGA	Mahatma Gandhi National Rural Employment Guarantee Act
MLS	Multilateral System
MNCs	Multi-National Companies
MoEF	Ministry of Environment and Forests
MoEFCC	Ministry of Environment, Forest and Climate Change
MoFSC	Ministry of Forests and Soil Conservation
MOP	Meeting of the Parties
MOU	Memorandum of Understanding
MPCAs	Medicinal Plant Conservation Areas
MPEDA	Marine Products Export Development Authority
MSSRF	M S Swaminathan Research Foundation
NAGS	National Active Germplasm Sites
NAM	National Ayush Mission
NBA	National Biodiversity Authority
NBAIR	National Bureau of Agricultural Insects Resources
NBAP	National Biodiversity Action Plan
NBC	National Biodiversity Centre
NBF	National Biodiversity Fund
NBPGR	National Bureau of Plant Genetic Resources
NBSAP	Nepal Biodiversity Strategy and Action Plan
NDVI	Normalized Difference Vegetation Index
NEFEJ	Nepal Forum for Environmental Journalist
NERAMAC	North Eastern Regional Agricultural Marketing Corporation Limited
NGT	National Green Tribunal
NIF	National Innovation Foundation
NINA	Norwegian Institute for Nature Research
NLSIU	National Law School of India University
NMPB	National Medicinal Plants Board
NPV	Net present value
NRIs	Non-Resident Indians
NTC	Normally Traded Commodities
NTFPs	Non-timber Forest Products
NWFPs	European non-wood forest products
OECM	Other Effective Area-based Conservation measures
OTC	Over-the-counter
PBR	Peoples BiodiversityRegister
PBRs	People's Biodiversity Registers
PGFRA	Plant Genetic Resources for Food and Agriculture
PGR	Plant Genetic Resources
PIC	Prior Informed Consent
PPVFRA	Protection of Plant Varieties and Farmers' Rights Authority
PVP	Plant Variety Protection
RBEF	RitnandBalved Education Foundation

RET	Rare Endangered and Threatening
SBBs	State Biodiversity Boards
SBFs	State Biodiversity Funds
SCBD	Secretariat of the Convention on Biological Diversity
SCP	Standing Committee on Patents
SDGs	Sustainable Development Goals
SDM	Species Distribution Modeling
SMTA	Standard Material Transfer Agreement
SR	Scarcity Rent
TDU	Trans-Disciplinary University
TEK	Traditional Environmental Knowledge
TK	Traditional knowledge
TKDL	Traditional Knowledge Digital Library
TKRC	Traditional Knowledge Resource Classification
TL	Truthfully Labelled
TM	Traditional Medicine
TMT	Thousand Metric Tonnes
TNAU	Tamil Nadu Agricultural University
TRIFED	Tribal Cooperative Marketing Development Federation of India
TWCs	Third World Countries
UEBT	Union for Ethical BioTrade
UGC	University Grants Commission
UNCLOS	U N Convention on the Law of the Sea
UNCTAD	United Nations Conference on Trade and Development
UNDP	United Nations Development Programme
UNDRIP	United Nations Declaration on the Rights of Indigenous Peoples
UNEP	United Nations Environment Programme
UNEP/DELC	United Nations Environment Programme – Division of Environmental Law and Conventions
UNESCO	United Nations Educational, Scientific and Cultural Organization
UNPFIL	UN Permanent Forum on Indigenous Issues
UNSR	UN Special Rapporteur on the Rights
UNU-IAS	United Nations University – Institute of Advanced Studies
UPOV	Union for the Protection of Plant Varieties
USDA-APHIS	U.S. Department Of Agriculture Animal and Plant Health Inspection Service
USPTO	United States Patent and Trademark Office
WHO	World Health Organization
WIPO	World Intellectual Property Organization
WTO-TRIPS	World Trade Organization agreement on Trade-Related Aspects of Intellectual Property Rights
ZSI	Zoological Survey of India

Part I
Premise of Access and Benefit Sharing (ABS)

Chapter 1
Convention on Biological Diversity, Access and Benefit Sharing: A Glance

Prakash Nelliyat, Balakrishna Pisupati, K. P. Laladhas, and Oommen V. Oommen

Abstract The advent of the Convention on Biological Diversity (CBD) made a paradigm shift in the management of biodiversity. The common property (global public good) nature of biodiversity or genetic/biological resources and associated traditional knowledge has been transformed into a sovereign right of the State, where Access and Benefit Sharing (ABS) attain the universal acceptance. This compelled the nations, including India, to introduce the legal, administrative and institutional mechanisms for operationalizing the ABS as an innovative financial and incentive mechanism for biodiversity conservation and its sustainable use with the support of the local communities, who are the custodian of the resources. Even if, ABS has enormous scope in safeguarding the biodiversity from its current crisis, its implementation imposes huge challenges which should be taken into account with proactive steps. This chapter addresses the above issues with an emphasis on: (a) the growth of biotechnology, bio-prospecting and bio-piracy, (b) emergence of CBD and the ABS philosophy in the context of benefit sharing and equity principles, (c) role of Nagoya Protocol in the ease of implementation of the ABS mechanism, (d) Intellectual Property Rights (IPR) and the emerging ABS issues including the tracking and monitoring of biological resources based patenting, (e) financing for biodiversity and the role of ABS as an innovative financial option, (f) valuation of biological/genetic resources for determining the benefit sharing norms/amount, (g) environmental laws vs. biodiversity laws, and (h) as a pioneering country in implementing the ABS, India's experiences, challenges and the way forward.

Keywords Benefit sharing · Sovereign right · Financing for biodiversity · Intellectual property rights · Valuation of biological resources

---

P. Nelliyat (✉)
Former Fellow (ABS), Centre for Biodiversity Policy and Law, National Biodiversity Authority, 5th Floor, TICEL Biopark, Taramani, Chennai 600113, India
e-mail: prakashnelliyat@gmail.com

B. Pisupati
Law Division, United Nations Environment Programme (UNEP), Nairobi, Kenya

K. P. Laladhas · O. V. Oommen
Department of Computational Biology and Bioinformatics, University of Kerala, Kariavattom North campus, Thiruvananthapuram, Kerala 695581, India

O. V. Oommen et al. (eds.), *Biodiversity Conservation Through Access and Benefit Sharing (ABS)*, https://doi.org/10.1007/978-3-031-16186-5_1

1.1 Introduction

Compared with developed countries, developing countries are rich in biological resources and associated traditional knowledge. However, these resources were largely under the common-pool or open access system of management for centuries. Hence, over-exploitation and biopiracy were widely prevalent on these resources. Further, colonial governments' rule in developing countries was targeted primarily on revenue maximization from biodiversity hotspots like forests and viewed as the custodians of biological resources. Besides, they considered indigenous and local peoples (ILPs) as "encroachers" and threats to wildlife conservation and marginalized them by denying ownership and/ or management rights over these resources. These factors ultimately not only reduced the livelihood opportunities of the ILPs, but also contributed to biodiversity loss.

However, the UN Convention on Biological Diversity (CBD) is a landmark event that contradicted these thoughts. The convention says it is aware of the intrinsic value of biological diversity and of the ecological, genetic, social, economic, scientific, educational, cultural, recreational and aesthetic values of biological diversity and its components' (CBD 1992 Preamble). It reaffirms that States have sovereign rights over their biological resources.

The ABS regime provides an opportunity to providers of biological resources as well as the possessors of biological resources based traditional knowledge (TK) holders to explore new markets for their products and TK and thereby enhance their livelihoods. This mechanism is feasible with the government's support in the form of guaranteeing the individual and community rights of the ILPs and facilitating opportunities to them for regeneration and conservation of biodiversity. In this regard appropriate legal, administrative, and institutional mechanisms need to be placed in each country or Parties of the Convention. Besides, proper cooperation and transparent approach in accessing the biological resources and benefit sharing are also significant pre-requisites.

However, fixing the benefit sharing criterion is a challenge. In this regard, the valuation of bio-resources is important as it ensures equitable benefits to be transferred to the custodians of biological resources and associated knowledge that are used in the development of new products. In practice, when we access the valuation of biological resources for ABS one should not consider it just for the market value of the product, but the availability (richness/scarcity) of a species, its associated traditional knowledge, cultural and intrinsic values related to the use of the biological resource, etc.

1.2 Emergence of Convention on Biological Diversity and Access and Benefit Sharing (ABS) Philosophy

Biodiversity has provided a number of ecosystem services including sequester atmospheric carbon, moderate weather, and facilitate hydrological and nutrient cycles, soil and water quality maintenance etc. Besides, the spiritual, religious, cultural and existence values of biodiversity are also significant, especially to certain ethnic communities located in different parts of the world. Biodiversity is the basis for health, food, employment, and livelihood security. Besides, biodiversity provides significant contribution to the Nation's economy, primarily due to its commercial importance. Biological/genetic resources are the source materials for different industries and provide services worth of US $ 125 trillion per annum globally.

There is no doubt that globally, during the last few decades, biodiversity loss has been unprecedented and shocking. The rapid population explosion along with economic growth has led to the unsustainable harvest of natural resources, particularly species (animals and plants), from different ecosystems including forests, agriculture, marine and coastal, and fresh waters. The over-extraction of these renewable resources leads to the loss of habitats. Further, the climate change impacts (frequent draughts and floods) widespread invasion of alien species in different ecosystems, and environmental pollution also hastened the degradation of our precious biodiversity, a universal asset. These global challenges need to be addressed in a collaborative manner with the support of all countries, where the universal treaty like the Convention on Biological Diversity (CBD) has to play a critical role.

The CBD emerged as an International treaty in 1993 and its article 15 on fair and equitable sharing of benefits arising from the utilization of genetic resources and associated traditional knowledge is a landmark initiative in the history of biodiversity conservation. The CBD elevated Access and Benefit Sharing (ABS), one of its objectives, as an exclusive mechanism for the conservation and sustainable use of biodiversity. ABS can be a facilitating regime for countries and biodiversity stakeholders across the world, to share genetic resources and associated knowledge in a legally bound as well as mutually agreed upon manner.

Prior to the CBD, access to genetic/biological resources and local/traditional knowledge remained free for anyone and the major share of the benefits from it went to the traders. In certain cases, if any benefits were shared by the entrepreneurs, they went to the Government and never reached the primary stakeholders or communities, or the custodians of the resources. In those years there was no provision for a fair and equitable sharing of benefits. In this context, the CBD and its primary objective ABS are instrumental as they argue for: (a) the equity principle through benefit sharing, (b) local people as custodians of biodiversity and biological resources, (c) empowerment of the local community through obtaining a free share of biological resources' business, (d) resource (biodiversity) conservation through community participation.

1.3 Growth of Biotechnology, Genetic Resources Drain and ABS Philosophy

The technological interventions on biological or genetic resources through biotechnology in the late 1980s is the fundamental aspect of the access regulations as well as the fair and equitable sharing of benefits arising from the utilization of genetic resources and associated traditional knowledge.

Biotechnology advancement has emerged in the western developed world. However, the major share of biodiversity is possessed by the eastern developing nations. This unequal distribution of bio-wealth compelled the developed (technologically advanced) nations towards indiscriminate extraction or exploitation of the resources for their research and product development. As biodiversity and genetic resources and the knowledge associated with them were considered as global public good, huge volumes of biological/genetic resources were drained from developing nations to the developed nations.

The overexploitation of biological resources and the emerging bio-piracy becomes a threat to global biodiversity. This fact was realized and debated upon by the nations around the world and it was decided to streamline biological resources access regulations through administrative, legal and institutional norms. In reality, this mechanism has been materialised through the advent of the CBD and its objectives: (a) the conservation of biological diversity, (b) the sustainable use of its components and, (c) the fair and equitable sharing of benefits arising from the utilization of biological or genetic resources.

The post CBD system guaranteed nations their sovereignty over the genetic or biological resources entitled by them or present in their geographical areas. In other words, CBD has established regulation over their biological resources to conserve and use sustainably through a strict legal framework. No doubt the ABS mechanism is a paradigm shift in the biotechnology sectors' research and its development resulted in different products from the sectors such as: pharmaceuticals, cosmetics, perfumes,, industrial enzymes, biocides, nutritionals, biotechnology and emerging synthetic biology. In the broader perspective, ABS was expected to be a dominant mechanism to achieve the other two objectives of CBD: sustainable use of biodiversity and its conservation.

In brief, with the rapid growth of biotechnology and increasing demand for nature based health remedies, access to biological resources and traditional knowledge has become a paramount necessity in the present context. In the absence of biological resources, access regulations, providing countries really lost their bio-wealth or natural capital. The CBD and the Nagoya Protocol on ABS have ascertained legally binding obligations to their parties.

1.4 Why ABS is Experiencing Distortion?

Even if the Convention has universal acceptability (196 countries as Parties), the practicability of ABS among the Parties is not promising but getting more complex, which compels the policy makers and scholars to re-examine the underlying principles of ABS philosophy from practical as well as operational perspectives.

Ideally, the ABS mechanism gives the ownership rights of biological resources to the local communities and considers them as the custodians of the resources. However, the operationalization of the ABS requires the commitment and cooperation of all the players (stakeholders) as well as an effective and functional institutional system in place. Further, multi-tiered responsibilities from the governance angle are also required at the global, national, state and local levels. Unfortunately, most of the countries are struggling in establishing the above requirements. Besides, those nations which established backgrounds for the enforcement of the legal provisions on biological resources' access and the achievements of ABS mechanisms, are struggling as the enforcement becomes a cumbersome task.

Facilitating the ABS process is a huge challenge, as convincing the importance of the different components to the stakeholders is not so easy. For example, in the Hindu-Kush Himalayan region, the International Centre for Integrated Mountain Development (ICIMOD) has been engaged in facilitating the ABS process for developing a common framework. Despite the existence of the legislative, administrative and official support for the implementation of ABS in India, Nepal and Bhutan, the ground reality is still discouraging and the indigenous and local communities (ILCs) are yet to benefit from the ABS arrangements. Biological resources in these areas are over-exploited and huge volumes are siphoned off to different parts of the world.

Further, awareness regarding TK protection has not received enough attention despite Acts, Rules and Guidelines brought about by the governments. The people in the Himalayan region wanted to protect their knowledge by law. People in the region feared misappropriation, bio-piracy and degradation of their knowledge and resources.

No doubt that, owing to the enormously rich biological/genetic resources and associated TK in the Himalayan region, their increased market demand could offer new opportunities for generating benefits and enhanced incomes for ILCs. However, the current reality seems to be quite far from achieving this. The involvement and engagement of local, indigenous, and marginalized people in the ABS process is still unclear.

In the ABS regime, restrictions are imposed on the free exchange and use of biological/genetic resources even for academic research, which is controversial among the scientists and is widely debated. For example, in the case of insects, even the dead specimens' exchange between taxonomists in different countries was debated. They argue that "internationally the accepted taxonomic practice involves the exchange of dead specimens between taxonomists for comparison, which is imperative not only to arrive at the correct identity of the insect species, but also for precise documentation and recommendation of ideal management/conservation strategies". They also claim that the ABS regime is seriously hampering the future of classical biological control (CBC) initiatives to manage alien invasive insect pests. CBC is applied world-wide and is dependent on the exchange of exotic natural enemies. The main beneficiaries of CBC are the global farmers, as their pest problems can be managed through reduced use of pesticides, resulting in lower pesticide residues in food.

However, when Law prevails and access regulation exists, it may be applicable to all types of biological resources' access including research and the users should abide by it. Whether it is with good or bad intention, it does not matter. No doubt that, through the implementation of ABS regulations, our rich species' illegal access might be restricted. In this regard, all stakeholders' support is essential rather than the justifications for non-compliance of the legislation.

1.5 Nagoya Protocol for More ABS Visibility

After realising the complexity in enforcing the legal norms for materializing the ABS, the Convention paid special attention to the ABS and it was widely discussed in various Conferences of Parties (CoP). Subsequently, the Nagoya Protocol on 'access to genetic resources and the fair and equitable sharing of benefits arising from their utilization' came in to effect in October 2014 (CBD 2011). It provided a transparent international legal framework to govern and access and utilise genetic resources and associated traditional knowledge.

The countries, which ratified the Protocol (till date 136 Parties), would have regulations necessary for access to biological resources from their land, predetermined with Prior Informed Consent (PIC) from the biological resources' providers and Mutually Agreed Terms (MAT) from the biological resource users for benefit sharing. The Nagoya Protocol reaffirms the sovereign rights of States over their natural resources and calls for fair and equitable sharing of benefits when such resources and associated traditional knowledge are used.

The World Trade Organization agreement on Trade-Related Aspects of Intellectual Property Rights, the Food and Agriculture Organization's - International Treaty on Plant Genetic Resources for Food and Agriculture (FAO-ITPGRFA 2009), and the Bonn guidelines (2002) provide a broad framework for ABS procedures along with the Nagoya protocol.

The Protocol clearly discourages anyone from using such resources and associated TK without PIC and MAT for such use. With 133 countries ratifying the Protocol, the Protocol receives much prominence in realizing the objective of the CBD and its Article 15. Several countries, both developing and developed, are currently working on the national implementation of the Protocol by designing administrative, regulatory and legal regimes. Parallel to this, recent advances in science and technology, especially biotechnology and synthetic biology, have been evolving with new ways of using the genetic sequence information which, in a number of instances, is displacing the need to have physical access to any genetic resource.

Developments in the use of digital sequence information (DSI) over the past decade make it possible to synthesize new organisms and products using DSI that are currently available publicly and are also held in private. This has now resulted in debates within the CBD and the Nagoya Protocol, as well as other forums including the United Nations Food and Agricultural Organization (FAO) and the discussions under the United Nations Convention on the Law of the Seas (UNCLOS) on issues of Areas beyond National Jurisdiction (ABNJ).

The Nagoya protocol exemplifies a trend toward using legal measures to clarify rights and drive sustainable development through market mechanisms such as contracts, intellectual property rights, and mutually supportive institutions at the national, regional, and community levels, with a substantive focus on the traditional knowledge of Indigenous communities. However, the major challenge is how the Parties implement the provisions of the Protocol through legal, regulatory and administrative measures which contributes to fulfill the CBD objectives as well as the scale-up of BioTrade.

1.6 Intellectual Property Rights (IPR) and ABS

The Intellectual Property Rights (IPR) aspects of biological/genetic resources related innovation and their future commercialization and ABS is a highly challenging and debated area. It demands the need for universal accountability and monitoring on biological resources- based invention, product manufacturing and patents. The utilization of genetic/biological resources and associated knowledge has grown manifold. The CBD mandates that utilization, prospecting of biological resources and deriving IP from them be necessarily linked to ABS. However, its practicability and operationalization is the most challenging aspect.

Historically, biological resources-based trade has a significant role in global value addition and enhancing economic growth and this was well established during mercantilism. The expansion of IPR in relation to biological resources has been at the center of global debate and the attributing factors include bio-piracy and overexploitation of biological/genetic resources.

The expansion of the work program on ABS under the CBD led to the adoption of voluntary guidelines in relation to prior informed consent, code of ethics, respecting dwellings and cultural aspects of indigenous people and repatriation of TK. Under the guidance of the intergovernmental committee on genetic resources under the WIPO, the expansion of the context of TK to develop an international understanding on TK is ongoing.

Post TRIPS, the expansion of the development dimension at the request of the developing countries was taken up by the WIPO. While the pressure to negotiate ABS mechanisms increases, developing countries face a challenge of varied implementation of ABS as well as biodiversity monitoring. Advancements in technology have facilitated greater access of biological resources, their commercialization and increase in IP filings. There is a need for a balanced approach in relation to ABS and IP rights.

From the ABS perspective, tracking and close monitoring the use of biological resources for patenting is important. For example, Section 18(4) of the Biological Diversity Act empowers the National Biodiversity Authority to oppose IPR applications that are filed outside the country on any biological resource or associated knowledge which is derived from India. As bio-piracy has emerged as a serious threat, its volume and extent needs to be properly ascertained for developing combative measures. Fulfillment of this task requires access to patent applications filed in various countries around the world.

1.7 Financing Biodiversity and Valuation of Biological Resources

Considering the rate of degradation of biodiversity, restoration requires huge funds. The BIOFIN assessment reveals that the gaps in meeting biodiversity strategies and targets are large and filling them through public financing alone is extremely difficult. We need a shift towards a new investment and policy paradigm incorporating the economic value and financial benefits of biodiversity. In this regard, companies and the financial sectors should come forward for biodiversity management. After the advent of the CBD, we have seen a quick transition of biodiversity from a global public good/heritage to a resource of those countries where the biodiversity occurs (sovereign right). In this context, innovative financial mechanisms like ABS have a good scope.

In the ABS process, fixing the criteria for benefit sharing is important, as also the valuation of biological resources and associated traditional knowledge. The CBD urges its member countries and other stakeholders to engage in plural values' approach to implementing ABS, where the commercial use of genetic resources and their value chain is under consideration.

Therefore, the forefront of valuation for ABS requires ensuring equity amongst the different stakeholders involved in the value chain. Besides, valuation is also to be considered of factors such as knowledge associated with the use of a species, the contribution or probability of this knowledge in developing a successful product and/or in reducing costs of development, and potential income that may be generated over a reasonable period of time. All these factors are dependent/vary on the type of sector the species utilizes. For example: the same species or biological resources may generate different values in different sectors. Valuation of ABS has to follow more plural approaches as it captures economic values of intangibles (eg. associated traditional knowledge). Besides, issues of species' population (status and their endemicity) also need to be considered.

In brief, ABS is an economic instrument that has evolved with techno-legal issues. Hence, its proper understanding and enforcement require special attention with a more scientific approach. No doubt, in a mega bio-diverse country like India, which possesses substantial bio-wealth in its rich ecosystems, the scope of the ABS potential is huge and its effective administration will be a reliable source for mobilizing the finances for the conservation of biodiversity.

1.8 India's ABS Track

India is a mega-diverse country having great strength in knowledge associated with biological resources. India possesses 7 to 8 per cent of the recorded species of the world. Even if India accounts for only 2.4 per cent of the world's total land areas, the country possesses 7 to 8 per cent of the recorded species of the world with around 45,000 species of plants and 91,000 species of animals. Of the 34 global biodiversity hotspots, four are in India: the Himalayas, the Western Ghats, the North-east region and the Nicobar Islands (MoEFCC 2014). The rich biological/genetic resources and the heritage of traditional knowledge in the country obtained global attention centuries ago.

The different ecosystems (agriculture, forests, marine and coastal, freshwater etc.,) of India and the biological resources arising out of them are significant from the ABS perspective. India's location, especially the confluence of three major bio-geographic dominions, contributes to its rich biodiversity.

From the agro-biodiversity perspective, a majority of the farming population in the country is the small and marginal land holders. Over the years, Indian farming communities have been developing and conserving varieties of crops that are suitable to different habitats and ecological conditions in a successful manner. India had developed its own legal measures to protect plant varieties and farmer's rights under the International Treaty on Plant Genetic Resources for Food and Agriculture (ITPGRFA). However, the scope in access to the agro-biodiversity heritage (genetic resources) of the country and its benefit sharing to the farming communities need to be explored. As the farmers and breeders are working for sustainable food security, their rights must be preserved. Farmers' Rights and ABS need to be synthesized in an amicable manner for the conservation and sustainable use of the rich agro-biodiversity of the country.

Certain species (plants and animals) conservation perspectives of ABS also need to be explored. For example, globally, insects are playing crucial roles in pollination, decomposition, dung burial, pest control and supporting animal and human life. But, people are unaware of these innumerable services provided by insects and their significance in enhancing human welfare. However, the fact is that insect biodiversity faces threats which should be addressed from the ABS perspective. Through the ABS mechanisms, the country's sovereign right over its biological resources, especially the endangered and vulnerable species, including insects, needs to be protected. But, the lack of complete scientific data on insect species in the country is a major constraint along with the ABS implementation challenges.

Marine resources are another example. The blue economy concept and the 2030 United Nations Sustainable Development Goals (SDGs), particularly Goal 2 (end hunger, achieve food security and improved nutrition, and promote sustainable agriculture) and Goal 14 (conserve and sustainably use the oceans, seas and marine resources for sustainable development) are highly interlinked and marine and coastal resources (especially fish) play a major role. In a country like India, having a coast line of about 7,500 km, and millions of fishermen depending on fishing, marine resources' extraction and commercialization is an important economic activity. Fish and fisheries play a significant role in India's exports and the income and employment generation in the coastal states of the country.

The ABS has a great scope in accessing the fisheries and aquaculture sector, as most of the resources and associated traditional knowledge are under commercial utilization. The open access nature of fishery resources in India has a long history of utilizing biodiversity for multiple purposes. There is a need to assess the genetic resources, research trends and further commercialization in both fisheries and aquaculture. There is a global focus on bio-prospecting of marine organisms as part of the "Drugs from the Ocean" programme, and the focal point would be marine microbes. The terrestrial genetic resources–those within the Exclusive Economic Zones (EEZs) of nations and the issues regarding traceability of marine genetic resources collected from Areas Beyond National Jurisdictions (ABNJs), used in research studies, and the process and product development, should be brought under the ambit of the ABS. In this regard, all fisheries research organizations including the National Bureau of Fish Genetic Resources (ICAR-NBFGR) and export agencies like Marine Products Export Development Authority's (MPEDA) cooperation with the Biological Diversity Act enforcement Agencies (NBA and SBBs) are required.

Medicinal plants in India are another unique bio-wealth, attractive globally. Its varied topography and climatic conditions has led India to flourish in rich biodiversity, especially the floral wealth of the country with its associated traditional knowledge and cultural heritage. Since time immemorial, besides daily life support, the floral wealth of the country has been used in traditional medicine for local health care. But over time, these medicinal plants had been explored/exploited by many foreign entities, especially during the colonial periods, leading to the rapid depletion of these resources, promoting bio-piracy along with erosion of associated traditional knowledge. No doubt that through the effective implementation of the BD Act, the medicinal plants and their associated knowledge in the country can be conserved and sustainably used for ever.

From the above discussion, it is clear that the protection of genetic resources is a multidimensional aspect and it is required for the environment and the ecosystem where the plant/animal in which the genetic component survives. Hence, the protection of the environment (including the water bodies, soil quality, forests, agriculture and the entire ecosystem) and its sustainable management is required. The emerging thought is that the benefit sharing should not be limited to the genetic resources' custodians alone, but to the environment (ecosystem) as well. In the case of aquatic biological resources, riparian vegetation for bank protection, managing water quantity and water quality, conservation of breeding populations, genetic resources' protection, regulated extraction etc. are the primary steps for their conservation.

India is one of the pioneering parties in promoting the ABS mechanism through the enactment of the Biological Diversity Act (BD Act) 2002 and Rules (2004) to regulate the ABS process to conserve biodiversity and to protect the interests of traditional knowledge holders. Further, the Guidelines on Access to Biological Resources and Associated Knowledge and Benefit Sharing Regulations, 2014, number of Notifications and the State specific Biodiversity Rules also exist for enforcing the ABS at their jurisdiction. The BD Act is implemented through a three-tier institutional mechanism: the National Biodiversity Authority (NBA), State Biodiversity Boards (SBBs) and the Biodiversity Management Committees (BMCs) at the local level. However, enforcement of the provisions of the BD Act needs a fresh look to cope with the modern day requirements of the law (NBA 2010).

India is a party in the Nagoya Protocol on Access and Benefit Sharing. As a party to the protocol, India submitted the interim national report on the implementation of the Nagoya Protocol (2017) to CBD, which explains India's achievements and experiences with respect to the various Articles of the Protocol.

In brief, as a party to the CBD, India initiated legal and institutional measures for the implementation of the ABS and mobilized an amount of Rs. 1.3 billion for biodiversity conservation. However, the ABS process in the country faces huge challenges, as different stakeholders are not well aware about the ABS principles and their operation. At present, only very few cases come under the ambit of ABS and the amount collected has also not been effectively utilized for biodiversity conservation.

The National Biodiversity Authority with the support of the State Biodiversity Boards and the Biodiversity Management Committee made efforts to implement the BD Act since its inception. However, its overall achievements are not commendable as many users of India's biological resources with commercial intent have not complied with the Act. Further, many criticisms emerged from different angles against many provisions of the law. The reasons for criticism include:: (a) lack of legislative acumen in drafting the law in the first place that led to vagueness in some provisions of the BD Act, (b) requirement of formal prior approval for all sorts of research activities, and (c) benefit sharing obligations for cultivated bio-resources.

In this context, the Biological Diversity (Amendment) Bill, 2021 was introduced in the Lok Sabha on 16 December 2021. The Bill was referred to the Joint Parliamentary Committee on 20 December 2021. The Bill seeks to introduce fundamental changes in the law by introducing many new provisions. However, this also may trigger several enforcement issues. One has to wait and see how many of these provisions prevail when the bill comes into effect as an amendment.

1.9 Environmental Versus Biodiversity Law and Policy Shifts from the Benefit Sharing Perspective

An evaluation of different environmental laws and biodiversity law in India is interesting. The background of the law, provocations in the making of the law, the legal frame, form, substance, institutional arrangements and the working of each law may differ. Broadly, most of the environment laws may primarily focus on the pollution control regime. However, these laws have a good connection with biodiversity, as its existence depends upon the complex bio-physical environment.

The laws concerning Forests and Wildlife have a closer nexus to the Biodiversity Law (as both deal with life and life-forms and the eco-systems that support them). These laws emphasize on the conservation and protection of the plants and animals. On the other hand, the pollution control laws give more thrust to the protective and preventive safeguards against abuse and destruction of environmental (natural) resources. There is no doubt that the pollution control laws have a strong indirect connection with the wildlife and biodiversity laws as they facilitate in protecting the system or environment where wildlife/biodiversity survives.

Another interesting aspect in the Biodiversity law is its application of the resource (biological or genetic) use, either for research or commerce. Besides, the biodiversity law has drawn inspiration from the pollution control laws, in substance, penal actions and accessing justice. Forests and Wildlife laws have a State centric approach to environmental management, with a clear vision of the exercise of sovereign right over the resources by the agencies of the State themselves. It is also true, that in the pollution control and biodiversity legal regimes, the State control and action dominates the discourse. However, the feel, texture and focus of attention and action concerning the Biodiversity Act is to bring in the individual and local and indigenous community into the management regime, as participants, partners and right holders.

No doubt that the important role of indigenous and local peoples (ILPs) in the conservation and sustainable use of biodiversity is getting greater recognition in recent years. Globally, a substantial policy shift from the exclusion of the stakeholders to their inclusion and integration in biodiversity conservation and development is emerging. This dynamism or innovative approach would correct not only the 'historical injustice' to the ILPs but also enhance their livelihood opportunities and increase biodiversity. However, the success of this policy shift would depend on legal measures, rights over biological resources, lowering of transaction costs of exchange of biological resources and creation of the enabling institutions and incentives for efficient participation in the ABS regime of the CBD. In brief, ABS gives the local communities a sense of belonging and ownership of the natural resources around them.

1.10 Conclusion

Biodiversity plays a critical role in enhancing human well-being and various economic activities. Basic needs like clean air, food security and fresh water, which are necessary for a healthy and productive human life, are part of biodiversity. Nearly half the human beings, particularly the most vulnerable people depend directly on biodiversity for their livelihood. However, the unprecedented degradation and depletion of biodiversity are a global concern. Even the recent climate change induced frequent floods and droughts and pandemics like Covid-19 have strong connections with biodiversity degradation. In this regard, the role of CBD and its members in maintaining and restoring biodiversity through an innovative approach like ABS is critical. By preserving biodiversity and ecosystems, we are retaining the ability of the planet to sustain our prosperity (UNDP 2018).

For biodiversity management, adequate finance is a pre-requisite, which is a struggle for developing countries like India. At present, the share of the Government budget allocation for biodiversity in these countries is inadequate in meeting the funding for biodiversity conservation, where the private and corporate sectors have a role. As developing nations, but biodiversity rich countries, and providers of biological/genetic resources to the biotechnologically advanced development nations, ABS should be an ideal mechanism for conserving global biodiversity. We should elevate ABS for biodiversity management.

The role of the Nagoya Protocol on Access to Genetic Resources and the Fair and Equitable Sharing of Benefits Arising from their Utilization, adopted by the CBD at the tenth meeting of the CoP on 29th October 2010, in Nagoya, Japan, in operating the ABS provision is important. The scope and challenges in operating the Nagoya Protocol on ABS is discussed in detail by the chapters presented in this book. No doubt that through the successful implementation of the Nagoya Protocol on the ground, ABS can be elevated as a successful innovative financial (incentive) mechanism for biodiversity conservation along with empowering the ILCs and their livelihood.

Conflict of Interest The views and opinions expressed in this article are those of the author in his personal capacity. None of the observations, opinions, or conclusions expressed herein are to be attributed to or construed as the views or positions of any organization, association, or company. No conflict of interest.

References

Convention on Biological Diversity (1992) Convention on biological diversity preamble. United Nations

Convention on Biological Diversity (2011) Nagoya Protocol on access to genetic resources and the fair and equitable sharing of benefits arising from their utilization to the convention on biological diversity. Convention on Biological Diversity (CBD), United Nations

FAO-ITPGRFA (2009) International treaty on plant genetic resources for food and agriculture. A Global Treaty for Food Security and Sustainable Agriculture, Food and Agriculture Organization (FAO) of the United Nations

Ministry of Environment Forests and Climate Change (2014) India's fifth national report to the Convention on Biological Diversity (CBD). Ministry of Environment and Forests, Government of India

National Biodiversity Authority (2010) Biological diversity act, 2002. NBA, Chennai.

UNDP (2018) Biodiversity Finance Initiative (BIOFIN) workbook. United Nations Development Programme, New York

Chapter 2
Implementation of the Nagoya Protocol in India: Experiences, Challenges and Learnings

Prakash Nelliyat

Abstract India is one of the 17 mega biodiversity countries of the world. The rich biological/genetic resources and the heritage of traditional knowledge in the country attracted global attention centuries ago. As a party to the Convention on Biological Diversity (CBD), India's initiatives in fulfilling its objectives were highly appreciable. India ratified the Nagoya Protocol on Access and Benefit Sharing (ABS) on 12th October 2014.This chapter examines India's initiatives in implementing the Nagoya Protocol through various legal, administrative, policy and institutional measures. The Articles of the Protocol were examined in detail with respect to the measures India has taken (especially the compliance provisions implied in the domestic legislations), the practical achievements, major barriers and challenges, and lessons learnt. Even if India enacted the Biological Diversity Act in 2002 and Biological Diversity Rules in 2004, concurrent with the CBD objectives, its provisions for meeting most of the requirements of the Nagoya Protocol on ABS (which existed in 2010) were instrumental. Further, the Guidelines on Access to Biological Resources and Associated Knowledge and Benefit Sharing Regulations, 2014 also came in to action. Besides, state-level legislation contributes significantly to Nagoya Protocol implementation in India. However, considerable efforts at the ground level with the support of the different ABS stakeholders are required. India is yet to establish check points and user country measures. Lack of understanding and support from the ABS stakeholders, Ministries and Departments associated with natural resources and biodiversity, slow processing and enforcing the Biological Diversity Act are the major constraints in implementing the Nogoya Protocol in India. To protect the country's rich biodiversity and genetic resources, as well as to control bio-piracy, an immediate effort for the implementation of Nogoya Protocol is required.

Keywords Nagoya protocol · Biodiversity · Access and benefit sharing · Biological/genetic resources · Traditional knowledge

---

P. Nelliyat (✉)
Former Fellow (ABS) Centre for Biodiversity Policy and Law, National Biodiversity Authority, 5th Floor, TICEL Biopark, Taramani, Chennai 600113, India
e-mail: prakashnelliyat@gmail.com

O. V. Oommen et al. (eds.), *Biodiversity Conservation Through Access and Benefit Sharing (ABS)*, https://doi.org/10.1007/978-3-031-16186-5_2

2.1 Introduction

The divergent ecosystems of India contain rich species diversity. Hence, the bio-prospective scope of its genetic resources is huge. However, for anthropogenic and natural reasons, India's biodiversity is under threat. India is a party to the Convention on Biological Diversity (CBD). Various legislative and institutional measures have been taken to meet the Conventions objectives: (a) the conservation of biological diversity, (b) the sustainable use of its components and, (c) the fair and equitable sharing of benefits arising from the utilisation of genetic resources. The Nagoya Protocol on Access and Benefit Sharing (ABS) ratified by India on 12th October 2014.

This chapter broadly discusses the Nagoya Protocol and its implementation in India based on: the Interim National Report on the Implementation of the Nagoya Protocol, various ABS literatures, and the inferences obtained from the discussions with experts who are primarily the researchers and officials who enforce the BD Act in the country. The background of the Interim Report is given in the following Box. India's initiatives in implementing various Articles of the Protocol, practical achievements, and implementation challenges are examined in detail. Further, appropriate measures for the Protocol's effective implementation are also suggested.

Box
Interim National Report on the Implementation of the Nagoya Protocol: Background.

The Article 29 of the Nagoya Protocol (Monitoring and Reporting) mandates parties to submit a report to the CBD for monitoring the implementation of its obligation under the protocol. In decision NP-1/3, paragraph 4, the Conference of the Parties serving as the meeting of the Parties to the Nagoya Protocol (COP-MOP) had requested the Parties to the Protocol to submit an interim national report on the implementation of the Nagoya Protocol, on November 2017. Accordingly, the Ministry of Environment, Forest and Climate Change (MoEFCC) with the support of the National Biodiversity Authority (NBA) prepared the report through a series of expert/stakeholders consultations and submitted the same to the CBD. The report is a helpful tool for nations to assess the level of implementation of the Protocol, as well as gaps, and needs in terms of capacity. In this regard, the Interim National Report is a valuable tool for building and developing the capacity to implement the Nagoya Protocol and for designing the capacity-building activities more effectively.

2.2 Nagoya Protocol on Access and Benefit Sharing

The market potential of genetic resources and associated TK is huge and their role in the global economy is considerable. Globally, bio-prospective and value-adding depend on the richness of biodiversity and bio-genetic resource potential. Products, including pharmaceuticals and nutraceuticals, are sourced from biological resources (Pushpangadan and Nair 2005). Laird and Kate (2002) estimated the global markets for biological resources based products as USD 500 billion. No doubt with the advent of modern biotechnology, the share could be much higher in recent years.

In brief, biodiversity provides enormous role for improving human welfare. Biological diversity is considered as a global asset, with tremendous value to the present and future generations. But, the ecosystem and species and are under greater threat in recent years than ever before. The recent Intergovernmental Science-Policy Platform on Biodiversity and Ecosystem Services (IPBES) Global Assessment presents a warning picture on the deterioration of the health of ecosystems on which human and all other species depend (United Nation 2019).Other estimate indicates that biodiversity losses have become a global concern and the species loss is from 45–250 per day. But biodiversity once lost is lost forever and is likely to cause serious consequences to the ecosystem, environment, and human wellbeing (Nelliyat 2017). Considering this fact, arresting the decline of biodiversity (species) and ecosystems' degradation is the major objective of environmental policy at different levels: global, national, regional/state and local.

With this objective, the Convention of Biological Diversity (CBD) was established on 5th June 1992 at the United Nations Conference on Environment and Development (the Rio "Earth Summit") and it came into action on December, 1993 as an international instrument for comprehensively addressing all issues related to biological diversity. The Convention's three objectives (indicated earlier) need to operate in a continuous and cyclical manner (as indicated in Fig. 2.1) towards the successful ecological or biodiversity functions which facilitate life and enhance human welfare.

In a realistic sense, the fair and equitable sharing of benefits arising from the utilisation of biological or genetic resources (the third objective of the Convention) is more instrumental for achieving the other objectives: the conservation of biological diversity and the sustainable use of its components of CBD. Thus, at the global level, further advance of the implementation of the third objective was essential. The World Summit on Sustainable Development at Johannesburg, (September 2002) called for the negotiation of an international regime, to promote and safeguard the fair and equitable sharing of benefits arising from the utilisation of genetic resources. The Convention's Conference of the Parties (CoP) responded at its seventh meeting, in 2004, by mandating its Ad Hoc Open-ended Working Group. After six years of negotiation, the Nagoya Protocol on Access to Genetic Resources and the Fair and Equitable Sharing of Benefits Arising from their Utilization, to the CBD was adopted at the tenth meeting of the CoP on 29th October 2010, in Nagoya, Japan (CBD 2011).

This clearly reveals the complexity of the ABS principles and its operationalisation envisaged by the CBD. During the negotiations of the CBD, the ABS issues

Fig. 2.1 Convention on Biological Diversity (CBD) objectives

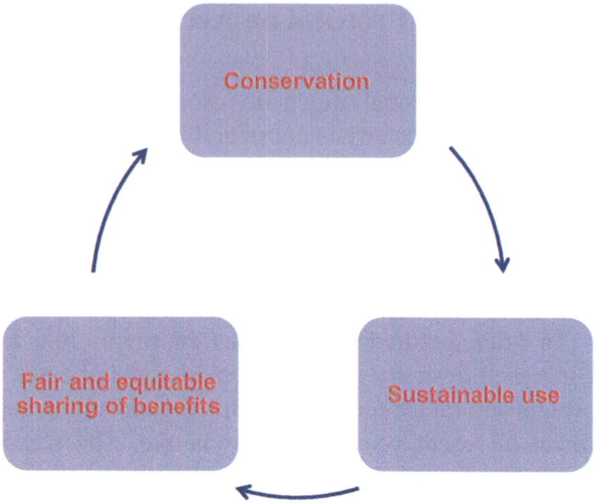

were discussed substantially. Further, other/related topics, such as State sovereignty, economic development, indigenous and local communities, scientific research, the industries dependent on genetic resources and associated TK, and the conservation and sustainable use of biological diversity also under debate. The lack of awareness regarding ABS, lack of clarity on its scope and legal principles, and gaps in the States' policies and legislation have hampered the efficient and effective implementation of ABS in practice (IUCN 2012), where the Nagoya Protocol is instrumental.

The Protocol significantly advances the Convention's third objective in the following ways:

- Providing a strong basis for greater legal certainty and transparency for both providers and users of biological or genetic resources.
- Executing specific obligations to support compliance with domestic legislation or regulatory requirements of the Party providing genetic resources.
- Contractual obligations reflected in Mutually Agreed Terms (MAT).
- Facilitating compliance provisions.
- Making more predictable conditions for access to genetic resources.
- Sharing the benefits arising from genetic resources to the community.
- Emphasizing the provisions of access to traditional knowledge (associated with genetic resources) owned by indigenous and local communities
- Benefit sharing to the community, when a company makes use of its knowledge, innovations and practices.
- Clearly demarcating various monetary and non-monetary benefit sharing strategies.

According to Bavikatte and Morten (2015), fleshing out the CBD, the Nagoya Protocol details the rights and obligations of countries of origin of genetic or biological resources and associated TK (provider countries) and countries in whose jurisdiction these genetic or biological resources and associated TK are used (user countries).

The Protocol is also aware with the rights of indigenous and local communities over genetic resources and associated TK in accordance with the domestic laws of countries where these communities are located. Therefore, the Nagoya Protocol is an unprecedented step in public international law.

Another uniqueness of the Protocol is that it makes obligatory for parties to ensure that users of genetic or biological resources and associated TK (bio prospectors) within their jurisdiction comply with the domestic ABS regulatory requirements of the provider countries. This national/state obligation on bio prospectors demands them to not only comply with domestic ABS frameworks of provider countries, but where required by such laws to enter into ABS agreements with legitimate providers of such resources and knowledge (Bavikatte and Morten 2015).

In brief, based on the principle of State sovereignty and equity principle, the CBD establishes that benefits from using genetic resources shall be shared fairly and equitably with the provider of the resources, in return for providing access (Prip and Charlotte 2016). The above indicated provisions are the unique and significant innovation of the Protocol. By promoting the use of biological resources and associated traditional knowledge (TK), and by strengthening the opportunities for fair and equitable sharing of benefits from their use, the Protocol will create incentives to conserve biological diversity, and sustainably use its components. In this context, ABS is considered as an innovative financial mechanism for biodiversity conservation. Through this one can enhance the contribution of biological diversity to sustainable development and human well-being. No doubt that biodiversity has a significant role in achieving most of the Millennium Development Goals (MDGs). The protocol has 36 Articles and an annexure.

Out of the 36 Articles of the Nagoya Protocol, some are instrumental in achieving the philosophy of ABS, and these are examined below.

2.3 Nagoya Protocol Implementation in India

2.3.1 Legislative, Institutional, Administrative and Policy Measures on ABS

Introduction of appropriate legislative, institutional, administrative and policy measures on ABS is an important prerequisite for implementing the Nagoya Protocol. In this regard, India enacted the Biological Diversity Act in 2002 (here after BD Act) and Biological Diversity Rules 2004 (here after Rules, 2004), Guidelines on Access to Biological Resources and Associated Knowledge and Benefit Sharing Regulations, 2014 (here after Guidelines, 2014) and number of Notifications. Besides, different States have also come up with State specific Biodiversity Rules and number of Notifications for enforcing the ABS at their jurisdiction.

The Biological Diversity Act is implemented through a three-tier institutional mechanism. This includes: the National Biodiversity Authority (NBA) at the national level; State Biodiversity Boards (SBBs) at the State or provincial level; and the Biodiversity Management Committees (BMCs) set up by the elected bodies at the local level. At present all the States have established SBBs. Forming of BMCs is an ongoing process. So far, in India 2,76,690 BMCs have been constituted by the local bodies (NBA 2022). However, the total number of BMCs, considering the total number of local bodies in place in the country, is supposed to be 3, 17,519 (MoEFCC 2019).

To administer the Nagoya Protocol in the country, the Ministry of Environment Forests and Climate Change (MoEFCC) is nominated as the National Focal Point and NBA has been designated as the Competent National Authority (CNA) as provided in Article 13 of the Protocol. India was the first country to upload permits issued by CNA on ABS-CH, to constitute the Internationally Recognized Certificate of Compliance (IRCC). So far 2,385 IRCCs has been generated (CBD 2022).

India has not yet established the checkpoints as mentioned in Article 17 of the Protocol for monitoring the utilization of the genetic resources. Though, the efforts in this regard is progressing. In brief, regarding the administrative, legislative, institutional, and policy measures on ABS, India has made significant progress.

2.3.2 *Access to Genetic Resources (Article 6)*

The Protocol envisages various settings for accessing the biological or genetic resources. However, the providers and users of such resources should follow certain norms.

2.3.2.1 Significance of PIC and Fair and Non-arbitrary Rules and Procedures

Access to the biological resources in India, as specified in the BD Act, operates on the principles of Prior Informed Consent (PIC)as provided in Article 6.1 of the Nagoya Protocol. The permission/approval of NBA is required for carrying out research/bio-survey and bio-utilization/commercial utilization/obtaining Intellectual Property Rights (IPRs), transfer of results of research over biological resources and third party transfer of the already accessed biological resources. PIC is accorded with the involvement of local people and communities following a process of consultation with them through the established institutional mechanism of SBBs, which in turn, consult the BMCs at the local (village) level.

Further, the Rules (2004) and various notifications issued there under provide for fair and non-arbitrary rules and procedures for accessing the biological resources (includes genetic resources) as provided in Article 6.3 (b) (CBD 2017).Information on how to apply for PIC (Article 6.3c) is mentioned in Rules (2004). Further, the ABS Guidelines, 2014 provide a transparent procedure and information on how to apply for PIC. NBA has provided adequate information such as: Interactive Videos,

User Guidelines, Online ABS application format and process, and FAQs to facilitate applying for PIC.

2.3.2.2 Provision of a Clear and Transparent Written Decision and Permits

As indicated in the Article 6.3 (d) of the Protocol, the decision of the CNA to grant approval is in the form of an agreement to be signed between the CNA and the applicant (user). In situations where access is not allowed, the applicant is given an opportunity to present his case before a final decision is taken, which is communicated to the applicant with recorded reasons specified there in, in writing. The NBA (as CAN), upon approval of access, enters into an agreement with the applicant. The agreement, which is an evidence of PIC and MAT, is equivalent to the permit as provided in Article 6.3(e). Rule 14(5) of the Biological Diversity Rules, 2004 mandates the signing of an agreement between the applicant (user) and the NBA as evidence of PIC and MAT as provided in Article 6.3 (g). More over, Rule 14 (6) details the minimum conditions to be included in such agreements (CBD 2017).

In brief, it is clear that when accessing the genetic resources, the significance of PIC and Fair and Non-arbitrary Rules and Procedures and Clear and Transparent Written Decision and Permits are the key pre conditions the Protocol emphasised on. The Indian legislations have the transparent provisions to fulfil the above requirements and conditions.

2.3.3 Fair and Equitable Benefit-Sharing (Article 5)

Fair and equitable benefit-sharing is one of the key principles of the Nagoya Protocol. As per the BD Act the benefits arising from the utilization of biological resources for certain activities are to be shared with the identified benefit claimers in a fair and equitable manner as mentioned in Articles 5.3, 5.2 and 5.5 of the Protocol. In case the benefit claimers are not identifiable, the benefits accrued are to be used for the conservation of biological resources or biodiversity, and development, including socio-economic development, of the area from where such resources are accessed. The ABS Guidelines 2014 prescribe the scheme of processing applications along with the template and terms of benefit sharing to be followed by NBA/SBBs for access to biological resources and/or associated TK. Further, the Rules, 2004 provide for procedure for access to biological resources and associated TK (CBD 2017).

2.3.4 Regulatory Requirements on ABS (Article 15 and 16) and Monitoring the Utilization of Genetic Resources (Article 17)

In India, steps are being taken to develop appropriate user country measures as required under Article 15. In this regard, inter-ministerial discussions are in progress. Further, check points are yet to be designated. However, the Indian Patent Office is performing a similar function, as provided for in the Patent Office Manual and Guidelines (Controller General of Patents, Designs & Trademarks 2011). Patents for inventions based on biological resources and associated TK obtained from India can be granted only on submission of necessary approvals under the BD Act. India has taken measures to encourage users and providers to include provisions in MAT to share information on the implementation of such terms as provided in Article 17.1(b). The approval granted under the BD Act is in the form of an agreement, which is executed between the user and the CNA on MAT. The agreement inter alia contains provision for periodical reporting on the use of biological resources and/or associated knowledge accessed (CBD 2017).

2.3.5 Cost-effective Communication Tools for ABS

Cost-effective communication tools for enforcing ABS are strongly recommended by the Protocol. India is encouraging the use of cost-effective communication tools and systems as provided in Article 17.1 (c).India has launched an online ABS application filing system "ABS E-filing", and the use of this tool is being encouraged. Further, India is also encouraging correspondence with the applicants and others through cost-effective communication tools, such as email and fax.

2.3.6 Compliance with Mutually Agreed Terms (Article 18)

According to the Article 18.1(a), (b) and (c) Parties are to encourage the inclusion of provisions in MAT to cover dispute resolution. In India's case the ABS agreement signed between NBA/SBB and the intended user (applicant) on MAT, contains clauses on the governing law, jurisdiction and arbitration. In cases of disputes arising from MAT, recourse is available under the Indian law enabling the applicant to initiate civil litigation. As per Section 52A of the BD Act any person aggrieved with the determination of benefit sharing by the NBA or SBB can file an appeal to the National Green Tribunal (NGT) (Article 18.2 of the Protocol) (CBD 2017).

As per Article 18.3, India has taken measures on the following issues. In the case of access to justice, the BD Act (under Section 52A) includes measures for ensuring access to justice. As per Section 52A, any person aggrieved with the determination of benefit sharing by the NBA or SBB, can file an appeal to the NGT. On the utilization of mechanisms regarding mutual recognition and enforcement of foreign judgements and arbitral awards: the Civil Procedure Code, 1908 enables the enforcement of foreign judgements and arbitral awards in India. Additionally, the Foreign Award (Recognition and Enforcement) Act, 1961 gives effect to the New York Convention on the Recognition and Enforcement of Foreign Arbitral Awards, 1958, to which India is a Party (CBD 2017).

2.3.7 Special Considerations (Article 8)

Article 8 (a) of the Protocol envisages creating conditions to promote and encourage research which contributes to the conservation and sustainable use of biodiversity including simplified measures on access for non-commercial research purposes, taking into account the need to address a change of intent for such research. In this regard, the ABS Guidelines, 2014 provide a separate provision (Form B application), which can be used by Indian researchers/Government Institutions, who intend to carry or send biological resources outside India for basic non-commercial research. The declaration to be signed by the user and the recipient (non-Indian) has to be submitted to the NBA. The applications filed under this provision are considered and disposed off expeditiously within a specified timeframe through simplified measures. Further, no fee is charged for such applications (CBD 2017).

Paying due regard to cases of present or imminent emergencies that threaten or damage human, animal or plant health has been emphasised in Article 8(b). In this regard, Regulation 13 of the ABS Guidelines, 2014 through Form B also facilitates transfer of biological resources by governmental institutions in India to carry out urgent studies to avert emergencies like epidemics etc. through an expeditious and simplified process.

India also taken into consideration the need for expeditious access to genetic resources and fair and equitable sharing of benefits arising out of the use of such genetic resources, including access to affordable treatments by those in need, especially, in developing countries, as provided in Article 8(b). Regarding the consideration of the importance of genetic resources for food and agriculture and their special role for food security as provided in Article 8 (c), the Government of India has issued a notification under Section 40 of the BD Act, 2002 for exempting crops listed in Annexure–1 of the International Treaty on Plant Genetic Resources for Food and Agriculture (ITPGRFA) from the provisions of the BD Act, if they are utilized for the purpose of breeding, research and training (CBD 2017).

2.3.8 Provisions Related to Indigenous and Local Communities (Articles 6, 7 and 12)

India does not have indigenous communities. India has local communities, who have customary rights and entitlements. These communities do not have any separate set of rules and regulations governing them. Regarding the local communities' established right to grant access to genetic resources (Article 6.2), the BD Act, provides for mandatory consultation with the concerned BMCs by NBA and SBBs before granting access to biological resources. Further, the BD Act also provides the power to BMCs to levy a fee from the users who collect biological resources from areas under their jurisdiction, thus further empowering the communities.

Again for ensuring theset out criteria and/or process forPIC or approval and involvement of local communities (Article 6.2 and 6.3f), the BD Act, provides for the involvement of local communities through BMCs in the decision making process relating to access to biological resources and associated TK through consultation with the NBA.

Article 7 of the Protocol discusses the measures taken in accordance with domestic law, with the aim of ensuring that the TK associated with genetic resources held by indigenous and local communities within the country is accessed by the PIC or with the approval and involvement of these indigenous and local communities and establishment of that MAT. India's BD Act, provides for mandatory consultation with the BMCs by NBA and SBBs before granting access to biological resources. Hence, the decisions taken at the national and provincial levels relating to access and utilization of biological resources are taken with the involvement of the local people/communities through consultations.

Chronicling of knowledge relating to biological diversity and preparation of People's Biodiversity Registers (PBRs) in consultation with local people is the main function of the BMC. The Register contains comprehensive information on the availability and knowledge of local biological resources, their medicinal or any other use or any other TK associated with them. Further, Section 36(5) of the BD Act provides for respecting and protecting the knowledge of the local people relating to biological diversity, through measures, such as registration of such knowledge at the local, state or national levels, and other measures for protection, including the sui generis system.

The BD Act takes into consideration customary practices and procedures with respect to the associated TK (Article 12.1).The obligations upon the potential users of biological resources and the knowledge associated therewith (Article 12.2) are contained in the BD Act, Rules, 2004, ABS Guidelines, 2014 and various state level Rules. Besides NBA is initiating various awareness generation programmes for different stakeholders.

2.3.9 Community Protocols and Model Contractual Clauses

India is supporting the development of community protocols, minimum requirements for mutually agreed terms (MAT), model contractual clauses by local communities. The BD Act mandates the preparation of PBRs by BMCs through chronicling of knowledge related to local biodiversity. The PBR has information on existing knowledge of local biological resources, their medicinal values, different uses, and trade or any other TK associated with them. The minimum requirements for mutually agreed terms are granted in the agreements signed by the NBA with the applicant for access to bio-resources/TK as provided in Rule 14(6) of the Rules, 2004.NBA, in consultation with stakeholders has formulated model contractual clauses incorporating the minimum requirements of MAT in the agreement which is entered into with the users.

The BD Act provides for certain objective exemptions so as not to restrict the customary use and exchange of biological resources and associated TK within and among local people and communities (Article 12.4). The conventional breeding or traditional practices in use in agriculture, horticulture, poultry, dairy farming, animal husbandry or bee keeping are not included in the purview of 'commercial utilisation'. Further, Section 7 of the BD Act which requires prior intimation to be given to the SBB for obtaining biological resources excludes the local people and communities of the area, including growers and cultivators of biodiversity, and *vaids and hakims*, who have been practicing indigenous medicine.

2.3.10 Contribution to Conservation and Sustainable Use (Article 9)

Article 9 of the Nagoya Protocol mentioned about benefits arising from the utilization of genetic resources towards the conservation of biological diversity and sustainable use of its components. Section 27(2) of the BD Act, governs the utilization of the accrued benefits with the benefit claimers for the purposes of conservation of biodiversity. This includes the conservation and promotion of biological resources of the areas from where the resources are accessed and also for the socio-economic development of those areas. The NBA and the SBBs together collected around Rs.130 crores through ABS. Out of it around Rs.100 crores were obtained through the red sanders (*Pterocarpussantalinus*) sale (NBA 2020). NBA has distributed a certain percentage of the amount (through Andhra Pradesh Biodiversity Board) to the Andhra Pradesh Forest Department. The Andhra Pradesh Forest Department is using this amount for the red sanders (an endemic species in India) conservation and sustainable use in the pockets where it exists. Similarly, a portion of the ABS amount accrued at the state level has been distributed to the BMCs for biodiversity conservation.

2.3.11 TransboundaryCooperation (Article 11)

India is cooperating with the neighbouring countries of the region with a view to implementing the Nagoya Protocol in instances where the same genetic resources are found in situ within the territory of more than one Party (Article 11.1), through various projects including the Kailash Landscape Project (KSLCDI) and Bay of Bengal Large Marine Ecosystem (BOBLME) Project (CBD 2017).

2.3.12 Model Contractual Agreements

In India to cater to the requirements of different activities covered under the BD Act four different types of model contractual agreements have been developed in consultation with the concerned stakeholders. These agreements relate to: (a) Research or commercial utilisation or bio-survey and bio-utilisation (b) Transfer of research results (c) Seeking an intellectual property right (d) Third party transfer. Although the model contractual clauses in these agreements are not sector specific, there are common terms and conditions that exist in these four agreements, which are simple and easy to understand. Some of them are: written notice in case of non-compliance, administrative sanctions in case of breach of contractual clauses, termination and revocation, liabilities of the user, confidentiality, and scope for making amendments. There are other clauses which are case specific, and depend on the purpose for which access is sought. These include: period of collection, obligations of the applicant during the existence of the agreement, annual status reports, benefit sharing component, and details of biological resources.

2.3.13 Codes of Conduct, Guidelines and Best Practices

India is encouraging the development, update and use of codes of conduct, guidelines and best practices or standards as mentioned in Article 20. According to the Interim Report, there are many experiences/best practices across the country on codes of conduct and guidelines on ABS. A few of these are summarised below (CBD 2017):

- Kovel Foundation, an NGO in Vishakhapatnam, Andhra Pradesh works with 253 Gum picker Associations for conservation of Gum Karaya plants and other non-timber forest products (NTFP), including wild honey, amla (*Phyllanthusemblica*), and kalmegh (*Adrographispaniculate*). Kovel Foundation's other activities include:: creation of awareness on the conservation of biodiversity, resource mapping and biological resources inventory preparation, imparting technical knowledge to primary collectors on sustainable harvesting, post-harvesting value addition and marketing of NTFPs. The Gum Karaya harvested through sustainable methods fetches a good price in the market due to the brand image of the Foundation. Further, the NGO also promotes sustainable agriculture through organic farming in 15 tribal districts of AP.

- BMC of the village Piprai Morena in Madhya Pradeshis located in the Chambal ravines which are dreaded dacoit territory with a degraded ecosystem and thorny shrub and grasses. Water is very scarce and vegetation is meager. Wildlife hunting, trade of prohibited species and animals and many illegal activities are common. Another backdrop in the village was Guggal, a critically endangered plant species, that experienced over exploitation for its gum. Destructive extraction methods like debarking and branch cuttings even of young trees pose a risk to their existence. Besides, satavar (*Asparagus racemosus*) and gokhuru (*Tribulusterrestris*), important medicinal plants, also harvested destructively. But the timely action of the Piprai Morena BMC has been helpful. The BMC with the support of the Sujagriti Samaj Sansthan, an NGO, has taken up projects for the reclamation of ravine areas and in-situ conservation of local biological resources. This includes preparation of PBR, conservation of existing plants of Guggal and planting of around 10,000 Guggal plants in the village. Training for improved techniques for safe harvesting of gum has also been provided to the local community and the produce is sold to Dabur. For water conservation, three thousand metres of bund has been constructed. In brief, the BMC restored the ecosystem and biodiversity of the village.
- Gram Mooligai Company Ltd. is a collective community enterprise that works in drought affected areas around Madurai in Tamil Nadu, for medicinal plant collection and trade. Several common medicinal plants grow in the fallow farm lands and waste lands, and these are collected by landless women and sold in the market. However, unsustainable harvesting of medicinal plants was done from common lands, leading to habitat destruction and reduction in availability. But, the company has given training to women for sustainable harvesting of medicinal plants and has also initiated a kitchen garden program. A nursery of 10,000 saplings of important medicinal plants of the area has been set up to provide saplings to the kitchen gardens and schools. Two threatened taxa, i.e. Gudmar (*Gymnemasylvestre*) and guggal (*Commiphorawightii*) have been replanted and protected. The work has now been extended to other States; Vidarbhain Maharashtra, Chambal in Madhya Pradesh, Balangirin Odisha and villages in Chhattisgarh. Studies revealed that the sustainable harvesting of about 20 medicinal plants in the area has facilitated an increase in the annual income of the community of around 10%.
- India Business & Biodiversity Initiative (IBBI), was launched in 2014 at the invitation of MOEFCC, and is hosted by the Confederation of Indian Industry (CII) with the support of German Development Cooperation through GIZ. The vision of IBBI is to sensitize, guide and mentor business organizations in India on the conservation and sustainable use of biodiversity related to their operations across their value chain, and documenting and sharing of local and global best practices in biodiversity conservation and its sustainable use. As a business-led initiative, IBBI serves as a national platform of business and its stakeholders for dialogue

to promote sharing and learning, ultimately leading to mainstreaming sustainable management of biological diversity into business. IBBI has also brought out innovative case studies on biodiversity and business.

- Biodiversity Awards, is a joint initiative of the MOEFCC, NBA and the UNDP. The India Biodiversity Awards recognize the contribution of a range of stakeholders towards conservation of biodiversity. The aim is to sensitize the effective implementation of the BD Act, which ensures conservation, sustainable use of biodiversity and access and benefit sharing arising from the use of biological resources. A large number of good practices on biodiversity conservation, sustainable use, biodiversity governance and ABS have evolved through various rounds of the India Biodiversity Awards since 2012.

In brief, all the above mentioned initiatives and practices really helped in enhancing biodiversity through its conservation and sustainable use as well as through the management of ecosystems in various pockets in the country.

2.3.14 Awareness Raising and Capacity Building (Article 21 and 22)

Awareness raising and capacity building about the components of Nagoya Protocol is a real challenge in a huge country like India, which also possesses around 18% of the world population. India has taken several measures to raise awareness of the importance of genetic/biological resources and the TK associated with genetic resources and ABS issues (Article 21) through the support of different agencies. This includes seminars, trainings, and workshops organised for various stakeholders such as government officials, researchers and scientists, industrialists, local communities, media etc.

India has also taken many measures to build and develop institutional capacity and strengthening of human resources to effectively implement the Protocol as indicated in Article 22. The BD Act provides for a three tier institutional structure consisting of NBA, SBBs and BMCs for effectively implementing the Act. For organising the capacity building programmes, external support has been received from agencies such as UNDP, GEF, GIZ, and the Norwegian Government. Some of the major projects related to the Nagoya Protocol/ABS which facilitate in enhancing the awareness and capacity building in the country are:

- UNDP project on "Strengthening the Institutional Structure to Implement the Biological Diversity Act" (2009–2013).
- UNEP-GEF **project on "Strengthening the implementation of the Biological Diversity Act and Rules with focus on its Access and Benefit Sharing Provisions"** (2011–2017).

- Government of India and Norwegian Government technical cooperation project on "Centre for Biodiversity Policy and Law (CEBPOL)" (2014–2018).ABS is one of the thematic areas in this programme.
- GIZ supported ABS Partnership project (2016–2020).
- UNDP-GEF ABS Capacity Building Project on "Strengthening the Implementation of the Nagoya Protocol" (2017–2019) (CBD 2017).

Besides, State Governments also carried out a number of capacity building programmes on biodiversity and ABS for various stakeholders through the SBBs.

2.3.15 Support for Building and Developing Capacity to other Parties (External Support)

India provided external support for building and developing capacity for the implementation of the Nagoya Protocol through various ways.

2.3.16 Technology Transfer, Collaboration and Cooperation (Article 23)

India is collaborating and cooperating in technical and scientific research and development programmes as a means to achieve the objective of the Protocol (Article 23). Some of such major programmes are.

- The Centre for Biodiversity Policy and Law (CEBPOL) formed in collaboration with the Norwegian Government, with the intension to develop professional expertise on biodiversity policies and laws, including ABS.
- Nature Index initiative under CEBPOL is a collaborative technical and scientific research programme between The Norwegian Institute for Nature Research (NINA) and NBA.
- Indo-German Biodiversity Working Group has been established in 2017 under the Indo-German Environment Forum.

2.3.17 Budget Allocations (Funds) and Human Resources

India has established mechanism for budgetary (fund) allocations for the implementation of the Nagoya Protocol. As the CAN, NBA is fully funded by the Government of India. Similarly, the SBBs and BMCs are supported by the respective State Governments and NBA. Further, financial resources have been received from other countries/parties of the CBD, including Norway; financial institutions such as Global Environmental Facility, and from other sources including the UNDP and GIZ.

India also has specific staff to administer the functions directly related to the implementation of the Nagoya Protocol. As the National Focal Point for implementing the Protocol, MoEFCC has a dedicated division for dealing with issues related to the CBD, Nagoya Protocol and for supervising the implementation of the BD Act. Apart from the National Focal Point of the CBD at the Ministry of Environment and Climate Change, six scientists/staff are working in this division. NBA has 16 full-time staff, including the Chairperson and Secretary and around 60 contractual staff having technical and scientific knowledge on biodiversity. SBBs set up in the 29 States are also having full time staff, including the Chairman and the Member Secretary and contractual staff and they are fully involved in ABS and biodiversity issues (CBD 2017).

2.4 Major Challenges in Implementing the Nagoya Protocol

- Biodiversity is an interdisciplinary and complex subject, where different discipline's ideas to be integrated. Hence, its management requires extraordinary efforts with scientific, technical and participatory approach. Biodiversity management through an innovative mechanism like the ABS requires the involvement of different stakeholders. However, enhancing the awareness among them (who have mutually contridectory approach and interests) as well as proper coordination among all these different agencies is a challenge.
- Biodiversity related issues and problems are being dealt with by various Ministries or Departments at the centrl and State levels.
- Each Ministry and Department viewed biodiversity from their perspectives, which are different. Hence, establishing co-ordination among these agencies is a pre requiest, but a huge task.
- India is a large country with a high population which is heterogeneous in nature. People follow different cultures and speak different languages. Hence, communicating the significance of biodiversity and ABS issues is not an easy task.
- Key ABS stakeholders, especially the commercial users of biological or genetic resoures, providers of the resources and enforcement agencies are not in a amiable relationship. Hence, establishing a conducive environment for deriving the 'Mutually Agreed Term'are not existing.
- Accessing biological resources by the users with commercial intent is not a new phenomenon. Even centuries before India's biological resources had been recognised, and timber and spices had attracted the west. During colonisation, huge quantities of biological resources were traded from India to western countries. Similarly, the biological resources based entrepreneurs in India had also used various biological resources without any commitments from time immemorial. Therefore, a sudden acceptence of the provisions of the CBD/Nagoya Protocol/BD Act (which are completely contradictory to the norms that had been followed and the freedom enjoyed in accessing the biological resources) might be a challenge.

2.5 Key Lessons and the Way Foreword

The above analyses and discussions on the 'implementation on Nagoya Protocol' in India provide the following lessons and conclusions:

1. The Nagoya Protocol on Access to Genetic Resources and the Fair and Equitable Sharing of Benefits Arising from their Utilization, by the CBD was adopted at the tenth meeting of the CoP on 29th October 2010, in Nagoya, Japan. India ratified the Protocol on 12th October 2014.
2. India is primarily implementing the Nagoya Protocol on ABS through the BD Act and the Rules (2004). Further, several notifications including the Guidelines (2014) on ABS have been issued for activating the various provisions of the BD Act, and State specific biodiversity Rules also facilitate implementing the Nagoya Protocol in India.
3. The comparison of various Nagoya Protocol Articles with India's initiative on its implementation reveals that the administrative, legal, policy and institutional structure introduced in India on the umbrella of Biological Diversity Act is appropriate and instrumental in implementing most of the provision of the Protocol.
4. However, the paradox is, the BD Act was enacted by the parliament on 2002, eight years before the emergence of the Nagoya Protocol (2010)! In other words, it revealed that, the BD Act (designed concurrent with CBD objectives) gave top priority in meeting the CBD's third objective (the fair and equitable sharing of benefits arising from the utilization of genetic resources), which is the exclusive focus of the Nagoya Protocol.
5. This might be one of the reasons behind designating India as one of the pioneering countries of the world in implementing the principles of CBD.
6. However, the practical achievements in terms of compliance of ABS among the commercial users of India's biological/genetic resources are limited. This indicates that, India's Nagoya Protocol implementation should be strengthened at multiple layers.
7. India is yet to establish the user country measures (Article 15) and the check point (Article 17), which is considered as a major limitation in the journey of the Nagoya Protocol implementation.

For the successful implementation of the Nagoya Protocol in India, the following strategies have been proposed.

- The implementation of the Nagoya Protocol in the country has improved the knowledge and understanding of the ABS mechanism among the stakeholders.
- This has strengthened India's ongoing efforts for the management (conservation and sustainable use) of biodiversity.
- Biodiversity conservation and management is an universal responsibility. Hence, strict enforcement of all existing laws in the domain of environment, including the BD Act is urgent.

- Enforcement is needed at all levels of government (national, state and local) in a federal system like India.
- Stringent action should be taken against those who violate the legislation. According to Section 58 of the BD Act the offence under the Act shall be cognizable and non-bailable.
- Mass awareness on the ABS philosophy among all the stakeholders, not only the providers and users of biological/genetic resources, but the public who use/consume bio-resources based end products, is also important. Public preferences of ABS cleared products might have a big impact on the benefit sharing paradigm.
- ABS is generally perceived as a tax imposed by the enforcement agencies (NBA and SBBs) on the users of biological or genetic resources. This wrong notion should be rectified or re-conveyed. Normally, a taxpayer cannot anticipate any benefit from the Government on behalf of the tax he paid. But in the case of ABS, whoever shared the royalty can anticipate the biological resources' continuous availability (raw material security) for business, when the community conserves and sustainably uses the biological resources.
- Establishment of the check point (Article 17) and introduction of the user country measures (Article 15) is the key in implementing the Nagoya Protocol in any country. India can designate the following institutions as the checkpoints:

  (a) The National Bureaus of Genetic Resources (which include: Plant, Animal, Microbial, Insect, and Fisheries)
  (b) The Protection of Plant Varieties and Farmers' Rights Authority (PPVFRA)
  (c) The Controller General of Patents, Designs & Trade Marks
  (d) The Central Board of Excise and Customs
  (e) Ports Authorities (Airport, Sea Port)
  (f) Directorate General of Foreign Trade
  (g) Department of Biotechnology
  (h) Indian Chamber of Commerce
  (i) The Geographical Indications Registry, and
  (j) Key State Government's Line Departments connected to biodiversity,especially Environment and Forests.

For materialising this venture, ministerial level discussion and policy decision is urgently required. According to Narayanan and Pisupati (2015), the mandate of the checkpoints could include providing information on the activities, including movement of genetic resources and/or associated TK with details of the geographical location of access, the purpose for accessing with specifications of use, confirming whether the collection or utilization of biological resources and/or associated TK was in accordance with the BD Act and whether prior approval was obtained by the user from the NBA for access to biological resources for the purpose for which it has been accessed.

- For the Successful implementation of the Nagoya Protocol, Parties need to introduce mechanisms (user country measures) as countries and their citizens importing or accessing genetic resources and/or associated TK. According to Prip and Charlotte (2016), user-measure requirements are the cornerstone of the Nagoya Protocol. They have come about as the result of hard, persistent pressure from developing countries on the developed countries to take co-responsibility in making the ABS regime functional. The degree of national implementation of Articles 15 to 18 will be an important indicator of the success of the Nagoya Protocol.
- At present, only a very few countries have user country measures and Norway is one among them. The Government of India is obliged to issue appropriate notifications/orders to regulate the use of genetic resources and/or associated TK brought into the country. The check points identified above could be potential sources of such information. The user country measures need to mandate those, both Indian and non Indian, to declare the source of the resources, whether the material and associated TK was accessed according to national provisions in the accessed country, the nature of use of the resource along with a declaration that the user will inform both the NCA and the national focal point on the use of the resource and associated knowledge (Narayanan and Pisupati 2015).

Considering the rapid degradation of biodiversity, its conservation and sustainable use is critical. But, conservation of biodiversity requires huge funding and the finance gap (with respect to meeting the strategies and action plan on biodiversity) is huge. In developing countries like India, the government's budget allocation for biodiversity is insignificant. Hence, mobilizing adequate funds through corporates and the private sector is an alternative, where ABS plays a significant role. No doubt that through the successful implementation of the Nagoya Protocol on ground zero, ABS can be elevated as a successful innovative financial mechanism for biodiversity conservation in a mega biodiversity country like India.

Acknowledgements This paper has been prepared based on the insights from one of my assignments or thematic areas (Interim National Report on the Implementation of the Nagoya Protocol) in ABS. I have worked at the Centre for Biodiversity Policy and Law (CEBPOL), National Biodiversity Authority (NBA), under the supervision of Dr. Sujata Arora, Former Adviser/Scientist 'G' Ministry of Environment, Forest and Climate Change and Shri. T. Rabikumar IFS, Former Secretary, NBA. I am thankful to the technical staff in the NBA and the State Biodiversity Boards, who provided the required information and the ABS experts, who supported me in drafting the report. The views expressed in this chapter/paper are personal and do not reflect those of the CEBPOL/NBA.

Conflict of Interest The views and opinions expressed in this article are those of the author in his personal capacity. No conflict of interest.

References

Bavikatte KS, Morten WT (2015) Beyond the thumb rule approach: regulatory innovations for bioprospecting in India. Law Environ Dev J 11(1). http://www.lead-journal.org

CBD (2011) Nagoya Protocol on access to genetic resources and the fair and equitable sharing of benefits arising from their utilization to the Convention on Biological Diversity. Convention on Biological Diversity, United Nations

CBD (2017) Interim national report on the implementation of the Nagoya Protocol. https://absch.cbd.int/en/search?schema=absNationalReport&currentPage=3&rowsPerPage=25

CBD (2022) The Access and Benefit-sharing Clearing-House (ABSCH), Convention on Biological Diversity. https://absch.cbd.int/en/countries/in

Controller General of Patents (2011) Manual of patent office: practice and procedures. Office of the Controller General of Patents, Designs & Trademarks. http://ipindia.gov.in/writereaddata/Portal/IPOGuidelinesManuals/1_28_1_manual-of-patent-office-practice_and-procedure.pdf

IUCN (2012) An explanatory guide to the Nagoya Protocol on access and benefit-sharing. International Union for Conservation of Nature and Natural Resources (IUCN) Switzerland

Laird SA, Ten Kate K (2002) Biodiversity prospecting: the commercial use of genetic resources and best practices in benefit sharing. In: Laird SA (ed) Biodiversity and traditional knowledge equitable partnerships in practice. Earth Scan Publication Limited, London, pp 241–286.

Ministry of Environment Forests and Climate Change (2014) India's fifth national report to the Convention on Biological Diversity (CBD). Ministry of Environment and Forests, Government of India

Narayanan S, Pisupati B (2015) Reconciling implementation of the Nagoya Protocol in India. FLEDGE, India. http://fledge.in/wp-content/uploads/2015/09/FLEDGE-Policy-Brief3-2015.pdf. Accessed 1 Nov 15

NBA (2020) ABS collection in India. National Biodiversity Chennai

Nelliyat P (2017) Bio-resources valuation for ensuring equity in access and benefit sharing: issues and challenges. In: Laladhas KP, Nilayangode P, Oomen OV (eds) Biodiversity for sustainable development. Springer International Publishing, Switzerland

Prip C, Charlotte Kl (2016) The Nagoya Protocol on access to genetic resources and benefit sharing: User-country measures and implementation in India. FNI Report 2/2016, Fridtjof Nansen Institute

Pushpangadan P, NairK N (2005) Value addition and commercialization of biodiversity and associated traditional knowledge in the context of the intellectual property regime. J Intellect Prop Rights 10:441–453

United Nation (2019) UN Report: nature's dangerous decline 'Unprecedented'; species extinction rates 'Accelerating', available at UN report: nature's dangerous decline 'Unprecedented'; species extinction rates 'Accelerating'-United Nations Sustainable Development

Chapter 3
Achievements in India's ABS Mechanism

K. P. Laladhas, J. R. Rani, Joseph Mar Dionysius, Achuthsankar S. Nair, P. R. Sudhakaran, and Oommen V. Oommen

Abstract India is one of the first countries to enact National legislation for CBD. The India Biological Diversity Act was in place in 2002. Subsequently, federal legislations were framed to comply with the trio-cardinal principles of the convention such as conservation of biodiversity, sustainable utilisation of its resources and access and benefit-sharing of biological resources of its occurrence. A three-tier regulatory structure has been initiated in India and at the national level such as National Biodiversity Authority (NBA), at the federal level 29State Biodiversity Boards and at the local level including Union territories, we have 276,690 Biodiversity Management Committees (BMC). NBA is assisted in its functioning through the constitution of various subject-specific expert committees which are deemed necessary for the implementation of the Act in India. By this manner for the implementation of ABS mechanism in India, there were Expert Committee on ABS; who are well versed in the administration of biodiversity, trade, Law, bio-resources etc. It is the statutory responsibility of the NBA and the State Boards to consult the BMCs while taking any decision relating to permitting the access of biological resources and associated knowledge occurring within the respective territorial jurisdiction of the BMCs. A statutory Biodiversity Funds is envisaged in the Biodiversity Act for Sharing and Distribution of Benefits to the stakeholders. Development of biodiversity-related activities such as conservation of bio-resources, propagation of the demanded bio-resources, and eco-restoration of such precious ecosystems giving valid bio-resources is a duty of LSG and BMC. This task can be achieved by identifying specific resource holders and supporting their resource conservation and utilization. The technical and administrative support for the BMC will be received from State Biodiversity Boards.

K. P. Laladhas (✉) · J. R. Rani · A. S. Nair · P. R. Sudhakaran · O. V. Oommen
Department of Computational Biology and Bioinformatics, University of Kerala, Kariyavattom, Thiruvananthapuram 695581, Kerala, India
e-mail: laladhas@gmail.com

K. P. Laladhas · O. V. Oommen
CSEIBA: Centre for Sustainable Environment-India Biodiversity Action, Thiruvananthapuram, Kerala, India

J. M. Dionysius
St. Stephen's College, Pathanapuram, Kerala, India

O. V. Oommen et al. (eds.), *Biodiversity Conservation Through Access and Benefit Sharing (ABS)*, https://doi.org/10.1007/978-3-031-16186-5_3

At present, about 5793 applications have been received at the National Biodiversity Authority for accessing bio-resources. Progressive demand for access to Indian bio-resources is seen in the data from NBA. Demand for access to bio-resources from State biodiversity boards is also progressively gaining. The amount of funds received in terms of Access to bio-resources is also increasing at NBA for sharing to its custodians.

Keywords ABS · Kerala State Biodiversity Board · KSBB · Statutory Board · BMC · PBR · NBA · NBC · Biodiversity Conservation

3.1 Introduction

Access and benefit-sharing (ABS) is international legislation it aims at the effective transfer of benefits arising from genetic resources by the users and resource holder countries. Convention on Biological Diversity (CBD) (United Nations Convention on Biological Diversity, 1992) envisages the sovereign rights over its resources to a Country and emphatically providing the ownership of their bio/genetic resources available at their territory. Transaction of genetic resources streamlined further in the Nagoya Protocol on Access to Genetic Resources and the Fair and Equitable Sharing of Benefits Arising from their Utilization to the Convention on Biological Diversity (Nagoya Protocol 2011). Principles derived out of CBD and the Nagoya Protocol, together with the right over their bio-resources and freedom of transaction of the Parties (Countries) are the real players of International ABS mechanism.

3.2 The Nagoya Protocol

The Nagoya Protocol opens up to create a transparent international legal framework to govern access to and utilization of non-human genetic resources associated with traditional knowledge from any nation. There by access to its genetic resources is subject to prior informed consent in a transparent and acceptable manner and pursuant to fair and non-arbitrary rules. In exchange for allowing such bioresource access, the country with sovereignty over the genetic resources must require fair and equitable sharing of any monetary and non-monetary benefits arising from utilization of such

Table 3.1 Status of ABS Application in India as on 22.02.2022

Received applications	Approval granted and agreement signed	Pending with applicant for execution of agreement	Closed applications	Applications under process
5793	3082	1505	894	268

Table 3.2 Bioresource Claimers from different State Biodiversity Boards of India

Sl. No	Number of biological resources	Name of benefit claimers	Amount to benefit claimers	Administrative charges 2.5%
1	1	Central Institute of cotton Research-ICAR	85,500.00	2250.00
2	12	Maharashtra State Biodiversity Board	34,200.00	487.50
3	14	BMC-Mumbai, Municipal Corporation	247,000.00	6500.00
4	1	BMC-BhorandMurbad, Maharashtra	19,000.00	500.00
5	1	Maharashtra State Biodiversity Board	7600.00	500.00
6	1	BMC-Sarkarvadi, Kopargaom, Maharashtra	8550.00	225.00
7	5	Maharashtra State Biodiversity Board	15,437.00	406.25
8	30	Maharashtra State Biodiversity Board	2,137,500.00	32,812.50
9	9	Maharashtra State Biodiversity Board	116,850.00	1283.33
10	1	Maharashtra State Biodiversity Board	57,000.00	1500.00
11	1	Maharashtra State Biodiversity Board	10,450.00	275.00
12	1	Maharashtra State Biodiversity Board	10,450.00	275.00
13	1	Maharashtra State Biodiversity Board	10,450.00	25.00
14	11	Maharashtra State Biodiversity Board	71,060.00	1870.00
15	9	Maharashtra State Biodiversity Board	48,165.00	1267.00
16	8	Maharashtra State Biodiversity Board	69,160.00	1820.00
17	50	Maharashtra State Biodiversity Board	475,000.00	3687.50
18	4	Maharashtra State Biodiversity Board	394,162.40	6223.61
19	4	Maharashtra State Biodiversity Board	10,106.00	153.95
20	1	Maharashtra State Biodiversity Board	968,905.00	83,947.50

(continued)

Table 3.2 (continued)

Sl. No	Number of biological resources	Name of benefit claimers	Amount to benefit claimers	Administrative charges 2.5%
21	6	BMC-Pune Municipal Corporation BMC-Mumbai Municipal Corporation	17,100.00	450.00
22	13	National Collection of Industrial Microorganism (NCIM) Pune	33,250.00	541.67
23	1	BARC-Trombay Dr PDKV Akola, Maharashtra	11,400.00	573.21
24	1	Maharashtra State Biodiversity Board	541,010.43	14,237.12
25	5	BMC-ChandoliBudruk, Maharashtra Parag Milk Food-Maharashtra	13,300.00	3500.00
26	38	Maharashtra State Biodiversity Board	236,550.00	6225.00
27	1	BMC-Kopargaon, Maharashtra	8334.35	219.33
28	3	Maharashtra State Biodiversity Board	154,113.75	6199.38
29	3	Maharashtra State Biodiversity Board	31,666.67	2790.00
		Total	**5,962,971.20**	**180,995.35**

resources. This access and benefit-sharing ("ABS") is effectively achieved through an agreement spelled out with elaborative contractual terms between the provider country and the user or research entity (Tables 3.1, 3.2 and 3.3).

To implement an effective international ABS legal system, each Party have to adopt domestic-level legal protocols for legible provider measures and user measures that determine access to and utilization of genetic resources. "Provider measures" require those seeking to access genetic resources to do so only with prior informed consent of the country (and local indigenous community) and in accordance with mutually agreed terms for benefit-sharing. After implementing Nagoya Protocol, as various nations have adopted new provider and user measures, the Protocol's legal framework has begun to mingle into an international regulatory scheme governing access to and use of genetic resources for research and development. It is not clear that many Parties have not yet adopted laws to implement the Protocol, and others are relying on existing laws, which is rarely fit to implement the Protocol.

Table 3.3 Details of applications since the inception of NBA as on 22.09.2021

Sl. No	Year wise status of applications	Access to Bio-resources for research/commercial use (Form I)	Transfer of research results (Form II)	Approval for obtaining IPR (Form III)	Third party transfer (Form IV)	Form B	Total
1	2006–2007	4	1	0	2	0	7
2	2007–2008	5	3	12	6	0	26
3	2008–2009	4	4	21	6	0	35
4	2009–2010	2	1	9	1	0	13
5	2010–2011	3	1	4	1	0	9
6	2011–2012	1	2	6	0	0	9
7	2012–2013	1	0	8	7	0	16
8	2013–2014	1	0	14	2	0	17
9	2014–2015	19	0	22	1	0	42
10	2015–2016	31	1	51	2	7	92
11	2016–2017	36	4	127	0	15	182
12	2017–2018	37	2	245	1	31	316
13	2018–2019	62	1	209	1	17	290
14	2019–2020	29	1	258	0	40	328
15	2020–2021	76	8	947	0	23	1054
16	2021–2022	567	0	175	0	8	339
Total		**368**	**30**	**2111**	**30**	**141**	**2680**

3.3 ABS Implementation in India

The implementation of ABS mechanism in India isestablished through a three-tier institutional structure. National Biodiversity Authority (NBA)is the upper bodyat the National level, the State Biodiversity Boards (SBBs) at the federal level and at the local level, the Biodiversity Management Committees (BMCs). These three institutions are established as per Biodiversity Act (2002) as the independent statutory bodies which have multiple duties to carry out and performin a decentralised fashion.

3.3.1 Institutional Structures for ABS Implementation

3.3.1.1 National Biodiversity Authority (NBA)

In India the NBA has been established under the purview of Ministry of Environment, Forest and Climate change (MoEFCC) with its headquarters at Chennai. The Authority is constituted as per Biodiversity Act (2002) by MoEFCC and which consists of a chairperson, who is an expert in the area along with ten ex officio members from line departments dealing with the implementation of environment and biodiversity aspects and five non-official members who are experts from the fields of biodiversity and environment. The ex officio members represent the Ministries and Departments. NBA is assisted in its functioning through constitution of various subject-specific expert committees which deemed necessary for the implementation of the Act in India. By this manner for the implementation of ABS mechanism in India there are Expert Committee on ABS; Expert Committee on Normally Traded Commodities; the Expert Committee on Agro biodiversity; the NBA is assisted in its functioning through the constitution of various subject-specific expert committees.

Functions and Powers of NBA

Ref. 1: Sect. 8, BD Act

Ref. 2: Sect. 18, BD Act

As per Biodiversity Act, the NBA has been entrusted with certain powers and functions which include granting approvals to access biological resources for commercial utilization, research, bio-survey and bio-utilization.

Approval for obtaining IPRs requests, approval for already accessed bio-resources and grant of approval for transfer of research results.

NBA is entrusted to take necessary measures on behalf of the Central Government to defend the right of our bio-resources in other countries who are granting IPR on our biological resources or associated knowledge derived from India.

3.3.1.2 State Biodiversity Boards (SBBs)

Ref.3: Sect. 22 BD Act

Indian States are entrusted to constitute State Biodiversity Boards in all its federal divisions as per Biodiversity Act. Currently, 29 SBBs have been constituted in India. The structure of constitution of SBB consists of a chairperson, not more than five ex officio members from environment and biodiversity line departments and not more than five non-official members specialized on matters related to the three objectives of the BD Act. NBA discharges the duties of Union Territories and has to constitute Biodiversity Councils in all the 9 Union Territories vide a set of Office Memorandums dated 31st December 2019.

SBBs have the deemed functions to advise the State Government on matters related to the three cardinal objectives of the BD Act; regulate (a) commercial utilization and (b) bio-survey and bio-utilization for commercial utilization of biological resources within the revenue boundaries of the concerned State and other functions necessary for the conservation of biodiversity of the State to carry out the provisions of the Act.

3.3.1.3 Biodiversity Management Committees (BMCs)

Ref 4: Sect. 41 BD Act

Biodiversity Management Committees are established for conservation and management of biodiversity at decentralised bottom level through local self-governments in rural and urban areas. BMC is constituted with a chairperson and not more than six members who are nominated by the elected body concerned from the Local Self Governments. One third of such nominated members shall be women and not less than 18% of the BMC members should belong to the Scheduled Castes/Tribes.

The intention of the constitution of BMC is for the Conservation, sustainable use and documentation of biological diversity from the LSG, Preservation of habitats, Conservation of landraces, folk varieties and cultivars, domesticated stocks and breeds of animals and microorganisms and chronicling of knowledge associated with biological diversity.

It is the statutory responsibility of the NBA and the State Boards to consult the BMCs while taking any decision relating to the use of biological resources and associated knowledge occurring within the respective territorial jurisdiction of the BMCs. Decision making with regard to transfer or selling the biological resources and/or associated TK is through this consultation process and to ensure the participation or approval and involvement of the conservers for ABS prior to access.

Phase II & III—Benefit Sharing and benefit distribution

Sharing of the benefits acquired by the use of bio resource or knowledge associated, from the applicant having mutually agreed term with NBA or SBBs is the second phase of the ABS process. Benefits ought to be shared in respect of conservation and replication of harvested bio-resources and which is the mandated activity. The quantum of benefits to be shared would be mutually agreed upon before the access of bio-resources between the applicant, the benefit claimants and the local bodies concerned. It is the duty of the NBA to ensure the right and obvious advantages in mutually agreed terms and conditions to secure fair and equitable benefit sharing. This would also mean that when benefit claimants are not identified, NBA could negotiate for mutually agreed terms with the applicant.

The BD Act defines fair and equitable benefit sharing to mean "sharing of benefits as determined by NBA for the best interest of the holders in all or any of the following manners.

Ref 5: Sect. 2(g) r/w Sect. 21 BD Act

Ref. 6: As per the BD Rules, 2004

Ref 7: Rule 20 (5), BD Rules

Ref 8.; Regulation 14, ABS Guidelines, 2014

A statutory Biodiversity Funds is envisaged in the Biodiversity Act for Sharing and Distribution of Benefits to the stakeholders. Biodiversity fund is the facility to act upon the needs of the local level conservation, biodiversity enhancement, ecosystem conservation, ecosystem restoration, biodiversity enumeration and development of Biodiversity register, conservation of biodiversity Heritage sites etc. at the three levels of Biodiversity administration of the country. At the National level, we have the National biodiversity fund, and at the federal level we have the State Biodiversity Fund and at the local LSG level, we have the Biodiversity fund.

Biodiversity Funds under BD Act

National Biodiversity Fund (NBF)

State Biodiversity Funds (SBFs)

Local Biodiversity Funds (LBFs)

National Biodiversity Fund

The application fee collected from the resource seekers would be deposited in the NBF and also the amount of money received after mutually agreed-on share from the resource assessors and development of funds from the outcome of research results from TK associated bio-resources and any such permissible fund generated by the utilization of our bio-resources.

3.3.1.4 Channelling Benefits to the Identified Benefit Claimers

Development of biodiversity-related activities such as conservation of bio-resources, propagation of demanded bio-resources, eco-restoration of such precious ecosystems giving valid bio-resources is a bound duty of LSG and BMC. This task can be achieved by identifying specific resource holders and supporting his resource conservation and utilization. The technical and administrative support for the BMC will be received from State Biodiversity Boards.

For the socio-economic development of areas from where the concerned biological resources have been accessed in consultation with the concerned local bodies.

Ref 9: Sect. 27 BD Act

Ref 10: Sect. 21(3) BD Act24

Ref 11: Regulation 15, ABS Guidelines

State Biodiversity Fund (SBF)

The sources of State biodiversity funds are mainly planning funds received from the State Government (as in the case of Kerala SBB), funds from NBA, other funds generated as per project modes from different funding agencies and the main source expected is from the ABS mechanism to be channelized to the BMCs. The management and conservation of heritage sites, compensating and rehabilitating people economically affected by notification and declaration of biodiversity heritage sites and funds will be shared to promote livelihood for the biodiversity custodians who are holding bio-resources locally.

Local Biodiversity Fund

Local Biodiversity fund is formulated by the LSG to get the different funds and grants from the Government, State Biodiversity Board and also from NBA. ABS requests and it's fees will be received by the LBF for its jurisdiction. Management of LBF is by the BMC and the fund is envisaged for the conservation activities only.

3.4 India's First ABS Approval After the Inception of Biological Diversity Act 2002

A successful ABS approval given for bio-resources having TK associated by India after implementing Biological Diversity Act 2002 is for anti-venom tablet 'PINAK' by Maharashtra State Biodiversity Board. An application received for seeking prior approval of NBA for obtaining patent (IPR) for the invention titled "*an Ayurvedic anti snake venom comprises four medicinal plants*" in the treatment of snake bite. This anti-venom tablet acts as a temporary relief instantly before victim is taken to the hospital. After due consultations with the Maharashtra State Biodiversity Board and also expert committee constituted for evaluation of ABS applications as per regulation, the approval has been given. NBA fixed the benefit sharing as 2% of the gross sales or gross revenue of the product derived from the use of biological resources accessed. All formalities and approval have been executed by NBA in 2009. This was the first successful approval since the inception of Biological Diversity Act.

On commercialisation of the patent product and as per the conditions of the agreement, the applicant has paid Rs. 10, 638. 00, between April 2011 and April 2017 at six terms towards royalty as benefit sharing to NBA. From the amount 5% of the benefit has been shared equally between NBA and Maharashtra State Biodiversity Board as the administrative expense. The remaining 95% of the benefit shared amount has been transferred to the beneficiaries in the Karad and Patan Biodiversity Management Committees in Satara District through Maharashtra State Biodiversity Board.

3.5 Details of Benefit Sharing During 2021 by NBA

As per Sec 21 of Biological Diversity Act, 2002 with Biodiversity Rule 20, NBA has received a total of Rs.4,50,07,188.38 (Rupees Four Crores fifty lakhs seven thousand one hundred and eighty-eight and thirty-eight paise), as benefit sharing remitted to the National Biodiversity Fund (NBF). This amount was gained from 29 applications for accessing biological resources for Research. Benefit claimers were identified based on their geographical location. The amount of transfer of benefit was as per agreement that 2.5% each to the NBA and State Biodiversity Boards and the remaining 95% to the benefit claimers.

As per the decision of the NBA, the amount of benefit sharing shall be disbursed to the benefit claimers via the concerned State Biodiversity Boards. The share in respect of State Biodiversity Boards will be deposited in the State Biodiversity Funds (SBF) for undertaking activities enumerated in the Biological Diversity Act as a Rule for conservation and promotion of their precious bio-resources. Shared amount has to be transferred to the Benefit claimers of the respective BMCs not later than a year for the conservation activities of the BMC.

Till August, 2020 a total of 191 companies transacting bio-resources from Kerala have been entered into ABS agreement with Kerala State Biodiversity Board and four companies paid their share of benefits. They are, Kakkanat Ayurveda Ashramam, ShornurPalakad; Peermade Development Society, Idukki; N P Ayurveda Pharmacy, Maranadu, Kollam and Kannikkat Pharmaceuticals, Mallappally, Pathanamthitta. A total of Rupees Ninety-Six thousand six hundred and fifty-seven (Rs.96657/-) has been received as benefit sharing.

3.5.1 *Status of ABS Applications in India*

There were heavy demands for accessing Indian bio-resources. As per data available from NBA web site as on 31 March 2021, 5073 applications have been received by the Authority. Approvals were granted and agreement signed to 2436 cases. Due to formality irregularities 1580 applications are pending. About 839 cases have been closed due to errors in application and other formality objections.

http://nbaindia.org/content/333/25/1/applicationstatus.html.

3.5.2 *Details of Applications Since the Inception of NBA*

http://nbaindia.org/content/683/61/1/approvals.html.

Number of applications of Access and benefit sharing has been granted by the NBA as on 31 March 2021 was 2436. A total of Rs. 100.45 crores of money has been gained as benefit sharing component from the transaction of Bio-resources from India. An amount Rs.48.45 crores have been transferred to the benefit claimers. Applications for Intellectual property right (IPR) to the findings from Indian bio-resources have been approved for different cases by NBA. IPR approvals to 251 cases have been given with regards to knowledge associated with the Indian bio-resources. A total of 102 approvals have been given to agencies having association or collaboration or share with Multi-National Companies (MNCs). From the IPR approvals given about 13 cases have been turned successful for ABS. There are no approvals for ABS given to bio-resources and associated traditional knowledge from tribal population from India. There are no litigations at present against international IPR violations with regard to Indian bio-resources after the enactment of the biological diversity Act.

3.6 Discussion

3.6.1 Trust-Building Between Users and Providers Must Be Necessary

States have their potential and sovereign rights over natural resources under their jurisdiction. They are obligated to put in place right conditions that facilitate access to these precious resources for environmentally sound uses. Providers agree terms for their transaction, which include PIC and MAT, for granting access and sharing benefits equitably. Users are responsible for sharing the benefits derived from genetic resources with the real benefit claimers. They seek access to genetic/biological resources for a wide range of purposes, from basic research to the development of new commercial products. The agreement aimed at creating transparent and legally secured relations that are appropriate to the needs and intentions of both the parties for the sustenance. The suggested terms and clauses are intended to meet the needs of both the providers of the genetic resources and the researchers seeking access. There are plenty of international litigations going on with regard to procuring biological or genetic resources illegally, especially when certain multinational companies procured bio-resources from third world nations and developed new pharmaceutical and other products. We need to bridge gaps for more partnership between these key players. There must be a trust developed with the real players with mutual agreement, which will really help the development of research on such bio-resources and the development of new products for the betterment of mankind https://abs-sustainabled evelopment.net/component/two/.

3.6.2 ABS Informations Are More Intractable

More user supportive and transparent exhibition of voluntary public information is very essential for the accessor who wishes to partnership with a resource holder. The display of users' guidelines or code of conduct to access to genetic resources is very important to build up a climate of trust between users and providers of the genetic or biological resources excluding any hidden adversaries. Sharing of acquired benefits from accessed bio resource is the cardinal dogma for the accessor and this should be in all way linked to the conservation of acquiring resources and the livelihood of holder or custodian of the bio-resource. A true interception of available legal and accessing protocol at national level is necessary to see whether the present ABS legal frame work is attracting access to our bio-resources readily and the subsequent progress in the generation of funds for the conservation of our precious bio-resources (Manuel Ruiz Mullere 2016). The role of NBA is inevitably paramount to see the existing and ongoing progress in this respect and see that whether it needs any policy shift or modification.

3.6.3 Access to Bio-resources is High in India but Less Benefit Sharing

As we are aware that India is a mega diverse country having big treasury of bio-resources capable enough to gain ABS opportunities. We are blessed with great potency of bio-resources with traditional knowledge. From the data available from NBA, we have great number of applicants for accessing Indian bio-resources (http://nbaindia.org/content/683/61/1/approvals.html) but which is not developing to a level of benefit sharing from the value-added products. The main deficit is the systematic research in our bio-resources specifically for identifying bioactive molecules. A strong policy shift is necessary for developing indigenous research institutions capable enough to do research in bio-prospecting using high end research methodology for molecular characterisation and also facility for bioassay testing. Huge number of bio-resources are used in India for livelihood maintenance and trade in rural world without assessing its potent bio-capability and benefits (Laladhas et al. 2020).

3.6.4 Decentralised Research for Bio Capability

We have large numbers of Government with research institutions with around 1043 Universities, public funded 42,343 colleges many with research facilities, 9995 standalone research and development centres across India doing academic research in fundamental as well as high end quality research as per all India survey of higher education 2020 (Ref). These institutions have their potential role in identifying credible bio-resources having capability to do progressive research and subsequent trade as per ABS mechanism via their preliminary research. Systematic and scientific research is very essential for our bio-resources to compete with international ABS arena. Bio-resources with bio capabilities are necessary for giving clues to the technology empowered accessors for doing high end research and subsequent development for benefit gaining.

Ref. https://www.education.gov.in/sites/upload_files/mhrd/files/statistics-new/aishe_eng.pdf.

3.6.5 Documentation of Best ABS Practices Across the Globe Will Boost More Transactions

For a well-established ABS mechanism, a compilation of best practices on contractual clauses established in other member nations collated to build a library for user-provider cooperation. It is necessary to support the elaboration of sectoral standard clauses for MAT with a view to increase cooperation and feasibility. There is a need for continued interaction among ABS actors for source of information on access requirements as it is not effectively displayed and sufficiently updated in a regular interval.

A thrust is needed to be put on Benefit Sharing because it is the key to trust biological transaction. There should be a predominant thrust given for the Research and Development of bio-resources or genetic resources to address its bio-capability for better trade and transaction. A culture of promotion of non-monitory benefits for access of bio-resources to society is highly essential for the overall development and progress.

3.6.6 *Roobios Reference*

The ABS Capacity development initiatives, Access and Benefit Sharing of Roobios, associated traditional knowledge. https://rooibos.abs-biotrade.info/#/genetic-resource.

3.6.7 *An Easy Mechanism for Accessing Bio-resources Traditional Knowledge Necessary*

Traditional knowledge associated with bio-resource is a potent source for research leading to viable commercial outcome. Need for protecting traditional knowledge is a hot debate globally for past several decades. An acceptable global protocol may be a distant reality. Anarchy and mistrust in accessing traditional knowledge is evident in several cases globally due to the unfair access and its subsequent litigations. This mindset triggered the global community holding knowledge associated with bio-resources to prevent the access to biological resources in the name of protecting associated traditional knowledge. Implementation of the Convention on Biological Diversity in India proposed more formalities which was envisaged for a transparent and judicious access has created substantial uncertainty in collaborative scientific research between Indian and foreign entities. It is mainly with regard to filing application for intellectual property rights. A reality persists is that right now CBD provided the States the sovereign right over its resources and the bio-resources and knowledge associated with bio-resources are protected under the regulation is clearer globally. States are very keen to see and declare the origin of bio-resources and its subsequent research outcome globally by any governmental or private incorporation. Any breaching or violation in this regard can be readily caught and be booked. An easy mechanism has to be developed in India to access our bio-resources linked with traditional knowledge. A transparent user-friendly online application facility to facilitate the communication and exchange of information in a swift time and simplified and standardised procedure for ABS transaction should be adopted.

3.6.8 Sustainability is More Evident in Community Linked Access to Bio-resources

When a bulk quantity of bio-resource is accessed by an entity for extraction or for industrial use, from the existing experience it is better to collect from community or cooperative firms that helps to implement a protocol of sustainable harvest and propagation of harvesting bio-resource which will sustain the livelihood enhancement of rural poor and entrust conservation in a local participatory model (Sirakaya 2019). Community protocols can be useful in the establishment of clear procedures to follow with local communities and indigenous peoples. When the direct involvement of communities for accessing bio-resource as per National frame work obtaining prior informed consent (PIC) and establishing mutually agreed terms (MAT) for access to genetic resources or associated traditional knowledge is more stringently adopted.

3.6.9 Intermediary or Neutral Facilitator for Local Level Capacity Building

Capacity building at local level is a main blockade in the access of bio-resources in India. BMC is the statutory organ at local level entrusted with great responsibility of conserving its biodiversity and its sustainable utilisation at local level, is unsuccessfully established in India. Local level empowerment is very necessary with the help of available voluntary experts, academicians, retired officers and staff from Medicinal plant board, fishery, forest, agriculture and animal husbandry that can support as an intermediary or neutral facilitator to support the identification of bio-resources suitable for trade, its quantification, wild access, necessity for cultivation or conservation etc. It is essential to explain the ABS rules and facilitate the process to the different stakeholders including the benefit of our bio-resource, endemicity, rarity. It will be helpful to support with practical aspects such as translating documents into local languages or submitting requested application documents to the relevant authorities. The ongoing need for awareness-raising of various international and national trends in ABS implementation is also very essential at local level (Andreas 2012).

3.6.10 Multiple Actors for a Successful ABS

ABS implementation often involves multiple actors along the much essential value chain, which may include government representatives, communities, TK providers, research institutions and companies. Communication and dialogue between the various actors involved are of key importance in establishing this trust and acceptance. Research involving access of bio-resource can contribute to the establishment of very long-term partnerships. Long-term partnerships involving various scientifically committed actors along the value chain have to contribute to the valorisation of biological resources and traditional knowledge.

3.6.11 Non Monitory Benefits Are Essential for Conservation, Infrastructure and Capacity Building

Access to bio-resource is aimed to conserve biodiversity and its sustainable utilisation being the cardinal principle. It is evident that high emphasis is given for livelihood empowerment, which is quite essential for the existence of the fundamental stakeholder. But general implementation of infrastructural facilities for conservation, buildings instrumentation for research, technology transfer, development of new skills in the communities and job creation, IPR, research publications, livelihood enhancing infrastructure such as starting of schools and colleges etc. in any level of demands accepted by the accessor. It can also be focused to building the capacity of local students, scientists and to establishing even independent research facility in provider countries.

3.6.12 Benefit-Sharing, a Trustful Mutualism

Benefit-sharing mechanism is more successful when the parties involve in a mutualism, principle for the betterment of both parties. In the absence of pre-established measures and parameters for benefit-sharing in the ABS experience, there is flexibility in making more pluralistic demands and decisions which are useful to parties. This process may lead to lengthy and complex negotiations when more numbers of actors are involved. Benefits can be shared for the better progress of the resource holder in various fashion. In some cases, up-front payments have been considered useful to initiate scientific research as they provide legal certainty to users and enable to start with research and development. Useful requirements have to be implemented for further authorisations and benefit-sharing if commercialisation is to take place. The principal challenge of establishing fair and equitable benefit-sharing terms which are trustworthy and no loser's principle proportional to profits at the outset. This sharing process can be suggested to review the terms once greater visibility and progress are achieved with respect to sales and turnover of the product using the genetic resources for long term better contract and trust. The establishment of a benefit-sharing fund or biodiversity fund is the useful mechanism for the involvement of providers in the progressive decision making regarding the sustainable use of such funds, in relation to conservation and livelihood in cases where multiple communities are involved.

3.6.13 Innovative Research Linked Policy is More Inevitable

Mega biodiverse nations can establish national strategy for systematic assessment and valorisation the bio-resources associated with traditional knowledge and the development of National research facilities for better management of such precious bio-resources. A systematic collection, assessment, availability, benefits, viable associated knowledge, testing are necessary for showcasing our biodiversity before global entity for ABS transaction. There should be a policy for bio survey of our resources by Government owned national institutions for keeping the available resource intact for ABS.

Conflict of Interest The views and opinions expressed in this article are those of the author in his personal capacity. None of the observations, opinions, or conclusions expressed herein are to be attributed to or construed as the views or positions of any organization, association, or company. No conflict of interest.

References

AISHE (All India Survey of Higher Education 2019–20). Ministry of education, government of India. Available at https://www.education.gov.in/sites/upload_files/mhrd/files/statistics-new/aishe_eng.pdf

Andreas D (2012) Access and benefit sharing, strategic communication for ABS: a conceptual guide and toolkit for practitioners. ABS capacity development initiative division 47 environment and climate change deutsche GesellschaftfürInternationaleZusammenarbeit, (GIZ) GmbH Postfach 5180 65726, Eschborn Germany. Available at https://www.cbd.int/cepa/toolkit/other-toolkits/cepa-abs.pdf

Building trust between users and providers of genetic resources to facilitate the identification of bio-discovery efforts. Available at https://abs-sustainabledevelopment.net/component/two/

CBD (1993) Convention on Biological Diversity (with annexes). Concluded at Rio de Janeiro on 5, June 1992CBD (1993) Convention on biological diversity (with annexes). Concluded at Rio de Janeiro on 5 June 1992. United Nations, Treaty Series 1760(30619):142–382

Laladhas KP, Roy John Fr, Saraswathy V, Bharucha E, Nair AS, Oommen OV (2020)Successful livelihood models through BMCs in India, 212–223. In: Oommen OV, Laladhas KP, Bharucha E (eds)Biodiversity and livelihood: lessons from community research in India, Bentham Science Publishers ISBN (Online): 978–981–14–8215–1 ISBN (Print): 978–981–14–8307–3. Available at https://benthambooks.com/book/9789811482151/

Manuel Ruiz Mullere (2016) Access and benefit-sharing policy and legal framework in central America and the Dominican Republic deliberations in the "First Regional Conference on ABS issues" of the member states of the central American Commission on environment and development (CCAD), 7th and 8th of July 2015, San Salvador/El Salvador, in the context of the programme ABS/CCAD-GIZ of the German cooperation. Available athttps://absch.cbd.int/api/v2013/documents/76AAE394-EB57-FB19-64F6-F2FB9CDD07DB/attachments/ABS%20Policy%20and%20Legal%20Framework%20Central%20America%20and%20Dom.Rep_english.pdf

Nagoya Protocol on Access to Genetic Resources and the Fair and Equitable Sharing of Benefits arising from their Utilization to the Convention on Biodiversity (2011). Available at https://www.cbd.int/abs/doc/protocol/nagoya-protocol-en.pdf

Sirakaya A (2019) Balanced options for access and benefit-sharing: stakeholder insights on provider country legislation. Front Plant Sci 10:1175. https://doi.org/10.3389/fpls.2019.01175. Available at https://doi.org/10.3389/fpls.2019.01175/full

The ABS Capacity development initiatives, Access and Benefit Sharing of Rooibois, associated traditional knowledge. Available at https://rooibos.abs-biotrade.info/#/genetic-resource

Chapter 4
Statutory and Obligatory Responsibilities of State Biodiversity Boards for the Conservation of Indigenous Biodiversity and ABS

K. P. Laladhas, J. R. Rani, A. S. Vijaya Sree, Achuthsankar S. Nair, and Oommen V. Oommen

Abstract CBD implemented an international legal instrument for "the conservation of biological diversity" by providing the nation's sovereign right over its Biodiversity. The sovereign right of biodiversity indebted a country to fully assert its exclusive right over the biological diversity found within its geographical boundaries. It addresses different departments of conservation of its biological diversity comprising genetic resources, species diversity, and ecosystems. India is a signatory and under the Convention, Governments undertake to conserve and sustainably use biodiversity. Global biodiversity loss is unprecedented and alarming which was addressed in the CBD 1992 and the key drivers involved in the crisis progressively shooting. The unsustainable natural resource harvest, habitat loss, climate change, population explosion, invasion of alien species and environmental pollution etc. causes degradation of its precious biodiversity. India's National legislation to oblige the CBD came into force in 2002 as Biological Diversity Act. A National, State and regional management mechanism was established as per the Act and gave an array of conservation protocols. Kerala State Biodiversity Board is the federal mechanism to conserve and manage the biodiversity of Kerala. Kerala Board is well established and showed good model conservation and regulatory mechanisms in the State as per the Kerala State Biodiversity Rule 2008. Kerala Board is the pioneer in India and displayed marvellously well in the constitution of Biodiversity Management committees, Making of Peoples Biodiversity Registers and also proposed model conservation Programmes. The detailed model regulatory, managerial and conservation programmes are discussed well in this manuscript.

K. P. Laladhas (✉) · J. R. Rani · A. S. Nair · O. V. Oommen
Department of Computational Biology and Bioinformatics, University of Kerala, Kariyavattom, Thiruvananthapuram 695581, Kerala, India
e-mail: laladhas@gmail.com

K. P. Laladhas
CSEIBA: Centre for Sustainable Environment-India Biodiversity Action, Thiruvananthapuram, Kerala, India

A. S. Vijaya Sree
Department of Zoology, Fathima Matha National College, Kollam, Kerala, India

O. V. Oommen et al. (eds.), *Biodiversity Conservation Through Access and Benefit Sharing (ABS)*, https://doi.org/10.1007/978-3-031-16186-5_4

Keywords Kerala State biodiversity board · National biodiversity authority · KSBB · BMC · PBR · NBA · NBC · Biodiversity conservation

4.1 Biodiversity Conservation—Global Enactment

National Biodiversity Authority and State Biodiversity Boards are established based on the recitals to the Indian Biological Diversity Act 2002; which is due to the obligatory responsibility, where India is a party to the United Nations Convention on Biological Diversity (CBD) in 1992. The convention that came in to force on the29th December 1993 declared that the sovereign rights of the State over their biological resources and the responsibility of conserving such biological resources, sustainable utilization of its components, fair and equitable sharing of benefits arising out of use of genetic resources are also vested with the State.

4.2 Kerala State Biodiversity Board

Kerala State Biodiversity Board (KSBB) is a statutory autonomous body falls under the provision of the Indian Biological Diversity Act 2002, Rules 2004 and Kerala State Biological Diversity Rules 2008 governed under the Environment Department of Government of Kerala. KSBB is envisaged for the conservation of Kerala's biodiversity, sustainable utilisation of its bio resources and fair & equitable sharing of benefits. A decentralised management of Biodiversity throughout the Nation was the cardinal aim of Biological Diversity Act 2002 and which was promulgated to meet the objectives of United Nations Convention on Biological Diversity (CBD). Kerala has established KSBB in 2005 as per the provisions of BD Act 2002 (GO (Ms) No. 1/2005/STED dated 28 February 2005). State Rule during 2008 was made based on the powers conferred by section 63 Of BD Act 2002 vide GO (P) No 1/2008/Envt. Dated, 10th June 2008.

4.2.1 Statutory Role of KSBB in Regulating and Conserving Kerala's Biodiversity

The section 12 of Kerala State Biodiversity Rule gives directions regarding the functions of the Board. The prominent responsibilities of KSBB are to advise the State Government, subject to different guidelines issued by the Central Government on the issues connecting to the conservation of biodiversity, sustainable utilization of its components and access and benefit sharing of biological and genetic resources from their territory. Give technical assistance and expertise to various departments of Government for implementing biodiversity governance, update and implement

Biodiversity Strategy and Action Plan in the State, commission research on our biodiversity, Preservation and strategic conservation of every genetic resource of the State. Collect, inventories, monitor, compile, analyse and document the data connected with biodiversity of the State. Establish a database on the biodiversity resources of the State, natural habitats, threatened taxa, traditional knowledge associated with bio-resources and develop strategy for conservation of such resources. Ensure protection of IPR over bio-resources and TK associated with this, ensure biodiversity dependent livelihood. Biodiversity governance, regulatory roles, mass media campaign, propose advanced research and introduce infrastructural inputs for biodiversity conservation, sustainable utilisation and access and benefit sharing are other responsibilities. Coordination of activities of Biodiversity Management Committees and give directions to the effective implementation of Act and its responsible conservation activities are the obligations of KSBB. Further report the various developments, technical information's, research outcome and all relevant developments to the Government and National Biodiversity authority. Decentralised governance and its model in Kerala has been reported by Mithrammbika et al. (2017).

4.2.2 Kerala is the First State to Constitute BMCs in All LSGs as Per BD Act (2002) in India

In pursuance of the Order of Hon'ble NGT, Principal Bench, New Delhi, NBA directed to constitute BMCs in all local bodies of the State including the three tiers of Panchayati Raj system. Accordingly, KSBB initiated constitution of BMCs in the remaining upper two tiers of Panchayati Raj systems in the State viz., Block Panchayats and District Panchayats. Principal Secretary of LSGD also gave special direction to all the 152 Block Panchayats and 14 District Panchayats to constitute BMCs in their jurisdiction by the end of December 2019. For expediting the process of BMC constitution in Block and District Panchayats and also for providing necessary guidelines in this regard, KSBB held video conferencing with all the 152 Block Panchayat Secretaries on 09–12-2019 in association with Rural Development Commissionerate. Similarly KSBB directly held Video Conference with all the 14 District Panchayat Secretaries on 18.12.2019. Thus, by constituting BMCs in the Block and District Panchayats, BMC constitution has been completed in all the 1200 local bodies of Kerala viz., 941 Grama Panchayats, 152 Block Panchayats, 14 District Panchayats, 87 Municipalities and 6 Corporations.[1]

[1] KSBB Annual Report: AR. 2019–2020 Page 21.

4.2.3 Constitution of Biodiversity Management Committees (BMCs) in Block and District Panchayats of Kerala

Completion of PBR is the long pending obligation of Board to document the biodiversity of all local self-governments of Kerala by the involvement of BMC, technical support of KSBB and Technical Support Group. In Kerala a three tire LSG governance system is implemented covering the whole State having 941 Grama Panchayats, 87 Municipalities and 6 Corporations.[2]

4.2.4 Asramam Mangrove and Wetland System (Kollam) as First Biodiversity Heritage Site of Kerala

Asramam Biodiversity Heritage site is located at Kollam East village of Kollam Corporation and demarcated in survey Block No 67, 160, 161 & 162 (Ashtamudi Lake) having geographical coordinates 76° 34′46″-76° 35′50″ E Longitude and 8° 53′ 27″-8°53′58″ N Latitude was declared open by Hon. Chief Minister under the notification, G.O. (P) No. 1/2019/Envt. On 3rd June 2019.[3]

4.2.5 Biodiversity Clubs

KSBB established Biodiversity clubs in educational institutions of the State including colleges and higher secondary schools to generate environment stewardship and conservation heritage to young generation for edifying and to involve them in the protection and conservation of rich biodiversity of Kerala. Biodiversity Clubs have been constituted in Kerala educational institutions for the said implementation of conservation programmes at local level. Large numbers of conservation programmes have been taken up by the biodiversity clubs at the campus level and the important one is the establishment of Shanthi Sthall, a conservation spot for RET species and wild animals to live safe. The major warriors involved in the propagation of conservation messages, organising seminars for the prominent environment and biodiversity important days, local level environmental assessments, awareness for pollution free campuses, programmes to regulate plastic menace at campus, biodiversity assessment in the campus, water conservation programmes and clean campus programmes. More greening programmes at campus and surrounding areas are also initiated by biodiversity clubs.[4]

---

[2] KSBB Annual Report: AR. 2019–2020 Page 20.

[3] KSBB Annual Report: AR. 2019–2020 Page 27.

[4] KSBB Annual Report: AR. 2016–2017 Page 40.

150 Biodiversity clubs were newly registered during the annual year 2019–20. Totally there are 1715 registered biodiversity clubs constituted in various institutions of Kerala. Biodiversity club is an arm very effectively extended throughout Kerala by KSBB for initiating strategies for biodiversity awareness and conservation programmes.[5]

4.3 Biodiversity Conservation Programmes of KSBB

4.3.1 Aquatic and Marine Biodiversity Conservation

4.3.1.1 Marine Biodiversity Register

Kerala State Biodiversity Board started a unique initiative to document the traditional knowledge of fishermen community together with their fishing, livelihood experience, and knowledge associated with sea. The proposed project was named as "Marine Biodiversity Register", depicting the vast diversity of fishes they catch, edible and non-edible, predatory and poisonous fishes, its coast line as per fisherman experiences, details of terrains in fishing sea bed, seasonal variation of fishes they catch along with the concerns of traditional fishery. It was a project carried out in association with NGOs, traditional fisher folks, common people, by gathering informations and various conservation measures useful for their livelihood along with the concerns of the traditional hook and line fishers in the coast.

Marine Biodiversity Register has documented the under-water ecology, terrain, diversity of fishes and other lives in the inshore areas of the fishing community. The sea bed and terrain enlisting the important reefs, rocky beds, sandy dunes and muddy flaps are also documented. The availability of fishes and seasonal occurrence, demersal fishes, bottom configuration of the sea bed, ocean current, wind pattern, seasonal sea wave occurrence and status of fish resource are also displayed in details.[6]

4.3.1.2 Turtle Conservation at, Neeleswaram—Thaikadappuram

Ecologically significant Olive Ridley Turtles frequently visit Neeleswaram—Thaikadappuram beach of Kasargod District, Kerala during rainy season for egg laying. IUCN classified the Olive Ridley turtle (*Lepidochelysolivacea*) as vulnerable as per IUCN Red Data Book, and is listed in Appendix I of CITES. A NGO from the area along with the local conservation-oriented volunteers proposed various steps to protect the turtle nesting area from predators. This was a participatory programme

[5] KSBB Annual Report: AR. 2019–2020 Page 68.

[6] KSBB Annual Report: AR. 2012–2013, Page 14.

(See Footnote 6). Laladhas et al. (2017a, b) has reported a community conservation model in Kerala.

4.3.2 *Conservation of Traditional Breeds and Varieties*

4.3.2.1 Tuber Conservancy in Wayanad

Tropical climate and its blessings are paramount to Kerala with respect to tuber diversity and is documented well by the Board. This diversity was preciously conserved for the generations and propagation to other regions are envisaged in this project. Wayanad is blessed with great diversity of 20 Aroids, 21 species of wild Dioscorea and 19 traditional Dioscoria cultivated varieties, 20 types of yams, 16 varieties of taro, 6 types of sweet potatoes, wide varieties of cassavas, arrowroots, yam beans, many other edible tuber varieties and more than one hundred such varieties are documented. The main threat of this biodiversity is that majority of such tubers are not used by local people and the rate of extinction from the cultivating field is more imminent due to plantation culture and change of food pattern.

A massive awareness campaign has been proposed by the Board to stabilise this diversity in the region and conserve it for tomorrow's use. The merit of such diversities was documented using electronic and print formats to spread the message with school students, public, stakeholders and farmers. Also, tried to spread the diversity to LSGs having agricultural and farming possibilities.[7]

4.3.3 *Conservation of Agro Biodiversity*

4.3.3.1 Propagative Conservation of Tuber Diversity at Onattukara Farming Area

This project was mainly for the propagative conservation of genetic diversities of neglected and underutilised root and tuber crops germplasm and popularise among farmers and public for decentralised conservation. Diversity of food crops including tubers, major and minor crops are very important for conservation especially at the time of scarcity or famine.

The Onattukara tuber conservation and dissemination project was a propagation project of rare indigenous diversities of tubers conserved in a decentralised farming field. An appropriate and active step is necessary to keep such germplasms which are edible and highly nutritious but are now underutilized due to availability of other sources of easy diet and lack of popularity.

[7] KSBB Annual Report: AR. 2013–2014 Page 14.

Planting materials of tubers of ten varieties of such crops were procured from the traditional farmers of Wayanad district, was transported to the BMC at Onattukara farming area for distribution to the conserving farmers. 127 Farmers were selected based on their conservation capability from five LSGs namely, Chennithala, Thazhakara, Mannar, Chettikulangara, Thekkekara Grama Panchayat and Mavelikara Municipality, Onattukara for this project. Supportive measures were given to the farmers for the successful conservation at fields and also entrusted the farmers to propagate the diversity in the other parts of the region in a participatory mode, as a chain of action for decentralised participatory conservation.[8]

4.3.3.2 Conservation of Traditional Cattle Breeds Such as Cheruvally and Vechur

Vechur and Cheruvally are two of the traditional cattle varieties of Kerala, long been conserved by farmers. Vechur cattle is considered to be the smallest cattle in the world. Since its numbers fall drastically due to lack of conservation, non interest of traditional farmers who are after hybrid breeds and lack of facilities for scientific propagation of its germ plasm, a project was proposed by the Board to generate semen bank from pure bulls of Vechur and cheruvally cattle bulls, which were scientifically reared and its semen was collected as per veterinary guidance and preserved and stored for long term use for the interested cattle breeders. Semen is delivered to farmers who are rearing vechur and cheruvally cattles. A systematic registration system has been formulated for this by the expertise of Dr. Sosamma Iype, formerly Deputy Director of Veterinary Department. Farmers receive the stored semen on need. This project is aimed at the genetic improvement and germplasm purity of the Vechur and Cheruvally cattle breeds of Kerala.[9]

4.3.3.3 Conservation of Native Inland Fishes

Native inland fishes in Pampa River have drastically reduced in numbers and many of them are likely to face extinction in the near future. Most fish species use flood plain (Mithrambika et al. 2017a, b) canal system as corridors to spawning grounds during high water flow and juvenile fishes utilise nutrient rich flood plains and their associated complex habits as nursery grounds. Excessive fishing takes place during migration of fishes from the river and low land to the upland paddy fields for pairing and egg laying. KSBB project aims to conserve the migrating fishes during monsoon flood plain fishery thereby enhancing fish population in Pampa River, its tributaries and allied water bodies. The project is implemented in Pandanad and Mannar Grama

[8] KSBB Annual Report: AR. 2014–2015 Page 17.

[9] KSBB Annual Report: AR. 2013–14 Page 15.

panchayats of Alappuzha district and Koippuram Grama Panchayat of Pathanamthitta district.[10]

4.3.3.4 Conservation of Indigenous Fish Diversity of Vellayani Lake

Kerala State Biodiversity Board has initiated Conservation of Indigenous fresh water fish species to ensure food security as well as indigenous fish diversity in the water bodies of the state. Vellayani, the largest fresh water lake of the capital city which revealed decline in the total biomass of indigenous fishes in the recent surveys due to heavy fishing pressures on the lake. So as to enhance the biomass as well as the fish diversity of the lake, seedlings of native varieties of prawn (Aattukonchu-*Macrobrachiumrosenbergii*), 3500 nos. and Pear spot (Karimeen-*Etroplussuratensis*) 20,000 nos. were released in to the lake. A model native fish conservation and livelihood enhancement by Sujith et al. (2017) has been reported.[11]

4.3.3.5 Afforestation Programmes

A tree planting campaign was organised by KSBB in Karunapuram Panchayat of Idukki district during August 2012. 20,000 tree saplings were planted by NSS volunteers of Mahatma Gandhi University. The participation of local BMC and panchayat administration under the guidance of Kerala State Biodiversity Board were involved.

4.3.4 Ecosystem Conservation

4.3.4.1 Action Plan for Conservation of Biodiversity of Pampa River jBMC

Conservation of natural ecosystem is a priority objective of KSBB. Kerala is blessed with diversity of ecosystems all along between its mountain and shore line terrain. There are critical ecosystems which are under pressure from anthropological pressure, pollution and state of rapid destruction. One among them is Pampa River, for this an action plan was proposed for the Conservation of Pampa River at LSG level. A detailed programme with 8 modules encompassing Base line data collection which includes the collection of information on Flora, Fauna, Water quality, Habitat improvement, Conservation of biodiversity, Environmental protection, Capacity building along with a special module focussed on Equitable sharing of benefits derived from use of biological resources was envisaged. The rate of degradation and

[10] KSBB Annual Report: AR. 2013–14 Page 13.

[11] KSBB Annual Report: AR. 2014–15 Page 16.

threats identified from the data collection and the degradation rate was calculated to generate multitask, conservation programmes for the river Pampa.[12]

4.3.4.2 Cabomba: An Aquatic Invasive Alien Species Management in Kerala

KSBB has been giving important thrust to conserve Kerala's biodiversity as its obligatory responsibility. Invasive alien menace is a major threat even to Kerala also. An ornamental plant turned to a destructive weed in Kerala's wetland area was managed by Board. Cabomba (*Limnocharis, Cyperus)* grows over the Pampa River tributaries and periphery fields. Weed is densely packed and grown thickly in the water bodies preventing the safe life of endemic and traditional aquatic diversity. Also obstruct livelihood possibilities of the traditional stakeholders and poses threats to many unique ecosystems and threatened taxa. A scientific and systematic study was proposed to understand the grave threat of the invasive menace in the ecosystem with help of experts and NGOs for a concrete management of the spread of the invasive weed. Awareness was given to the BMCs of LSGs having such imminent invasive threat.[13]

4.3.4.3 Korapuzha Conservation of Mangroves

Korapuzha river bank in Atholi Grama Panchayat of Kozhikode district was an area enriched with mangrove diversities. Due to anthropogenic invasion, the mangrove diversity was completely damaged and subsequently the fish catch also reduced a lot. Fisher Stakeholders have very limited yield and cry for their survival crisis. Board supported a project to enrich the mangroves diversity to plant more seedlings and also gave financial support to fish culture using penstock culture to the LSG by the support of scientific experts. This was a participatory conservation programme by the support of LSG, NGO, environmental experts, volunteers, stakeholders etc.[14]

4.3.4.4 Biodiversity Enhancement at Urban Area

Natural ecosystems in urban area are under severe threat due to anthropogenic activities. Majority of the natural ecosystems like ponds, rivers, other water bodies and free land areas are polluted, over exhausted, destroyed by various destructive activities. In Thiruvananthapuram corporation area we have several ponds with good water recharge and degraded and polluted by various activities. Board has initiated an urban

---

[12] KSBB Annual Report: AR. 2012–13 Page 15.

[13] KSBB Annual Report: AR. 2013–2014 Page 17.

[14] KSBB Annual Report: AR. 2012–2013 Page 17.

biodiversity enhancement programme to rejuvenate the damaged ponds by participatory method. The main objective of the programme was to conserve the ecosystem and enhance its biodiversity and quality of environment. Natural fencing facility promoted to conserve the habitat of natural and traditional fish fauna, enhance the riparian biodiversity, stop the polluting sources to the water body and also add more fish fauna and ornamental plant diversity to enhance its natural aesthetic look. This programme was implemented as participatory by involving residence associations, NGOs, volunteers and public (See footnote 14).

4.3.4.5 Joint Biodiversity Management Committee for Sasthamkotta Lake

KSBB constituted a joint biodiversity management committee (jBMC) to coordinate the activities of the BMCs in ecoregions or larger natural ecosystems covering many LSGs and BMCs and to give direction for effective implementation of all measures relating to conservation, sustainable use and equitable benefit sharing. Proposed conservation programmes for multi LSG linked natural ecosystems need a coordinated and systematically organised protocol to manage large ecosystems/ecoregions which are beyond the jurisdiction of one Local Self Government.

Sasthamkotta Lake, the State's largest freshwater lake and one among the 42 Ramsar sites in the country is undergoing severe degradation and biodiversity loss. As an internationally recognized Ramsar site, it is our responsibility to conserve the lake in its original state of existence. But it is now well understood that the indiscriminate usage of lake resources, various land constraints and unscientific waste management system have imposed serious environmental and socio-economic problems in the area. Hence, Kerala State Biodiversity Board has come forward with various measures to protect the lake through the constitution of jBMC.

BMCs have been constitutedat LSGs of Kerala for the management of biodiversity and ecosystem. But the Territory of Sasthamkotta Lake extends beyond a LSG revenue land area and abounds three LSGs. The section 12 (xxv and xxvi) of Kerala State Biodiversity Rule 2007, empowers Kerala State Biodiversity Board (KSBB) to coordinate the activities of the BMCs and give direction for effective implementation of the Act to facilitate their meaningful participation in all measures relating to conservation, sustainable use and equitable benefit sharing.

A jBMC consisting of BMC members of the three Grama panchayats of Sasthamkotta, Mynagapally and West Kallada was constituted with Sasthamkotta block panchayat president as the Chairman. The chairpersons of the three BMCs were added as Joint convenors and the members of the BMCs included as its executive members in the jBMC. A comprehensive management action plan was proposed to protect the watershed connectivity and pollution mitigation measures with conservation strategies for the floral/faunal biodiversity of the lake. Biodiversity documentation has been reported by John et al. (2017).

As part of the conservation programme, the KSBB made a Biodiversity register with detailed information on the flora and fauna of the wetland system with the participation of people, which was released in 2016.[15]

4.3.4.6 Conservation of Traditional Rice Varieties of Wayanad

Kerala is home to hundreds of diverse varieties of traditional rice which are cultivated since the beginning of existence of farming. Grow more food for eradicating poverty in India forced the Nation to propose policies like green revolution gave R & D involvement to develop hybrid seed, ploidy, fertilisers, pesticides, construction of dams etc., which pave the way for loss of traditional diversities and natural ecosystems. Decentralised field level conservation of such traditional rice varieties carried out by very few farmers in Wayanad was identified for this project. These rice varieties are highly beneficial with added resistance to disease and pests, rich in aroma, drought and flood resistance, medicinal value and also cost effective in cultivation especially organic mode of farming. A tribal farmer namely Cheruvayal Raman, was given financial support by the Board for conserving 41 varieties of traditional rice varieties in his field. More attractive traditional breeds such as Onamottan, Chennellu, Chomaala, Adukkan, Velumbala, Njavara etc. with potent tradable and medicinal properties are conserved in his farming area continuously.[16]

4.3.5 Marine Biodiversity

4.3.5.1 PAR-People's Artificial Reef Through Community Participation

Artificial reef making for the traditional fisher folks has emerged from Marine Biodiversity Register where the traditional knowledge that naturally established reef areas have rich swarming with fishes. PAR project was funded by Kerala State Biodiversity Board to see the progress of such knowledge in an artificial reef. Fisher man have the traditional way of methods to locate reefs or spots using visual triangulation method, an age-old traditional method. Two artificial reefs were constructed with the use of wooded planks and abandoned wooden boats by the financial support of Board and deposited at the inshore region of Kannanthura and Valiathura at 3,5 km distance and 20–30 fathoms depth from the coastal area of Thiruvanathapuram. The position of the PAR deposition as per the traditional knowledge of fisher folks. Artificial reefs will act as a natural ecosystem for providing space and shelter to marine diversities. A remarkable raise of fish catch was recorded in the artificial reef and fisher folks

[15] KSBB Annual Report: AR. 2015–2016 Page 7.

[16] KSBB Annual Report: AR. 2013–2014 Page 14.

have added income to their livelihood process. [17] Successful livelihood model has been reported by Laladhas et al. (2020).

4.3.5.2 Decentralised Conservation of Banana Varieties and Ecotypes Through Public Participation

Reintroduction of banana farm varieties by the local farmers was a measure of decentralised passive conservation with participatory concept. Hundreds of banana varieties are available in our country but farmers are only cultivating very limited varieties for commercial input. But a large group of varieties of banana have multiple strength in food quality, varied sugar strength, medicinal properties, aromatic and spiritual values. The main objective of this project is to conserve the diversity at local field level in a participatory and passive manner continuously. Also introduce the diverse variety with its significance to commercialisation and popularise the local varieties to the domestic consumers in Kerala. This will alternatively conserve the precious banana varieties in the local farmer's fields and homesteads, so as to stop their variety loss and extinction, a better message for conservation of biodiversity and be part of greater responsibility to species loss. Banana seedlings of 18 varieties conserved at Agricultural Banana Nursery at Peringamala were procured and distributed to 444 farmers of Local Self Government area, Kalliyoor. Adukkan, Krishnavazha, Matti, Perumpadaly, Karpooravally were the rare varieties distributed for conservation.[18]

4.3.5.3 Seed Bank for Conservation of Wild Crop Diversity

Biodiversity register is a repository of information regarding biodiversity of Kerala and documented with the local information in participatory modus. Huge data of information are available with respect to various types of information's about seeds of local land races, cultivars of agricultural crops, major and minor fruit crops, wild allies of food crops, stock crops, wild varieties capable enough to be used for food and leafy vegetables and medicinal plant stocks and seeds. Conserving all the varieties in field will be a gigantic exercise attracting maximum quantum of resources and may not be practically viable.

Biodiversity Board in association with expertise from the Department of Botany of University of Kerala established a state of the art seed bank for long term and medium-term storage of the diverse seeds for conservation which attracted varieties of our biodiversity. This seed germplasm is to be facilitated for research with respect to genetic diversity, conservation of wild relatives of essential and priority crops, RET plants, conservation of land races within species, wild crop relatives and farmers breeds are identified and stored. Farmers varieties form landraces of staple crops, such as cereals, domestic and local vegetables, indigenous fruit trees, are stored and

---

[17] KSBB Annual Report: AR. 2013–2014 Page 16.

[18] KSBB Annual Report: AR. 2014–2015 Page 18.

supported for propagation. This facility is a repository for wild flowering plants of State, repository for conserving plant diversity and wild crops relatives from Western Ghats, traditional, local and farm varieties. This is a translational project out of the intention of statutory responsibility of KSBB to conserve its vast and precious diversity.[19]

4.3.6 Eco Restoration

4.3.6.1 Restoration of Degraded and Abandoned Natural Ecosystems

As part of urbanisation, a huge area of rocky terrain has been quarried for rock needed for its construction purposes. An abandoned quarry was adopted having 4 Hectors of area at Ernakulam district in Kakkanad area near Palachuvadu called Ampalapara quarry. The Board to rejuvenated its ecosystem and cleaned the wetland area and converted it into a water body for drinking water purposes. A preliminary investigation was conducted at the water body to see the efficacy of water quality, availability, and quantity. A riparian diversity enhancement, addition of ornamental plants diversities and cleaning and modifying the according to the scenic and aesthetic beauty of the terrain were, to attract the public. Great improvement has been noticed after a duration of two years that more bird and wild native animals have been visiting the wetland and its worth conserving the area by a long-term protocol. This project was funded by Thrikkakara municipality and carried out by the monitoring of BMC, environmental experts and volunteers.[20]

4.3.6.2 Eco Restoration and Water Harvesting in Abandoned Quarries

The abandoned quarry at Ambalapara, Palachuvadu in Kakkand is a perennial source of water and is connected with Edapallythodu but the area is highly degraded. A preliminary study was conducted involving water quality, geographical details, quarry dimensions and details, and water availability. The project on Eco restoration of abandoned quarries envisages developing nature trail by planting recommended native species of grasses, shrubs, trees and riparian vegetation. The large water body attracts many birds, butterflies and serve as an important habitat for a range of amphibians, and other invertebrates. Hence development of this area is expected to enrich the biodiversity of this area and generate public awareness. The project is also beneficial to local communities that often face water scarcity. Water stored in the quarry can be utilized for household purposes and fish cultivation. The project will also help

[19] KSBB Annual Report: AR. 2014–15 Page 18.

[20] KSBB Annual Report: AR. 2012–2013 Page 17.

in recharging the underlying aquifer, raising the water table which will increase the vegetation in the surrounding area.[21]

4.3.6.3 Ecorestoration of Aduppukoottanpara Quarry

This was a project for rejuvenation of all abandoned, damaged and destructed ecosystem to be replenished to natural water body for the aesthetic significance, water conservation and aquatic biodiversity enhancement. Aduppukoottanpara quarry of Peroorkada in Thiruvananthapuram city, an abandoned and degraded natural ecosystem was a dumping ground of every sort of pollution and was a breeding ground for mosquitoes and factory of communicable diseases for several years.

The water logging area of the quarry is about 4 acres and depth is uneven and cleaning the pond area was a challenge. With the support of local volunteers, biodiversity Board took the initiative to clean and restore its natural health and beauty. A holistic terrain conservation, cleaning, pumping out polluted water, disposal of collected plastic waste to recycling centre, cleaning the road connecting to the nearby junction, levelling and clearing the entrance for visitors to view. This project had ample scope for recreational and tourism linked conservation activities in the degraded and damaged ecosystem. As a mark of understanding the tourism potential, a pilot boat was introduced to the water body as a recreational element. A potential addition to establish butterfly garden, fish culture, and ornamental garden for aesthetic and recreational value was proposed for its Eco restoration.[22,23,24]

4.3.6.4 Shanthi Sthal

Shanthi Sthal is a unique initiative of KSBB for conservation of wild ecosystem in urban and rural areas especially at the private land holds or conservation of wild ecosystem by common man with the participation and involvement of private lands. A common view that conservation of natural ecosystem is the duty of the Government or authorities not in private lands. A small financial support is being given to a private land holds for setting up of Shanthi Sthal for the conservation of natural ecosystem, where we may grow traditional, endemic, rare, threatened and endangered plant diversities for three continuous years. After that the area may be maintained as natural ecosystem without any further human intervention to attract natural wild animals to live without any disturbance, especially natural wild reptiles, birds, small mammals etc. This project was implemented with the support of Biodiversity clubs, Nature clubs, National Service Schemes etc.of educational institutions for its continuous maintenance. KSBB has established very important translation in Shanthi Sthal's

[21] KSBB Annual Report: AR. 2012–2013 Page 17.

[22] KSBB Annual Report: AR. 2014–15 Page 19.

[23] KSBB Annual Report: AR. 2015–16 Page 12.

[24] KSBB Annual Report: AR. 2016–17 Page 40.

project implementation in educational institutions and LSGs in Kerala covering **18. 6 acres of land for conserving such precious Rare, Endangered and threatened taxa by the participation of private entities**. Conservation of ecosystem through payment has been reported by Laladhas et al. (2017a, b) (KSBB-AR. 2015–16, pg. 12 & KSBB-AR. 2016–17, pg. 40).

4.3.6.5 Greening Palakkad Gap of Western Ghat

Kerala State Biodiversity Board implemented community afforestation programme in the rain shadow regions of Palakkad to improve the tree cover of the area. Palakkad Gap is a geological formation having major break in the mountain region of Western Ghat in South-western region of India. The project aims at the enhancement of green cover in the region and also stops the loss of biodiversity in the area. This was a participatory programme by incorporating BMCs of 17 LSGs, Biodiversity club members of Palakkad, Social Forestry division of Kerala Forest Department, farmers, volunteers and environmental experts. The projects was expected to deliver an enhanced green cover, improved hydrological parameters, sustainable water management, restricting soil erosion, restoration of habitats and enhance endemic and traditional diversities. A Hand Book on Greening Palakkad Gap was prepared by the Board and was circulated to the project members and stakeholders to explain the usefulness and procedure of implementation.[25]

4.3.7 Nature/Biodiversity Education

4.3.7.1 Kerala Biodiversity Museum

KSBB has set up a state of the art museum at Vallakadavu, in the coastal region of Thiruvananthapuram district of Kerala. A detailed display of biodiversity and ecosystems of India with special emphasis to Kerala are depicted with the aid of modern information and digital technology. The museum is housed at a heritage Vallakadavu Boat house, erstwhile the commercial boat jetty of the capital city used by traders and merchants who grabbed the traditional spices and agricultural crops from the State. The detailed component in the museum includes, the pictorial and interactive displays for awareness of Kerala's Biodiversity, endemic species, ecosystem diversity, conservation of Kerala's threatened biodiversity, and Biodiversity governance. Sustainable utilization of bio-resources is the cardinal concept and access to bio-resources and benefit sharing from its compounds from Kerala as per Nagoya Protocol are also keenly displayed. The attractive display of a variety of dioramas and models of species, ecosystems, pre historic animals, ecological significance of species, threatened taxa and its conservation, interactive kiosk, LCD displays are

---

[25] KSBB Annual Report: AR. 2014–15 Page 20.

the major themes of the museum. The dual storied museum was sorted with two sections, the first section was designed for the displays of biodiversity status of India and Kerala with different sorts of panels having interactive modes, backlit info panels, video stories of live species diversity and ecosystem significances of Kerala's great environment. The other section focuses on conservation of Biodiversity of Kerala and sustainable utilisation of our precious resources, agro biodiversity with interactive panels, video films and backlit panels. Conservation strategies and methods at global level and National level are well displayed. A three-dimensional theatre with 50 seating capacity is also erected in the museum for attracting students in live mode and involves show of rare and more scientifically oriented films. Science on sphere is another attractive feature of the museum by the help of computer and video projectors to project planetary data on to a sphere analogous to globe. Environmental and climatic changes affecting globally can be understood by this module by established programmes for the past data and designed models. More added features are like water fountain, aquarium, vertical garden, murals and facilitation centres. Boat house is constructed in wooden structure with an area of 5000sqft showing the charm of nineteenth century construction and Kerala heritage.[26]

4.3.7.2 Children's Ecological Congress

Kerala State Biodiversity Board gives greater importance to edify children's global concept of Biodiversity conservation, sustainable utilisation of its resources, access and benefit sharing of our precious bio resources for better management and survival for future generation. Children's Ecological Congress is an action oriented, fun involved, participatory workshop for School students organised every year, where innovative programmes, ideas and conservation initiatives for nature conservation are taken up. Loss of biodiversity is the irreparable loss of benefits to the future generation and its harm is directly affecting the children of today and non-replenishable process for tomorrow. Leaders of tomorrow are empowered to function as a catalyst to the process of conservation and a life practice of sustainable mode of harvest of natural resources and sustainable development in all dimensions of life. This annual congress is envisaged to promote the importance of biodiversity in our routine life and encourage the future generation to take action to better implementation and conservation of our natural resources and heritage.

4.3.7.3 National Biodiversity Exhibition

Kerala State Biodiversity Board frequently conducts biodiversity exhibitions at various levels to give better displayed, colourful, content oriented and pictorial displays of our heretic biodiversity. Large numbers of institutional level, LSG level, region wise, state level and national level exhibitions are conducted at different

[26] KSBB Annual Report: AR. 2018–2019 Page 35–36.

locations of the State. A ten days national Biodiversity expo was organised in the capital city of Thiruvananthapuram during December 21 to 30, 2012. A large mass of public participation was attracted in the capital city to view the great biodiversity and environmental displays. The veracity of displays from Critically endangered fauna and flora, myriad biodiversity of India, biodiversity heritage sites, displays from bio-geographical zones, traditional biodiversity, endemic biodiversity, domesticated biodiversity, traditional breeds and seeds, conservation measures, case studies, success stories of conservations, regulatory frameworks of Indian biodiversity, Access and benefit sharing information, natural resource management and best practices in resource management were displayed.

For each theme the Expo presented a series of technologically enabled display panels featuring text, images and graphics, as well as audio visual tools along with informative pamphlets. The picture gallery was a marvellous celebration of pristine nature, a gateway to crystalline display of stunning diversity of flora and fauna. The major attraction of the Expo was the aesthetic display on Western Ghats put up by the Central Ministry of Environment and Forests through successful display of the National Museum of Natural History with their supporting and interactive facilities.

It showcased diverse panels, each devoted to a specific aspect of the multifarious biodiversity that cloaks the Western Ghats. The stunning photographs were depicted with informative text that stresses on issues such as environmental crisis, over utilisation of resources, environmental pollution, encroachment that threatens the ecological treasures of the Western Ghats. Indigenous and traditional food festival from different ethnic groups from Kerala has added flavour to the expo. Kerala specific Cultural festival added more attractions to the visitors of other States and in all days different stage arts were performed for the Expo visitors.

It was a joint programme organised by the National Biodiversity Authority and KSBB at Kanakakunnu Venue of Thiruvananthapuram city, which attracted varied dignitaries from Chief Minister, Ministers, Member of Parliaments, Members of Legislative Assembly, Statutory officers from National and State Biodiversity Bodies, Scientists, Policy makers, Environmentalists, Researchers, Students, NGOs and public.[27]

4.3.7.4 National Biodiversity Conference 2012 (December 27–30, 2012)

National Biodiversity conference was a national level colloquium and get together of representatives from twenty State Biodiversity Boards, deliberations from more than 260 scientists and grass root level conservationists hosted by Kerala Sate Biodiversity Board under the auspices of National Biodiversity Authority from December 27–30. A systematic and stream lined scientific assessment was carried out by the participating experts in the scientific deliberations, reiterated the inevitable need for conservation and sustainable management of our precious biological resources and knowledge associated besides ensuring effective capacity building at various levels of

[27] KSBB Annual Report: AR. 2012–13 Page 25.

implementations for conservation and management. Updated awareness of advanced global research outcome in the area of biodiversity conservation, sustainable utilisation of its bio resources and access and benefit sharing should be translated to the empowerment of stakeholders of individual as well as institutional organo grams. The conference attracted various policy makers and ministers and actively initiated the establishment of a National Biodiversity Grid to empower National Biodiversity Authority in collating, analysing and processing of Nationwide biodiversity information and data for appropriate management of our rich biodiversity. This is very essential for the regulation and policy making of Indian Biodiversity by the National Statutory regulatory body. Another important delivered outcome was that a single window biodiversity information for the country to assess the national level quantification.

The conference was a platform for State Biodiversity Boards across India to discuss and deliberate issues and blockades on the process of implementation of Indian Biological diversity Act-2002 and Rules -2004. Difficulties faced by Boards in constitution of Biodiversity Management Committees, generation of Peoples Biodiversity Register, establishing Biodiversity Heritage cites, notifying threatened taxa of individual States, establishing State Biodiversity Fund, Access and benefit sharing of bio resources and its implications were effectively deliberated in the conference.[28] (KSBB-AR-2012–13, pg. 26).

A second National Biodiversity Congress has also been organised in 2015 by the Board at Thiruvananthapuram from 23rd to 27th February. Five day long National Biodiversity Congress has also had a colourful and contented show of Biodiversity Expo at Kanakakunnu Palace Venue. The focal theme of the Congress was 'Biodiversity: Access and Benefit Sharing'.[29] (KSBB-AR-2014–15, pg. 34–6).

4.3.7.5 Kerala State Biodiversity Congress

KSBB proposed a state level annual Biodiversity congress to be conducted every year to exchange the knowledge and acquired recent trends in conservation of biodiversity across the globe to the stakeholders and public of Kerala. First Kerala State Biodiversity Congress (KSBC) was organised at Government Brennen College, Thalassery, Kannur during 26th to 28th January 2019. Best practices adopted in the State and at National level for conservation of biodiversity, priorities for conservation of Kerala's biodiversity, BMC meet, Children's Biodiversity Congress and Biodiversity exhibition. The focal theme of first KSBC was "Biodiversity for Climate Resilient Kerala". Consultative workshops were held on the major issues faced in the State regarding the conservation of biodiversity incorporating BMC's, Scientists, academicians, farmers and other stakeholders, NGO's and conservationists. Biodiversity exhibition was an attractive event and about 98 institutions participated in the displaying of the vast diverse biodiversity progress and developments from the State. Traditional seeds

[28] KSBB Annual Report: AR. 2012–13 Page 26.

[29] KSBB Annual Report: AR. 2012–2013 Page 34–36.

from agro biodiversity and minor fruits were exchanged during the exhibition to the public. Dignitaries attended the congress includes Hon. Chief Minister, Ministers and LSG heads and implementing officers.[30]

4.3.8 *Sustainable Resource Utilization*

4.3.8.1 Conservation and Stainable Harvest of Medicinal Plant Resources of Wayanad at LSG Level

Sustainable utilisation of medicinal plant resources especially harvested from natural/wild ecosystem is an unregulated and unscientific process which destroys such resources irrecoverably. Biodiversity Board proposed a project to National Medicinal Plant Board for financial support to study the sustainable collection of medicinal plants and aromatic plant resources form Wayanad District for fair benefit sharing to the user groups. A modus operandi for sustainable harvest of aromatic and medicinal plant species from the forest should become an important and strategic initiative for long term harvest and survivability of medicinal plant resources. Guidelines have been given for sustainable collection practices especially plants which are used as whole or destructive collection. More such resources will be conserved in additional areas for better stock at wild. This project was implemented in the Wayanad district by the involvement of LSGs and Biodiversity Management committees.

A project funded by National Medicinal Plant Board entitled "Resource augmentation, sustainable harvesting and value addition of Medicinal Plant resources through BMCs at five LSGs namely Thanvinjal, Ambalavayal, Moopainad, Poothadi and Noolpuzha of Wayanad District, a sustainable harvesting of medicinal plant resources by the tribal stakeholders for their livelihood. Infra structure for drying ground and storing yard were added to this LSGs for a long-term utility and sustenance. Technical knowhow has been provided for identification of species, sustainable harvesting, post harvesting and processing, storage and sustainable trade has been given to the stakeholders.[31]

4.3.9 *Biodiversity Research Centre*

Kerala State Biodiversity Board has a research and development wing for giving added scientific inputs for the implementation of conservation strategy in the State. KSBB is an approved research centre affiliated to the University of Kerala. Systematic academic research can be carried out at this centre for providing biodiversity specific curricular implementation or research. Biodiversity data available in Peoples

[30] KSBB Annual Report: AR. 2018–2019 Page 36.

[31] KSBB Annual Report: AR. 2015–2016 Page 8.

Biodiversity Register is a mega data to be processed, collated, analysed, formulated or disseminated for decentralised LSG level conservation or for centralised State policy formulation at various statutory obligations carried out by this facility as a translation centre. Also, Board have the statutory obligation to advice the Government in Biodiversity conservation, sustainable utilisation and access and benefit sharing issues in the State.[32]

4.3.10 Advice the Government on Policy Issues

4.3.10.1 Conservation of Western Ghats

Ministry of Environment, Forest and Climate Change has proposed a draft regulatory notification for conserving the ailing Biodiversity and the degrading ecosystem of Western Ghat, which was under severe threats due to over exploitation of bio resources, urbanisation, anthropogenic intrusion. Western Ghats (Sahyadri), a 1600 km long mountain range, along six States Viz. Gujarat, Maharashtra, Goa, Karnataka, Kerala and Tamil Nadu. This is a world heritage site declared by UNESCO in 2012. Western Ghats is more critically important for monsoon and to give drinking water to 25 crores of people. It is estimated that natural ecosystem of Western Ghats neutralises around 4 million tonnes of carbon per year. This is the hot spot of biodiversity, contains great endemism and more than 325 globally threatened species.

Government constituted a committee and KSBB chairman was it convener to study the conservation issues and the apprehensions of people of Western Ghats Kerala. Board opened up discussions with Scientists, policy makers, implementing agencies, BMC's, LSG's, farmers, NGO's and other stakeholders for establishing modalities for policy dialogs with different players and stakeholders, by providing strategic environmental policy perspectives aimed at improved natural and biodiversity resource management and governance.

Ministry of Environment of Environment, Forest and Climate Change had constituted a High-Level working group (HLWG) chaired by Dr K Kasthurirangan on 17th August 2012 for a thorough assessment and make recommendations on how to conserve the precious biodiversity and environmental integrity of Sahyadri range. Recommendations of Dr Kasthurirangan was accepted in principle by MoEF & CC and a draft notification was issued declaring the regions identified in the Ghats are ecologically sensitive area and proposed certain regulatory and conservation measures.

A huge unrest and cry arose from the people residing in the Western Ghats regarding the regulations proposed by the MoEF &CC. Government of Kerala proposed a three-member Committee as per order no. 162/13/Envt. Dated 29. 10. 2013. Dr Oommen V. Oommen, Chairman, KSBB as the Convener of the committee, Prof. V. N. Rajasekharan Pillai, EVP, KSCST&E and former IAS officer Dr PC Syriac

[32] KSBB Annual Report: AR. 2013–2014 Page 21.

as members, and KSBB was the office coordinating the committee and submit the report in one month.

The committee had a tour across Western Ghats area and conducted sittings and public meetings and gathered inputs from more than 30, 000 people, also had discussions with Chief Minister, Ministers, MP's, MLA's, LSG representatives, NGO's and farmers and environmental organisations. KSBB translated the relevant part of Kasthurirangan report in Malayalam for public awareness and also uploaded in the website. More than 10, 000 copies of Malayalam report were distributed to public and more than one lakh persons visited the KSBB official website. Huge mass awareness campaign and 24 h help line was started at KSBB. The whole exercise was a mass movement and with the involvement of huge public participation. KSBB arranged wide publicity about the various issues raised by the public and the conversationalists, convening of meetings at different parts of Western Ghats, compilation of issues raised during meetings and supported the preparation of the final report.

A decentralised physical verification of cadastral maps of the 123 ESA villages were conducted to assess the exact ESA boundaries. Government appointed a committee comprising of LSG president, revenue officer, regional agriculture and forest officers to assess the cadastral maps at village level. Experts from KSBB, Kerala State Remote Sensing and Environment Centre, Forest department were involved in the refinement of cadastral maps. The final report of the expert committee was submitted to the MoEf & CC through the Government for the final notification.[33]

4.3.10.2 Impact of Floods/Landslides on Biodiversity

Kerala has experienced heavy rain, flood and landslides between 1st June and 18th August 2018 across the State. The rain devastated all walks of life, properties, and damaged domestic, agriculture and livelihood assets of people of the State.

KSBB proposed a rapid participatory assessment through the statutory Biodiversity Management Committees in flood prone 187 LSG's in 13 districts of Kerala on the impact of flood on Biodiversity. A 12-member core committee with senior experts from environment and biodiversity was nominated to monitor the rapid impact assessment along with100 regional experts who are having expertise in multidisciplinary areas like geology, environment, biodiversity, agriculture etc. The assessment was a joint process involving State, District and BMC levels through modalities of training, survey and assessment. A committee of four BMC members, two experts, five volunteers and representatives from KSBB was constituted to conduct a detailed rapid assessment of impact of flood and landslides on biodiversity of the 187 Panchayats covering 13 districts of Kerala. With the support from the experts of premier Environmental and biodiversity Research institutes, University Departments and KILA (Kerala Institute for Local Administration) developed a methodology manual for the assessment. Training for the volunteers were given by KILA and KSBB.

---

[33] KSBB Annual Report: AR. 2013–2014 Page 23–24.

Decentralised planning for environmental and biodiversity loss mitigation responsibilities of the LSGs were given through district level. A meeting comprising District Panchayat President as its Chairman along with representatives from LSG's district planning officer, Panchayat deputy director, Co-ordinators of Janakiyasoothranam, Haritha-Keralam, Suchithwa mission and other stakeholders in all 14 districts were organised for the awareness and implementation mitigation measures at local.

Local level assessment was made under the chairmanship of BMC President along with the BMC members, fisherman, farmers, tribal people, local experts and other stakeholders. The local implementing officers from agriculture, fisheries, forest, animal husbandry, were included to get the data from their department in this regard. Open data kit software was used to format the collected data from the affected area. The extent of damage, gravity and its nature were recorded from the field in the designed format. The expert team held deliberations with various line and implementing departments for their inputs and data. All relevant data and secondary inputs on biodiversity damage, livelihood loss, landscape destruction, wash away of fauna and flora were collected before the compilation of final report.

The report deliberated in detail how the flood and landslides disturbed the different ecological terrains, landscapes, biodiversity, traditional and agrobiodiversity, biocultural diversities and ecosystem diversities of Kerala. It was recorded that 771 different landscapes were affected by this flood and 86% soil erosion in this area. Animal diversity displacements, change in distribution pattern of floral diversity, loss of biodiversity at the riparian ecosystem etc.were the different impact observed during the study. A strong long term strategic policy is necessary for environment and biodiversity inclusive sustainable development is necessary for a green growth in Nava-Keralam. (AR 2018–19 Pg 32, AR 2019–20 Pg 43–7).

Conflict of Interest The views and opinions expressed in this article are those of the author in his personal capacity. None of the observations, opinions, or conclusions expressed herein are to be attributed to or construed as the views or positions of any organization, association, or company. No conflict of interest.

References

Gopan SV, John L, Laladhas KP, Oommen OV (2017) Sustainability of native fish brood and rural livelihood in "biodiversity for sustainable development. In: Laladhas KP, Nilayangod P, Oommen V (eds) Oommen for environment challenges and solutions series, 3. Springer International Publishing, Switzerland, pp 295–302.

Indian Biodiversity Act (2002). http://keralabiodiversity.org/images/act/bdact2002.pdf

John L, Laladhas KP, Kumar D, Gopalan SV, Cheruvat D, Oommen OV (2017) Preponderance of biodiversity at Vellayani worth to conserve as Biodiversity Heritage Site. Mainstreaming biodiversity for sustainable diversity published by KSBB (ISBN No. 978-81-934231-1-0), pp 181–201

John L, Laladhas KP, Oommen OV (2017) Restoring fish habitats for sustainable livelihood. In: Laladhas KP et al (eds) Biodiversity for sustainable development. Environment challenges and solution-3. Springer, Cham, pp 303–308. https://doi.org/10.1007/978-3-319-42162-9_4

Kerala State Biodiversity Board, Annual Report 2012–13 published by the Member Secretary, KSBB, Pallimukku, Petah PO, Thiruvananthapuram, Kerala. http://keralabiodiversity.org/images/pub/report_2012-2013_english.pdf

Kerala State Biodiversity Board, Annual Report 2013–14 published by the Member Secretary, KSBB, L-14, Jai Nagar, Medical College, PO, Thiruvananthapuram, Kerala. http://keralabiodiversity.org/images/2019/November/BOOKS/annual_report_2013-2014.pdf

Kerala State Biodiversity Board, Annual Report 2014–15 published by the Member Secretary, KSBB, L-14, Jai Nagar, Medical College, PO, Thiruvananthapuram, Kerala. http://keralabiodiversity.org/images/annual/annual_2014_15_opt_eng.pdf

Kerala State Biodiversity Board, Annual Report 2015–16 published by the Member Secretary, KSBB, L-14, Jai Nagar, Medical College, PO, Thiruvananthapuram, Kerala. http://keralabiodiversity.org/images/annual/annual_report_15_16_malayalamopt.pdf

Kerala State Biodiversity Board, Annual Report 2016–17published by the Member Secretary, KSBB, L-14, Jai Nagar, Medical College, PO, Thiruvananthapuram, Kerala. http://keralabiodiversity.org/images/annual/annual_report_malayalam_16-17.pdf

Kerala State Biodiversity Board, Annual Report 2017–18 published by the Member Secretary, KSBB, L-14, Jai Nagar, Medical College, PO, Thiruvananthapuram, Kerala. http://keralabiodiversity.org/images/2019/Annual_Report_17-18.pdf

Kerala State Biodiversity Board, Annual Report 2018–19published by the Member Secretary, KSBB, Kailasam, T.C.4/1679 (1), No. 43, Belhaven Gardens, Kowdiar P.O., Thiruvananthapuram 695 003, Kerala. http://keralabiodiversity.org/images/2020/february/books/annual_report_english_18_19.pdf

Kerala State Biodiversity Board, Annual Report 2019–20 published by the Member Secretary, KSBB, Kailasam, T.C.4/1679 (1), No. 43, Belhaven Gardens, Kowdiar P.O., Thiruvananthapuram 695 003, Kerala. http://keralabiodiversity.org/images/2021/august/Annual_Report_English_2019-2020_09_08.pdf

Laladhas KP, Roy John Fr, Saraswathy V, Bharucha E, Achuth Nair SS, Oommen OV (2020) Successful livelihood models through BMCs in India 2020, 212–223. In: Oommen OV, Laladhas KP, Bharucha E (eds) Biodiversity and livelihood: lessons from Ccmmunity research in India by All rights reserved-© 2020 Bentham Science Publishers ISBN (Online): 978-981-14-8215-1, ISBN (Print): 978-981-14-8307-3. https://benthambooks.com/book/9789811482151/

Laladhas KP, Nilayangod P, Gopan SV, Divya L, Oommen OV (2017a) Community conservation of marine turtles and sustainable ecosystem in "Biodiversity for Sustainable Development". In: Laladhas KP, Nilayangod P, Oommen V (eds) Oommen for environment challenges and solutions series, 3. Springer International Publishing, Switzerland, pp 177–84

Laladhas KP, Gopan SV, Nilayangod P, Oommen OV (2017b) Conservation through payment for conservation of ecosystems in "Biodiversity for Sustainable Development". In: Laladhas KP, Nilayangod P, Oommen V (eds) Oommen for environment challenges and solutions series, 3. Springer International Publishing, Switzerland, pp 193–224

Mithrambika NB, Laladhas KP, Nilayangod P, Oommen OV (2017a) Decentralised governance for sustainable development in "Biodiversity for Sustainable Development". In: Laladhas KP, Nilayangod P, Oommen V (eds) Oommen for environment challenges and solutions series, 3. Springer International Publishing, Switzerland, pp 165–76

Mithrambika NB, Laladhas KP, Baijulal B, Suresh Babu NB, Oommen OV (2017b) Monsoon flood plain fishery-conservation and management by biodiversity management committee. A Kerala Model. Mainstreaming Biodiversity for Sustainable Diversity Published by KSBB(ISBN No. 978-81-934231-1-0), pp 388–407

Part II
Technology Transfer

Chapter 5
Access and Benefit Sharing of Biological Resources and Associated TK: A Multi-tiered Responsibility

T. C. James and S. Rajasekharan

Abstract Access and Benefit Sharing (ABS) is a crucial objective of the Convention on Biological Diversity (CBD). It is this objective that gives an inclusive approach to the whole issue of conservation of biological resources. It also gives the local communities a sense of belonging and ownership of the natural resources around them. The implementation of the ABS mechanism since the signing of the CBD has raised many issues. With commitment and cooperation of all the players only can an effective system be in place. This chapter examines how ABS worked in the past in the Indian context and across other countries, what are the challenges that arose in the implementation and how they were overcome. It contains case studies of some important ABS instances in the country. Based on the past experience and legal provisions, it explores the multi-tiered responsibilities from a governance angle in the matter at global, national, state and local levels.

Keywords Benefit sharing · Community · Governance

5.1 Introduction

The Convention on Biological Diversity (CBD), 1992 has three main objectives, namely, "the conservation of biological diversity, the sustainable use of its components and the fair and equitable sharing of the benefits arising out of the utilization of genetic resources." In a telling statement in the preamble, the Convention stresses the importance of international, regional and global cooperation among States and intergovernmental organizations and the non-governmental sector in this regard. This

T. C. James (✉)
Visiting Fellow, Research and Information System for Developing Countries and President, NIPO, New Delhi, India
e-mail: tcjames@ris.org.in

S. Rajasekharan
Division of Ethnomedicine and Ethnopharmacology, Kerala State Biodiversity Board and Former Director Grade Scientist, Jawaharlal Nehru Tropical Botanic Garden and Research Institute, Thiruvananthapuram, India

O. V. Oommen et al. (eds.), *Biodiversity Conservation Through Access and Benefit Sharing (ABS)*, https://doi.org/10.1007/978-3-031-16186-5_5

brings to the fore the complexity of the responsibilities in ensuring proper implementation of the obligations laid down on the global community by the CBD. What it brings out is that there is a need for cooperation at all levels and across a wide spectrum of players. Consequently, the responsibilities also will get distributed. It is essentially because of the complications involved in the implementation, especially the provisions relating to Access and Benefit Sharing (ABS) that the 'Nagoya Protocol on Access to Genetic Resources and the Fair and Equitable Sharing of Benefits Arising from Their Utilization to the Convention on Biological Diversity (2010)' (Nagoya Protocol) had become necessary. This Protocol addressed issues relating to the international transfer of Genetic Resources (GR) and made recipient country governments also responsible for ensuring the ABS mechanism. International instruments and national laws can certainly introduce much-needed legal certainties on tackling situational problems. Bur a workable model of ABS system has both national and international dimensions and has to address many governance issues. In this chapter, we propose to explore these matters. It will, however, not be limited to the GR but will extend to the entire Biological Resources (BR) and Traditional Knowledge (TK), which are covered by India's domestic legislation.

5.2 ABS—A Global Governance Perspective

ABS mechanism involves a long chain of players from the person who wants to access, to the individual or institution who engages in research on the same and further to the enterprise who wants to market the product or process ultimately resulting from that access. Depending on the context and type and material accessed, the players and their institutional roles may vary. The accessor can be an individual who collects and sells products from the forest or a researcher who wants to do a survey or a scientist who wants to investigate the qualities of the product accessed and intend to come out with an invention. Depending on the result of the research it may lead to an invention that may or may not have commercial application. In some cases, the material collected may be used as raw material for the development of a commercial product. In some cases, like those of certain medicinal and aromatic plants (MAPs), the material itself can be the product. Apart from the accessor and exploiter, there are the owners or custodians of the BR who have to agree to the access. Depending on the credentials of the accessor, there may also be different authorities to permit or reject the access application. Sometimes approval may require a national authority, and a state or provincial authority.

5.2.1 International Dimension

The major important international legal instruments on BR are CBD and the different Protocols on the same. The CBD has 196 Parties, the Cartagena Protocol has 173

Parties, the Nagoya Protocol has 129 Parties, and the Nagoya-Kuala Lumpur Supplementary Protocol has 48 Parties. These statistics convey the different levels of commitment that countries of the world have made to these treaties. An important issue of global governance is to get all countries to ratify all the BR related treaties and fulfil their obligations, without which no ABS mechanism can work optimally as there will be countries who keep themselves outside the legal frameworks.

While these international instruments come with no strong implementation and enforcement provisions on the Parties who do not fulfil their obligations, the World Trade Organization (WTO) treaties which deal with trade, Intellectual Property Rights (IPRs), etc. have binding provisions on the Members. Besides the trade treaties, the Agreement on Trade-Related Aspects of Intellectual Property Rights (TRIPS Agreement), 1994, one of the treaties that form the basket of WTO treaties, has a salutary effect on ABS. This is principally because of the patents based on biological resources like the famous case of turmeric patent in the USA.

The International Treaty on Plant Genetic Resources for Food and Agriculture (2001) (PGFRA) has also established a multilateral system for ABS in the case of ex situ collections of plant genetic material. This is basically a global plant gene pool that can be shared cooperatively by all members. It operates around a Standard Material Transfer Agreement (SMTA). This document contains the terms of benefit sharing for accessing the material (Schroeder and Lucas 2013). Primarily, this is aimed at facilitating access for researchers. Several monetary and non-monetary benefit-sharing mechanisms have been provided in the Agreement. The coverage of the multilateral system is limited to 64 food crops. However, there are more than 7.4 million PGFRA accessions in the stocks in more than 1750 Gene Banks. There are 147 contracting states including India. The Protection of Plant Varieties and Farmers' Rights Act, 2001, which provides for plant variety protection has been enacted by India, however, in pursuance of the obligations under Article 27 of the TRIPS Agreement.

Along with the global governance issues, there are matters related to domestic administration of the national laws, particularly in the context of federal structures, where different levels of governments have to work in consonance with each other and involving the local people who are most affected by the law. The next section looks into these matters.

5.3 ABS in a Federal Structure

The Biological Diversity Act, 2002 (BDA), in its preambular statement has restated the three objectives of the CBD including "fair and equitable sharing of the benefits arising out of the use of biological resources, knowledge". In this respect, the scope of BDA is larger than that of CBD and is not limited to GR. However, India through a notification under section 40 of the BDA in 2016 excluded 385 commonly traded commodities like cereals, millets, pulses, green vegetables, etc. from the purview of the Act. The exclusion also covers large number of medicinal and aromatic plants and also spices like turmeric.

The Act has created a three-tiered structure for the implementation of the ABS provisions. These three tiers span across the three levels of government, namely, central, state and local levels. At the central level is the National Biodiversity Authority (NBA), at the state level is the State Biodiversity Board (SBB) and at the local level is the Biodiversity Management Committee (BMC). The duties and responsibilities of each level have been clearly defined in the Act.

5.3.1 National Biodiversity Authority

The NBA is the institution with the greatest role in the matter of ABS. It is the expert body to advise the Central Government on all matters relating to BR, including ABS, to regulate access to biological resources and "impose benefit sharing fee or royalty or both or impose conditions including the sharing of financial benefits arising out of the commercial exploitation of such (intellectual property) rights". Towards this, it has the responsibility to issue guidelines. It is, thus, the crucial institution in ABS. The Act and the Biological Diversity Rules, 2004 (BDR) have detailed provisions on how the NBA should perform this responsibility. These include, among others, providing technical assistance and guidance to the SBBs, "collect, compile and publish technical and statistical data, manuals, codes or guides", and mass media campaign. The Authority has the responsibility to prescribe the form of agreement for ABS.

What the legal provisions amount to is that the NBA has roles of an adviser, regulator, guide and executive, all rolled into one. Its advisory roles are mainly with reference to the Central Government and also with SBBs and BMCs. Its regulatory roles include prescribing guidelines and ensuring that ABS provisions are implemented. It has the onerous task of opposing the grant of IPR in other countries on biological resources obtained from India. In all the approvals that the Authority grants it has to ensure that ABS provisions are properly secured. This involves consultation with its own expert advisers and State Biodiversity Boards.

5.3.2 State Biodiversity Board

The BDA provides for the establishment of the State Biodiversity Board (SBB) in every State. Apart from the advisory function, the SBB has the power to grant approvals "for commercial utilization or bio-survey and bio-utilization of any biological resources by Indians". It also has the responsibility and powers to ensure equitable sharing of benefits in cases of access to biological resources by Indians. Towards this, the SBB has the power to enter into benefit sharing agreements with the applicants.

All the States have set up SBBs. It is the conduit for the NBA to consult with BMCs on matters of ABS.

5.3.3 Biodiversity Management Committee (BMC)

The lowest rung in the legal framework for ensuring ABS is the Biodiversity Management Committee (BMC) to be constituted by the local body. It has the power to levy collection fees from any person for accessing or collecting biological resources for commercial purposes. The NBA and SBBs are also required to consult the BMC while taking decisions on biological resources that occur in its territory. There are 2,73, 451 BMCs, as of 9 April 2021 (Table 5.1). This is as against a total of 2,65,161 g (village)*Panchayats* in the country. Constitution of the BMCs at an all-India scale also took a long time, understandably because the impetus had to come from the SBBs, whose constitution in many states was considerably delayed. The slow spread of awareness about the new legislation among the people also could be a reason for this lethargy.

Initially, the Union Territories were not to have any BMCs; their functions were to be performed by the NBA only. However, in December 2019, the NBA delegated those powers to the Union Territory Administrations, pursuant to a directive from the Government of India.

India has now a three-tiered institutional framework for the implementation of BDA provisions. This, by and large, correspond with the governance structure in India where we have governments at central and state levels and local self-governments. At the local level, there are three tiers to which representatives are elected by the people. These are the district, block-level and village level *Panchayats*. The BDA provides that BMCs be constituted by every local body. Local bodies are defined by the Act to include Panchayats and Municipalities, by whatever name called. The NBA Guidelines on constitution of BMCs cover the *Zilla* (District), Block, Taluk and *Mandal panchayat* levels also. Kerala has already constituted BMCs at district, block and village levels; Tamil Nadu and Tripura have set up block level BMCs. Wherever bioresources of a particular kind are spread over more than one panchayat, it would be appropriate to have BMCs that transcend village boundaries and covering a block or district as such. Beyond the district level, the SBBs can take care. This is especially so in the case of marine biodiversity. A draft policy paper on the blue economy suggests that in regard to protecting marine biological resources, the provisions of the CBD should be taken into account, but does not make any reference to the BDA. At the time of the drafting of the BDA, the marine biological resources had not attained the importance that it now receives under the concept of Blue Economy. The way it is emerging, it would be necessary to have an international treaty on resources beyond national boundaries.

However, as ABS is concerned, the Act and Rules and also the Regulations do not clearly lay down any role for SBBs and BMCs, except consultations by the NBA in IPR matters [sections 24(2) and 41(2)]. However, in actual practice, they came to

Table 5.1 State/UT-wise list of BMCs

Sl. No	State/UT	Number of BMCs
1	Andhra Pradesh	14,212
2	Arunachal Pradesh	1806
3	Assam	2549
4	Bihar	9099
5	Chhattisgarh	12,004
6	Goa	205
7	Gujarat	13,578
8	Haryana	6435
9	Himachal Pradesh	3371
10	Jharkhand	4684
11	Karnataka	6554
12	Kerala	1200
13	Madhya Pradesh	23,557
14	Maharashtra	28,649
15	Manipur	2260
16	Meghalaya	5451
17	Mizoram	894
18	Nagaland	1119
19	Odisha	7256
20	Punjab	13,599
21	Rajasthan	10,406
22	Sikkim	196
23	Tamil Nadu	13,604
24	Telangana	13,461
25	Tripura	1264
26	Uttarakhand	7991
27	Uttar Pradesh	59,407
28	West Bengal	3828
	Total	**268,639**
	Union territories	
29	Andaman & Nicobar Islands	71
30	Jammu & Kashmir	4666
31	Ladakh	65
32	Lakshadweep	10
	Total	**4812**
	Grand total	**273,451**

Source National Biodiversity Authority. (http://nbaindia.org/content/20/35/1/bmc.html)

play an important role as would be seen later. The role of the SBBs got much needed clarity and a boost in the case of *Divya Pharmacy vs. UoI & Others* (2018), wherein the court held that "SBB has got powers to demand Fair and Equitable Benefit Sharing from the petitioner, in view of its statutory function given under section 7 read with section 23 of the Act".

Besides this three-tiered structure, there are certain levels that also play a role in the ABS mechanism. These are the governments at central and state levels and local self-government institutions. The role and responsibilities of the Central Government are much larger compared to the other levels. The BDA itself in sections 36–40 details the duties and responsibilities of the Central and State Governments under the legislation. These relate to policies and strategies in the area of conservation and sustainable use of the BR and protection of TK and other related matters. Further there are 33 central legislations including the BDA and PV&FRA that have implications for the ABS mechanism. In addition to the 33, the Compensatory Afforestation Act, 2016 has also been enacted. The NBA has administrative responsibility only in respect of BDA. While some of the legislations are with the Ministry of Environment, Forest and Climate Change (MoEFCC), different Ministries are concerned with other legislations. In addition, there are good number of state laws and regulations. The Central Government has to ensure that there are no overlaps among the legislations and the institutions set up by the different laws work in harmony. In case of disputes between the NBA and one or more SBBs and among SBBs, with regard to any policy decision or implementation of any order or direction issued by the Central Government, an appeal can be preferred to the Secretary, MoEFCC. But in day-to-day activities and ordinarily in the matter of ABS mechanism, this level does not play any significant role. Regarding ABS, besides the NBA under BDA, there is the Plant Varieties and Farmers' Rights Authority under the PPV&FRA. That Authority is to decide benefit sharing in respect of registered plant varieties.

5.4 ABS Implementation in India

The BDA covers the following activities by Indians: commercial utilization, bio-survey and bio-utilization, research, transfer of research results, third party transfer of already accessed biological resources, sending/transfer of biological resources outside India, and applying for IPRs. Wherever the activity results in a commercial gain, directly or indirectly, there has to be fair and equitable benefit sharing. The ABS provisions in the BDA were brought into force on 1st July 2004. Since then, several guidelines were issued to ensure that the provisions are implemented smoothly. During the years from 2006–07, the NBA has granted 2,444 ABS approvals and agreements signed as presented in Table 5.2.

This was against a total of 5,219 applications received by the Authority, which includes 1,649 cases pending with applicants for the execution of the agreement and 842 closed cases.

Table 5.2 Approvals granted

Year	Form I access to Bioresources for research/commercial purpose	Form II transfer of research results	Form III approval for obtaining IPR	Form IV third party transfer	Form B Sending/carrying biological resources outside the country for research	Total
2006–07	4	1	0	2	0	7
2007–08	5	3	12	6	0	26
2008–09	4	4	21	6	0	35
2009–10	2	1	9	1	0	13
2010–11	3	1	4	1	0	9
2011–12	1	2	6	0	0	9
2012–13	1	0	8	7	0	16
2013–14	1	0	14	2	0	17
2014–15	19	0	22	1	0	42
2015–16	31	1	51	2	7	92
2016–17	36	4	127	0	15	182
2017–18	37	2	245	1	31	316
2018–19	62	1	209	1	17	290
2019–20	29	1	258	0	40	328
2020–21	76	8	949	0	23	1056
2021–22	2	0	0	0	3	5
Total	314	**29**	**1935**	**30**	**136**	**2444**

Source NBA updated as on 14.6.2021. (http://nbaindia.org/content/683/61/1/approvals.html)

What comes out is that the ABS implementation has a slow start in the country. The initiative was mostly with NBA. Perhaps, this might have something to do with a lack of clarity on the procedures. The issue of Guidelines in 2014 seems to have accelerated the implementation. Verdicts on certain cases concerning ABS pending with Tribunals and Courts such as the *Divya Pharmacy Ltd. Vs. UoI & Others* in the Uttarakhand High Court also might have accelerated the implementation as the SBBs are now in surer legal footing.

The 2014 Guidelines clarified that there can be both monetary and non-monetary options for benefit sharing. The monetary ones can be in the forms of (i) Up-front payment; (ii) One-time payment; (iii) Milestone payments; (iv) Share of the royalties and benefits accrued; (v) Share of the license fees; (vi) Contribution to National, State or Local Biodiversity Funds; (vii) Funding for research and development in India; (viii) Joint ventures with Indian institutions and companies; (ix) Joint ownership of relevant IPRs. The non-monetary options will include (i) Providing institutional capacity building, including training on sustainable use practices, creating infrastructure and undertaking the development of work related to conservation and sustainable use of biological resources; (ii) Transfer of technology or sharing of research and development (R&D) results with Indian institutions/individuals/entities; (iii) Strengthening of capacities for developing technologies and transfer of technology to India and/or collaborative R&D programmes with Indian institutions/individuals/entities; (iv) Contribution/collaboration related to education and training in India on conservation and sustainable use of BRs (v) Location of production, R&D units and measures for conservation and protection of species in the area from where the BR has been accessed, contributions to the local economy and income generation for the local communities; (vi) Sharing of scientific information relevant to conservation and sustainable use of biological diversity including biological inventories and taxonomic studies, (vii) Conducting research directed towards priority needs in India including food, health and livelihood security focusing on biological resources; (viii) Providing scholarships, bursaries and financial aid to Indian institutions/individuals preferably to regions, tribes/sects contributing to the delivery of BRs and subsequent profitability if any; (ix) Setting up of venture capital fund for aiding the cause of benefit claimers; and (x) Payment of monetary compensation and other non-monetary benefits to the benefit claimers as the NBA may deem fit. The Guidelines clearly indicate that the determination of benefit sharing is to be based on "considerations such as commercial utilization of the biological resource, stages of research and development, the potential market for the outcome of the research, amount of investment already made for research and development, nature of technology applied, timelines and milestones from initiation of research to the development of the product and risks involved in the commercialization of the product". The Guidelines also lay down certain graded norms for the benefit sharing which may range from 0.1 to 0.5% of the annual gross ex-factory sale of the product in case of access for commercial utilization. In case of transfer of research results, it shall be 3.0 to 5.0% of the monetary consideration, and in case of IPRs, it shall be 0.2 to 1.0% based on sectoral approach. The highest percentages for ABS are prescribed in case of transfer of biological resources, namely, between 2 to 5%.

As per the Act, the money received by way of benefit sharing is to be deposited in the National Biodiversity Fund, except where the biological resource was accessed from a specific individual or group of individuals or organisations, in which case the amount be paid directly to such individual or group of individuals or organisations, as directed by NBA. Out of the amount received by NBA, it shall retain 5% and the remaining 95% is to go to the BMC concerned or the benefit claimers. From the amount retained by the NBA, half will go to the SBB concerned. Thus, India has very detailed provisions covering various forms of access.

The realization of benefit-sharing amounts was not very high in the early years. As per a 2014 study on ABS Implementation in India (Dhar et al. 2014a) the actual ABS amount received by the NBA during the period from 2002 to 2014 is Rs. 43,39,698 only. Category-wise details are presented in Table 5.3.

While there has been some improvement in the matter since 2014, as a result of more vigorous steps by the SBBs, the actual ABS amounts received are not very high. It is interesting to note that the NBA has received more amounts through fees

Table 5.3 Category wise Details of Actual ABS received by the NBA (2002–2014)

Category	Amount in Rupees
Access and transfer of BR and associated TK (Forms II & IV)	39,09,765
Access of BR and associated TK (Form I)	4,25,993
IPR (Form III)	3,940
Total	**43,39,698**

Source National Study on ABS Implementation in India 2014

Table 5.4 Estimated Application Fee received by NBA

Category	Amount
Access to BR and associated TK	31,30,000
Transfer of Research results	2,50,000
Seeking 'No objection Certificate' for obtaining patent	5,94,500
Seeking approval for Third Party transfer of the accessed biological resources and associated TK	8,10,000
Total	**47,84,500**

Source Biodiversity Finance: Identification and Analysis of key Building Blocks of a Biodiversity Finance Plan in India. 2019

under the four categories of applications as brought out in a 2018 study by National Institute of Public Finance and Policy (Table 5.4) (Pandey et al. 2019).

Note: The estimate appears to be for the period from 2004–05 to 2016–17 and is against 1677 applications, which includes 31 applications for conducting non-commercial research and 13 incomplete applications for which no fee was received. The study has estimated the royalty received by the NBA during the period 2009–10 to 2016–17 as Rs 469,810,000 of which Rs. 465,805,000 was from one source, namely, Andhra Pradesh Forest Development Corporation Limited. The SBBs also receive various application fees, licence fees and processing fees as prescribed under state laws.

The Dhar et al. (2014a) study also brought out that there were cases in which although the NBA had directed payment of monetary benefits, actual payments had not been made. One instance presented was the case of a patent for a process for preparation of a herbal formulation of *Tejan, Bhootkeshi* and *Nilnirgundi* as a Bronchodilator, for which a patent was obtained and the patentee was directed by NBA to pay 5% of the net sales to NBA, but actual payments had not materialised. This case brings out the importance of central and state governments, the NBA and SBBs acting in concert to realize the ABS.

5.5 Case Studies

Learning from best practices is an ingredient of good governance. There are several cases in India and around the world which present valuable lessons in ABS for policymakers. One such case is the *Arogyapacha* case from India.

5.5.1 Arogyapacha Case

ABS in India predates the enactment of the legislation and the setting up of an institutional mechanism for the same. The case relates to the access to the biological resource *Arogyapacha (Trichopus* zeylanicus *ssp. travancoricus*) in the Western Ghats and the associated traditional knowledge (TK) of the Kani tribe of Kerala by a research institution. This is a case that has been documented very much. The case involved access to BRs, R&D, IPR and subsequent commercialization. The research was undertaken by a government institute, namely, the Jawaharlal Nehru Tropical Botanic Garden and Research Institute (JNTBGRI), Thiruvananthapuram, which collected the BRs from an area predominantly occupied by the Kani tribe, who assisted in the collection and who also shared their TK about the rejuvenating property of the plant with the scientists in 1987. In the subsequent years, the Institute separated 12 active ingredients, applied for and obtained three patents and developed a drug, *Jeevani.* Since the Indian Patents Act, 1970 had not been permitting product patents for pharmaceuticals those days, the JNTBGRI obtained process patent for the manufacture of Jeevani and licensed the Arya Vaidya Pharmacy (AVP), Coimbatore for making and marketing it. This was on a licence fee of Rs. 10 lakh and royalty @ 2% on ex-factory sales price. There was a tripartite agreement for benefit sharing involving all the three parties. Since the Kanis had then no common forum, a Trust (Kerala Kani Samudaya Kshema Trust) was established for transferring the royalty. Out of the royalty, a community building was built. Fifty per cent of the licence fee, i.e., Rs. 5 lakh and 50% of royalties, i.e., Rs. 1,41,172.50 were to go to Kanis. The Kani Trust was paid a total of Rs. 6,41,172.50 by the Institute, Rs. 1 lakh by Dr. Pushpangadhan, who had led the expedition and research and Rs. 5,000 by Shri. Anil Gupta. In addition, the Kani community could earn and add to the Trust Rs. 3 lakh through the cultivation of the plant. In 1997 three Kanis, who were part of the first expedition, were rewarded with cash by the Trust. Further, two of the guides were engaged as paid Consultants at a monthly remuneration of Rs. 3,000 for 7 years from 1993 to 1999 by the JNTBGRI. Their total payment amounted to Rs. 5,04,000. However, there were complaints that Kanis from other areas were ignored. The patent subsequently expired and no fresh agreement seems to have been made later. The case stands out for the reasons (1) that it is the first ABS case in India and predated BDA; (2) made when there was no legal requirement; (3) royalty was shared equally between the research institute and the holders of TK who are also the collectors of BR. It is a case of an accessor institution sincerely attempting to share benefits to the providers of the BRs and TK. The absence of the constraints of a system later introduced in the Act seems to have contributed to it. A take home lesson from the case is that it is the genuine human right and ethical considerations that can actually lead to benefit sharing and not always legal regulations. If in those days there were the 20-year product patents for pharmaceuticals in India, perhaps the benefit-sharing would have been for a longer period. However, it also throws open certain issues: (1) How should the royalty for access to BRs and TK be separated; and (2) How should the value of the material, i.e. biological resource, be separated from the fee required under BDA. Should not the manual labour involved in the collection or cultivation be considered separate from ABS?

One of the problems faced in this case was that of the difficulty and complexity of the determination of the amount of benefit sharing and its distribution (Anitha Ramanna-Pathak 2017). A lot of political and other pressures were there. The Kani community also is not a homogenous group, which also raised challenges in identifying the beneficiaries.

An issue of governance arises from the fact that the plant is spread over the territories of Kerala and Tamil Nadu. Since the collection of material was from the Kerala region only, the Kanis belonging to that State only were included in the Trust. This raises the issues as to (1) whether the community/tribe should be taken as one irrespective of the areas they are located, and (2) when the BRs are from one state only whether it would be fair to share benefits with residents of the other state who have not contributed. Though the Act was not in operation at the time of the access, similar issues can crop up in the future too. How and what kind of solutions can be devised for the same and how to ensure the cooperation of all the stakeholders are matters for consideration.

5.5.2 *Seaweed Case (2007)*

Among the early cases, the Seaweed access is the one which brought in high Benefit Sharing and, perhaps, the highest for a local community. The Seaweed, *Kappaphycus alvarezii*, cultivation was launched in the four southern districts of Tamil Nadu, namely, Ramanathapuram, Thoothukudi, Pudukkottai and Thanjavur districts by Pepsico in 2000 supported with technology from Bhavnagar-based Central Salt and Marine Chemicals Research Institute (CSMCRI). The cultivators belonged to one fishing community in the coastal areas. Pepsico exported 2000 MT seaweed to Indonesia, Malaysia and the Philippines. The collection, cleaning and baling was done by women self-help groups. For the export of the biological resource seaweed, NBA imposed a fee of 5% of the Freight on Board (FOB) value which amounted to Rs. 43, 57,176, ultimately netting an amount of Rs. 39,20,459 after-tax deduction (Jacob and Reddy 2015). The total fund available for the coastal communities for benefit sharing was Rs. 37,24,437 only after a 2.5% deduction by NBA and another 2.5% deduction for SBB. Even for this amount, there was no proper distribution to the community highlighting the problem at the bottom level of the ABS pyramid. To whom and how the amount should be passed on are issues in this case also as in other cases. There were 754 benefit claimers. Training was given to women to collect and dry the seaweed. Does that really amount to benefit-sharing or is it only a process in the exploitation of the resources optimally? Since the training does not have an application outside of the product, it cannot be said that it imbues the people with a sustainable or transferable skill. It also has not resulted in a technology transfer except rudimentary skills of efficient manual collection and drying. The non-financial ABS has to be such as to empower the community. In the absence of that, one can only presume that it is a normal part of the business of the company to speed up their raw material collection, which even without the NBA intervention they might have done in their own commercial interest. From a governance angle what matters is that while training, etc. are considered as benefit sharing, such training has to be in adaptable skills that will stand the recipients in good stead. That would require, involvement of many, administrators, academics, industry and the community as such.

5.5.3 Neem Export Case (2010)

BioIndia Biologicals paid an amount of Rs. 55,035 to the NBA for export of 2000 kg of *Neem* leaves (*Azadirachta indica*) to Japan. The leaves were collected, dried and bundled by the people of the village Amarchinta, Mahboob Nagar district, Andhra Pradesh under an agreement with the company. The case is similar to the sea-weed case in that the local community collected and dried the leaves before handing the same over to the company. The NBA, however, transferred an amount of Rs. 20,000 for "planting neem saplings and creation of awareness about biodiversity conservation." While the stated objective of the money-transfer would contribute to the cause of conservation, it really cannot be considered as a benefit for the community in the spirit of the Act. Their real gain seems to be in having gainful employment for a period comparable to that under the Mahatma Gandhi National Rural Employment Guarantee Scheme. That would have happened despite the Act. The ABS governance has to take into account the actual benefits that the community should get from a proper implementation of the BDA.

5.5.4 Habib Cosmetics Case (2015)

The Uttarakhand State Biodiversity Board and Habib Cosmetics Private Ltd had entered into an agreement on 15 April 2015 under which the company shared Rs. 3,22,991 as benefits of the year 2013–14 (NLSIU). This is the first case of an SBB entering into an ABS agreement after notification of the ABS Guidelines. It is also one of the early cases after the Guidelines on ABS with its detailed directions on how to decide ABS in different types of access had come into existence. As per the agreement the company accessed 39 different kinds of biological resources available in Uttarakhand for the manufacture of various cosmetics at its factory situated in the Dehradun district of that state. The calculation of the ABS amount was made on the basis of 0.5% of the Annual Gross ex-factory sale (Rs. 6,71,35,744) minus tax (Rs. 25,37,565) which was Rs 6,45,98,179. The amount that was to be paid to the SBB was Rs. 3,22,941. This amount was to be paid within 15 days of the signing of the agreement. In this case, the accessor belonged to the same state and operated a factory within the state. The resources did not have to be transported beyond the territory of the state though the headquarters of the firm was outside the state. The agreement does not refer to any consultation with local BMCs. It also does not say anything about the distribution of the ABS amount to any BMC. The case was practically between an SBB and an Indian accessor. Since there was no involvement of foreigners or application for IPR, the NBA was not in the picture.

5.5.5 Red Sanders Case (2015)

Andhra Pradesh government did a global e-auction of Red Sanders (*Pterocarpus santalinus*), one of the costly trees in Indian forests, which has also medicinal and aromatic properties. One of the conditions stipulated was for payment of 5% to NBA and 95% to BMCs. The export of Red Sanders netted an amount of Rs. 32 crore. (As per the Annual Report 2017–18 of the NBA, the amount received by the NBA during the year 2017–18 on this account was Rs. 18.38 crore and out of this an amount of Rs. 3 crore was distributed to the Andhra Pradesh Forest Department.) Quantity-wise, this has been so far the largest ABS. It displayed an interesting play of different stakeholders. It was the state government that invited the tender. It stipulated the ABS conditions in the bid document itself. This amount was to be paid to the three tiers, as indicated above. However, as per the NBA, it has distributed a portion of the amount received to the Andhra Pradesh Biodiversity Board for conservation and sustainable use of Red Sanders. The state has distributed it to BMCs for the same purpose. However, the conserved bioresources include bamboo, broom grass, cotton, wild cardamom, etc. The case brings out certain issues. The Red Sanders is a forest product and as per the law, the State Forest Department is the proprietor of the same. The NBA/SBB rightly enters the field for the ABS mechanism only. This depends on who the accessor is and what commercial exploitation is made of the forest produce.

5.5.6 Gram Mooligai Company Limited

The Gram Mooligai Company Limited (GMCL) is a case of people as cultivators and collectors of medicinal plants and herbs creating a system to secure better remuneration from biological resources provided by them to users. It is a public limited company formed in 2002 with collectors and cultivators holding the majority shares. It started as a collective community enterprise in Madurai, Tamil Nadu. Landless women were collecting the medicinal plants that were growing in the fallow farmlands and wastelands and were selling them in the market. The collection practices were not sustainable. The company launched training programme in harvesting and also cultivation. It now covers the states of Tamil Nadu, Madhya Pradesh, Odisha and Chhattisgarh. The objective of the company is to secure remunerative price from user companies accessing medicinal plants and herbs from primary cultivators and collectors. GMCL enters into agreements with companies like Dabur, Himalaya and Nature Remedies to provide them with quality components of medicinal plants and herbs as raw materials. GMCL has developed training packages combining traditional practices and modern techniques to train cultivators and collectors in best cultivation and sustainable harvesting practices. These practices include knowledge of the part of the plant or the herb to be harvested/collected, the manner and the extent of harvesting/collection, and the time and season of the year in which to be sown and harvested/collected. The initiative has resulted in direct monetary benefits to the people through better price for the resources provided to the industry; 70% of the returns go to the community. It has generated non-monetary benefits through capacity building of the people to negotiate with the markets and through sustainable and scientific harvesting and collection practices leading to conservation. It has led to an annual income growth of around 10%. This case shows how a community involved public company can bring benefits from biological resources to the community.

5.5.7 *Traditional Knowledge and ABS*

While most of the ABS cases in India involved access to substantial biological resources, there is one case, apart from the *Arogyapacha* case, in which the access to the associated traditional knowledge was very important. The applicant in this case also is the JNTBGRI. This relates to patent No. 357405 dated 4th September 2011 granted on 1st January 2021 for "A Novel Polyherbal Formulation with Multiple therapeutic effects as antidiabetic, antifatigue, hepatoprotective and antioxidant" using the biological resources (1) *Tetrastigma leucostaphylum*, (2) *Picroriza kurroa*, (3) *Terminalia chebula*, (4) *Phyllanthus emblica*, (5) *Zingiber officinale* and (6) *Piper longum*. As per the ABS agreement signed with the NBA on 17 September 2019 the JNTBGRI agreed to the ABS of (1) 0.4% of annual gross sales minus government taxes, and (2) 4% of licence fee and 4% of royalty. The actual amount of payment will depend on the income from the commercialisation of the product. However, the case deserves mention because of the special feature that the TK holder, namely, Thekkedathu Muhammedukunju Shahul Hameed Vaidyar, a traditional healer was named as an inventor along with the scientists. This approach (1) accepts that traditional knowledge is also valuable as modern scientific knowledge, (2) gives recognition to the traditional knowledge holder, and (3) encourages secret TK holders to come forward and share their knowledge for scientific research. This is in the right spirit of the CBD and BDA. The NBA and the research institution performed their responsibilities well in this case.

5.6 Gainers and Losers

As per the Annual Report 2017–18 (the latest report available on the NBA website), the NBA received a sum of about Rs. 18.38 crore as benefit-sharing during 2017–18. Out of this an amount of Rs. 3 crore was distributed to Andhra Pradesh Forest Department for the access to Red Sanders wood. No details are provided in the report about the balance. Under the ABS, earlier an amount of Rs. 32.00 lakh was released to the Tamil Nadu State Biodiversity Board for distribution to the beneficiaries in four districts on account of the seaweed collection. (NBA Annual Report). The Annual Report 2016–17 also refers to these two cases and amounts, though at that time the total receipt by NBA was Rs. 12.49 crore only. The Madhya Pradesh SBB during the period 2013–2021 signed 673 MoUs with traders and manufacturers for ABS against which it received an amount of Rs. 1.74 crore. The Maharashtra SBB reportedly received Rs. 10 lakh from 12 manufacturers as ABS royalty.

For a country of the size of India, the actual benefits that have flowed to the communities concerned over a period of two decades of implementation of the BDA are meagre, rather paltry. The NBA has estimated that USD 8852.3 million as the export of biodiversity related material in 2008, (NBA Fact Sheet) reflecting the economic potential of the biodiversity of the country. For a clear assessment, one has to separate the benefit-sharing from the cost of resources and wage payments. These are like the sale of timber from the forest or cultivated land.

The country has adopted a top-down approach in the matter of not only policy formulation but also in implementation. The NBA and to a lesser extent the SBBs are the decisive levels. The local communities have very little say in the matter. This approach is exemplified by the fact that the payments are mostly made to the NBA and the balance to the SBBs. There are no payments to the local communities directly except in the case of Natural Remedies Pvt. Ltd., Bengaluru who had purchased *Kalmegh* (*Andrographis paniculata*), a medicinal herb, from BMC of Malajkhand in the Balaghat district of MP and paid Rs. 21,000 directly to the BMC.

The local level has to await any payments that flow to them through the mercy of the national and state levels. A psychological consequence of this is the loss of sense of ownership of the biological resource by the local communities. It takes away their enthusiasm to conserve biological resources. There are criticisms that in benefit sharing in India, "the interests of the state are prioritised over community interests" (Anitha Ramanna-Pathak 2017). After all, as observed by the Hon'ble Uttarakhand High Court, biological resources while being the property of the state, "are also the property, in a manner of speaking, of the indigenous and local communities who have conserved it through centuries."

A country of the size of India with such rich biological resources and with a large number of manufacturers of traditional medicine, cosmetics, nutraceuticals and other pharmaceutical products are ordinarily expected to have a good number of ABS agreements contributing a significant amount of ABS. However, the total number of agreements and the total amount as reflected in the annual reports of the NBA are not very high. Uttarakhand is one of the more active states and the data on its site show a total of 139 applicants over a four-year period from 2014–15 to 2017–18, of which mutually agreed terms signed with 29 applicants only. The rest of the applications are shown as under process. Two points stand out in the Uttarakhand case: one, there is a significant delay in the processing of the applications, and two, the ABS seems to be between the SBB and the commercial entity only. Apart from the Habib Cosmetics case, no other case has been presented on the Uttarakhand SBB website.

5.7 ABS Implementation in Other Countries

ABS implementation has a mixed response across countries. While 195 countries have ratified the CBD, ABS implementation had not made much headway, because of a lack of clarity in the obligations of the receiving country. This issue was sorted out through the Nagoya Protocol, 2010. This made it obligatory for receiving countries to have mechanisms to ensure that the party had complied with the ABS procedures. There are 129 Parties to the NP. Many countries like Chile are grappling with issues relating to objections from indigenous communities this time (Heinrichet al. 2020). A pre-Nagoya Protocol study presents seven cases of ABS (Secretariat 2008). A 2014 study presented four cases in South Africa which involved bioprospecting permits issued by the Minister of Water and Environmental Affairs. The beneficiaries included various communities and in one case National Traditional Healers Committee. A similar 2014 study in Brazil in 2014 mentioned 1314 authorisation for access to GR and associated TK for scientific research, bioprospecting, technological development, etc. by the Genetic Heritage Management Council who has accredited various agencies to issue the authorizations. These include Brazilian Institute of the Environment and Renewable Natural Resources (IBAMA) to issue authorizations for access to GR with no associated TK for purpose of research, the Institute for National Historical and Artistic Heritage (IPHAN) to issue authorizations for access to associated TK without access to GR for the purpose of research, and the National Council for Scientific and Technological Development (CNPq) to issue authorizations for access to GR with no associated TK for the purposes of research, bioprospecting or technological development. Thus, Brazil has specialized agencies for granting access. While some of the approved agreements contain financial benefit sharing, data on the actual amounts are not available in the study.

In actual practice for a comprehensive and fool-proof benefit sharing regime, a good number of steps have to be undertaken; legislation is only one of them. An inventory of the biological resources is important. This is the one which enables both the regulator and the accessor to explore a proper ABS contract. The PBRs in India are envisaged from this angle. However, their preparation is still ongoing. At the same time, it must be appreciated that the SBBs have taken steps to digitise these registers, which in future will make handling of applications for access to biological resources, at various levels much easier and smoother.

It must be remembered that despite the lapse of more than three decades since the signing of the CBD, the systems are not yet in place and functioning in most of the countries. Consequently, the benefits are not reaching the communities concerned the world over, though access and utilization have been going on. A recent research paper published by the South Centre gives a list of 26 cases of plants worldwide involving patent applications numbering 73,388.

The concerns that were at the root of the CBD still persist in many ways. The 2007–2008 episode regarding the non-sharing of H1N1 data by Indonesia was a pointer. The data from developing countries in global databanks are accessed by commercial firms in industrialized countries who use the same to develop products for the market. Many of the products are medicines for viral infections which are caused by viruses whose data is depended upon. But the developing countries find it well-nigh impossible to obtain the drug because of the high prices. The access does not involve any benefit-sharing but *gratis* whereas in the medicine price there is no concession to the community who had the original virus.

The dimensions of the issue vary depending on the nature of exploitation. The contours of the problem and the roles of different players are generally clear in the cases of access to biological resources for direct commercialisation as in the case of the Seaweed or Red Sanders cases. When it comes to the case of ABS for IPRs, the matter poses many problems. Since the commercialisation of an IPR may not take place immediately after access or grant of the patent, it is difficult to assess the value of the IPR. As per the BDA, all cases of IPR applications based on access to biological resources, are to be routed through the NBA irrespective of the nationality of the applicant/accessor. It is the NBA that gets into the ABS agreement and also examines the PIC and ABS in other countries when patent applications based on BRs or associated TK of those countries have been the basis of patent applications in India.

Conceptual clarity is essential in implementing the provisions of ABS in the law. It is one thing to enforce the letters of the law and another to implement the spirit of the law. This applies to all the different tiers of the agencies that are engaged in this task. What is the rationale of the benefit-sharing? Is it just like another tax to raise finances for the government? One of the three objectives of the law is fair and equitable sharing of the benefits of the use of the BRs and the knowledge associated with that (Preamble). Three things stand out in this. First, ABS arises only when the resources are used, i.e. use is a must. Second, such access and use have to be by non-local people since the Act states unambiguously that section 7 shall not apply to the local people and communities in the area. Third, fair and equitable sharing of benefits arises only in the context of human beings, i.e., the beneficiaries are to be humans. The law has a clear definition of benefit claimers. They are conservers of biological resources, their by-products, creators and holders of knowledge and information relating to such biological resources, innovations and practices associated with such use and application. This is further clear from the fact that the law provides for the distribution of 95% of the benefits to the beneficiaries and allow the administrative bodies to retain 5% only for administrative expenses. The detailed procedure laid down in the Act, Rules and Guidelines also leads to this in that the benefit-sharing is to be based on Mutually Agreed Terms. This people-centric approach that ensures implementation of the provisions in the interest of the community or communities involved will have to imbue all the different levels. At a more philosophical level, one should also realise the larger ethical justification for benefit sharing. It is indeed a changing concept and very contextual (Dauda and Dierickx 2013).

Different institutions play different roles in the ABS mechanism. Seeking and granting approval is only one of the tasks. There are many issues linked with both access and exploitation of the material and the commercial application of the results where the institutions' performance matter. Benefit-sharing is closely linked with conservation and sustainability of the biological resources, since the depletion of the resources leads to drying up of benefits as was seen in the Kani case. In many instances, the lack of continuous availability of the raw material has led to the closure of the ABS mechanism. In the case of the *Arogyapacha* itself, after a certain period, the AVP found it difficult to get the raw material and they stopped the manufacture of *Jeevani*. A well-designed replantation of the plant, although the particular *Arogyapacha* species with the requisite qualities was available in Agastya Mala and Thenmala regions of the Western Ghats only being a habitat-specific plant species, could have ensured continuous availability and this would have led to regular ABS flowing to the beneficiary community. The NBA and the SBB could have taken anticipatory action once they were in place. Policy level intervention at the highest level also could have done many things. For example, when the process patent was to expire after a period of 7 years no action was taken to amend the law to make special provision for the patents based on Indian TK or BRs.

What one has to realize is that the case of ABS is quite different from that of royalty payment as per licensing terms for a patent. The parties in this case, i.e., the real beneficiaries are not well-organised groups. Their bargaining power is very limited. The law certainly has empowered them a lot in this and the NBA and the SBBs who guide the PIC and mutually agreed terms process are qualified to do their roles. But the manning of the SBBs is many a time as an additional responsibility for most in addition to their regular responsibilities as an officer of the state government. This is a handicap.

5.8 Way Forward

A fully effective ABS regime has to be both national and international. In a world where R&D and trade are global, materials and knowledge travel across borders quite frequently. As in the cases of turmeric and *Neem* patenting in the USPTO and European Patent Office respectively (NLSIU), the biological resource of one country was got researched and attempted to be patented in another country, though the effort ultimately failed. Unless the recipient country agrees to connect with the provider country no action is possible. An isolated national regime cannot ensure the benefits of access to the local communities.

It was the problems associated with the international regime that adversely affected benefit sharing that led to the Nagoya Protocol. Recipient countries did not have any responsibility earlier to ensure that the commercialization and other uses associated with biological resources are based on legal access and Mutually Agreed Terms (MAT) between the accessor and the benefit claimers. The Nagoya Protocol has made it obligatory for recipient states of GR to comply with the ABS regime. However, the state of implementation of the Protocol is very low. With the accession of Brazil on 4 March 2021, the number of ratifications has gone up to 130 only. Major industrialised countries like Australia, Canada, Israel, Italy, the Russian Federation and the United States of America have still not ratified the treaty. They are in fact major commercial users and recipients of a large number of patent applications. Many developing countries like Bangladesh, Sri Lanka and Thailand have also not become parties. Unless all major countries become parties to the Convention and introduce effective mechanisms in their countries, the ABS system will not be fully functional as non-ratifying countries will be havens for pirates.

As already stated above, there are a large number of players and tiers in the ABS system. There is the provider country, the user country and the international organisations. Apart from the institutions mentioned earlier, there are the Food and Agricultural Organisation, World Trade Organisation, Antarctic Treaty System, World Intellectual Property Organisation, and the United Nations Organisation (Schroder 2013). The countries and the international organisations have to work in harmony and with a commitment to ensure the success of the ABS mechanism.

From the experience with the CBD and the Nagoya Protocol and other related treaties and agreements what is clear is that international governance is rather weak in this area mainly because of the reluctance of the developed countries to fulfil their obligations. Governments and national authorities of developing countries will have to take a more proactive role in this matter and attempt to persuade the recipient countries of GR and TK to establish mechanisms and systems as per the Nagoya Protocol. They will have to use every available fora for this.

If one looks at the ABS mechanism from a holistic perspective of all the parties involved, we get three groups: The regulator group includes governments and international organisations. The accessor group includes researchers, their institutions and the industry. The provider group includes indigenous and local communities (Cape Town Dialogue 2014). What is required is that all the three groups work harmoniously and keeping in view the broader objective and rationale of ABS.

5.9 Conclusion

Harmonisation of procedures and regulations at international and national levels is necessary to have an effective system of fair and equitable sharing of benefits of access to biological resources and associated traditional knowledge. All agencies need to work in a cohort. Basic objectives of CBD need to be kept in view always.

BD conservation is a global responsibility and destruction of biological resources will be a loss not only for all humans but also for the entire living world.

An issue that needs to be flagged from a global governance angle is the new means of the utilisation of genetic resources without actual physical access. "The availability of the genetic sequence information and data of biological material and gene editing technologies have made it possible to avoid the need for physical access to the samples of biological or genetic material in order to utilize the same" (Syam and Romero 2021). This opens up new challenges as the CBD and Nagoya Protocol discuss physical access. How this issue will be sorted out in the future is likely to shape the way ABS emerges in the future. As of now, certainly, the best way to handle the challenges are strengthening all the tiers of stakeholders from the local level to the global level. The ways of doing this will differ depending on the tier. For example, at the local level, it will involve awareness-building, education in the intricacies of the process including valuation, development of databases on the local biological diversity on scientific lines and so on. At the state level, the role will involve coordination between the state body and local bodies and also among local bodies where more than one local body is involved. In the case of enforcement of the provisions, it would be better to handle them by the state body as they are in a better position in legal and technical expertise and to avoid contradiction in the stands. At the national level apart from coordination, the body will have to handle cases involving foreign nationals and applications for IPRs based on biological resources. It also will have to lead the negotiations at the international level to ensure the systems provided in the CBD and the Nagoya Protocol are in place at least in all major user countries of biological resources for commercial uses or acquisition of IPRs. It also will have to study and explore solutions to emerging challenges such as non-physical access to utilize GR.

In order to ensure that ABS effectively takes place and benefits the communities, a massive effort at empowerment of the lower tiers, particularly the BMCs, is required. Such empowerment involves capacity building, politically, socially, technologically and managerially. In the case of IPRs, a phrase often heard is the need for education and awareness building. This is often proposed by governments, international organisations and large industries. In the case of ABS also it is required. In this case, the education has to be of the industries and researchers since many of them are under the old belief that biological resources are free for anyone to access and the community knowledge can be freely appropriated without anyone's permission.

Conflict of Interest The views and opinions expressed in this article are those of the authors in their personal capacity. None of the observations, opinions, or conclusions expressed herein are to be attributed to or construed as the views or positions of any organization, association, or company. No conflict of interest.

References

Anitha Ramanna-Pathak (2017) Benefit sharing: reframing India's policy. FNI Report 1/2017

Cape Town Dialogue (2014) Dialogue on practical ways forward for the implementation of the Nagoya Protocol Cape Town, South Africa, 30–31 January 2014 Report. The ABS Capacity Building Initiative

CBD (Secretariat of the Convention on Biological Diversity) (2008) Access and benefit sharing in practice: trends in partnerships across sectors. Montreal, Technical Series No. 38, 140 p

Dauda B, Dierickx K (2013)Benefit sharing: an exploration on the contextual discourse of a changing concept. BMC Med Ethics 14:36. https://doi.org/10.1186/1472-6939-14-36.

Dhar B, James TC, Vinayak P (2014a) National study on ABS implementation in India. http://www.absinitiative.info/fileadmin/media/Knowledge_Center/Pulications/ABS_Dialogue_042014/National_study_on_ABS_implementation_in_India_20140716.pdf

Dhar B, James TC and Pandey V (2014b) Access and benefit sharing under the biodiversity act: towards a more effective regime. RIS policy brief No. 65. Research and information system for developing countries. ris.org.in

Jacob Thomson C, Reddy Achalendra C (2015) Implementation of the access and benefit sharing provisions of the biological diversity act, 2002: a case study on red seaweed (Kappaphycus alvarezii) in Asian Biotechnology and development review 17(3):19–31

Ministry of Environment, Forests and Climate Change. Annual Report 2017–18

NLSIU (National Law School of India University, Bengaluru, Centre for Environmental Law, Education, Research and Advocacy) (2019) Handbook on biological diversity and access and benefit sharing

Pandey R, Sane R, Yadav P, Agarwal S (2019) Biodiversity finance: identification and analysis of key building blocks of a biodiversity finance plan in India. National institute of public finance and policy, New Delhi, India

Schroeder D, Lucas Julie C (2013) Benefit sharing: from biodiversity to human genes. Springer

Secretariat of the Convention on Biological Diversity (2008) Access and benefit-sharing in practice: trends in partnerships across Sectors. Montreal, Technical Series No. 38

Syam N, Romero T (2021) Misappropriation of genetic resources and associated traditional knowledge: challenges posed by intellectual property and genetic sequence information. South Centre Research Paper 130. April 2021

Chapter 6
Intellectual Property Rights (IPR) and ABS: Need for Universal Accountability and Monitoring

Padmavati Manchikanti

Abstract The utilization of biological resources and associated knowledge has grown in several proportions. Trade in bio resources has significant global value and is an indicator for economic growth. The expansion of Intellectual Property Rights (IPR) in relation to bio resources has been at the centre of global debate due to issues of piracy, violation of human rights and overexploitation of bio resources. The Convention on Biological Diversity (CBD) mandates that utilization, prospecting of bio resources and deriving IP from them be necessarily linked to access and benefit sharing. The adoption and operationalization of the Nagoya Protocol on Access and Benefit Sharing (ABS) is the most interesting and challenging aspect of world biodiversity management. The expansion of the work program on ABS under the CBD led to the adoption of voluntary guidelines in relation to prior informed consent, code of ethics, respecting dwellings and cultural aspects of indigenous people and repatriation of TK. Under the aegis of the intergovernmental committee on genetic resources under the WIPO the expansion of the context of TK to develop an international understanding on TK is ongoing. Post TRIPS the expansion of the development dimension at the behest of the developing countries was taken up by the WIPO. While the pressure to negotiate ABS mechanisms increases, developing countries face a challenge of varied implementation of ABS as well as biodiversity monitoring. Advancements in technology have facilitated greater access of bio resources, their commercialization and increase in IP filings. The study examines the international developments in relation to IP and ABS and analyses the current perspectives. The statutory basis as well as the implementation of the ABS mechanism in relation to IP in India is analyzed from the standpoint of the Biological Diversity Act 2002. Further, whether benefit sharing is linked to biodiversity monitoring is analysed in the context of the implementation of the Biological Diversity Act 2002 drawing comparisons from a cross country context. In the mainstreaming of biodiversity for the global well-being, accountability is a key factor. The study advances a conservation centric approach to be imperative to build universal accountability and the need for biodiversity monitoring to be a shared goal. There is a need for a balanced approach in relation to ABS and IP rights.

P. Manchikanti (✉)
Rajiv Gandhi School of IP Law, IIT Kharagpur, Kharagpur 721302, India
e-mail: mpadma@rgsoipl.iitkgp.ac.in

O. V. Oommen et al. (eds.), *Biodiversity Conservation Through Access and Benefit Sharing (ABS)*, https://doi.org/10.1007/978-3-031-16186-5_6

Keywords Biodiversity conservation · Biotrade · Biodiversity monitoring · Access and benefit sharing (ABS) · Intellectual property rights · Bio resources · Patents

6.1 Introduction

Value of biodiversity differs based on the needs, cultural and religious beliefs, livelihood support, trade etc., The concept of traditional resource rights values biodiversity centric to livelihoods and, thereby, is a subset of human rights. This inextricable linkage perhaps sustained biodiversity for long years. Biodiversity is an important factor for achieving stability in the environment. The short term and long term effect of how anthropogenic factors affect biodiversity have been studied in some cases. It is observed that the changes in biodiversity due to anthropogenic environmental changes have similar effects on stability of ecosystem productivity as observed in biodiversity experiments done in a multiyear and involving multiple plant species. Changes in biodiversity cause consistent changes in ecosystem stability and decreased plant species numbers leads to lower ecosystem stability (Hautier et al. 2015). Alterations in structure and functional aspects of biodiversity affects species and their sustenance. The anthropogenic drivers have posed a global threat to biodiversity. Resilience of ecosystems have often been challenged due to human activities. Climate change has directly impacted agricultural production, survival of species and ecosystem function. The twenty Aichi targets that were part of the implementation for the Strategic Plan for Biodiversity 2011–2020 were not met. By 2020, as per the Aichi Target 11 at least 17% of the terrestrial and 10% of the marine regions respectively have to be conserved and managed. As can be observed from the National reports on biodiversity at each COP meeting, the Aichi Targets have been severely underperformed. The loss of habitat is known to lead to loss of species. Based on the analysis of the species/area ratios and congruence among taxa in 25 biodiversity hotspot regions, several aspects have been identified. 16 hotspots that are largely in the developing countries face the greatest threats with poor conservation resources. The analysis of the extent of endemism in terms of number of plants, vertebrates, the per area ratios for each and remaining primary vegetation as percentage of the original extent indicates the need to safeguard hotspots to prevent loss of biodiversity (Myers et al. 2000). A need for interdisciplinary collaborations for biodiversity conservation has been identified in studies that emphasise on the behavioral impacts on conservation targets. There are several factors such as citizens, cultivators, governments, businesses, school teachers, universities, religious groups who are variously graded in terms of their behavior towards conservation targets (Nielsen et al. 2021). Sustainable forest management and its role in mitigation of climate change has been discussed. Increasing tree biodiversity has positive effects on carbon sink and therefore can be an important facet for climate stabilization. Analysis of the extent of contribution of tree diversity through primary production in several biomes such as tundras, temperate grasslands, temperate coniferous forests, mangroves (fourteen biomes as defined by the World Wild Life Fund) has led to the identification of different Shared Socioeconomic Pathways that project different land-use changes. The pre-area loss of biodiversity-dependent productivity was observed to be less in warmer biomes. Regional heterogeneity estimates and country level estimates indicate the level of social cost of carbon. Countries with high level of social cost of carbon have greater incentives to mitigate climate change. The benefits of forest biodiversity as an important aspect of climate solutions has been emphasized (Mori et al. 2021). The efforts for creation of long term carbon sinks are vital and prioritizing biodiversity efforts for the same are necessary. Marine protected areas are also important carbon sinks and like forest biodiversity have a tremendous capacity to absorb carbon based emissions. It is projected that by the year 2030 at least 30% of the ocean biodiversity should be restored. Based on multi-objective prioritization, it has been observed that for marine protected areas there can be several cumulative benefit functions that include biodiversity protection, food provision and carbon storage. Strategic protection of 21% of the ocean would increase the average protection of endangered and critically endangered species; identify food provisioning outcomes in relation to these areas. Analysis of sediment carbon emissions reveal estimates in relation to the atmospheric CO_2 absorber per year by the oceans and how remineralisation can be prevented. There are several threats to the sediment carbon that need to be explored based on synergies and optimization. (Sala et al. 2021). Investment into biodiversity has been observed to be an important determinant and national level investment estimates derived from the BIOFIN (Biodiversity Finance Initiative) provide an account of the public biodiversity expenditure. An analysis of thirty countries for biodiversity expenditure measures for the decade of 2008–2018 revealed that GDP and biodiversity expenditure are correlated. Public biodiversity expenditures are higher when the number of threatened species increases by 1%. Developed countries are found to invest less than developing countries (Seidl et al. 2021). While there is already expanded research on mainstreaming biodiversity, funding initiatives, national, regional and local policies and on effects of climate change and other factors on several ecological systems, the globe faces competing interests to conserve biodiversity in the regions rich in diversity and also provide access to resource for the development of value added products. IPR have been necessarily in conflict with the conservation objectives in relation to bio researchers and it is not surprising. Balancing ABS and IPR is one of the toughest challenges that policy makers and implementing agencies face due to several sorts. There are fundamental differences in the rights paradigm in both the cases. While IPR are predominantly private rights, access to biodiversity is an essential component of human rights. Communities have since time immemorial have a shared basis of such rights. Biodiversity is not evenly distributed and national policies are subtly different. The contours of IPR are also different and have varied implementation across nations. Not unexpectedly the context of ABS and IPR takes a centre stage in international discussions.

This chapter begins with a description on the context of how innovations and commercial value of biodiversity encourage the pursuit of IPRs. The background of ABS and its elaboration from the CBD and post Nagoya perspective is discussed. Keeping in view the interface aspects of IP and ABS, the next section discusses the developments in this regard from a multilateral basis. The third section deals with the ABS implementation in the IPR process in India from the perspective of the Biological Diversity Act 2002 and the IPR legislations. The fourth section deals with the discussion on conservation centric approach to ABS and IPR. The last section identifies a framework on what would drive outcomes in relation to universal accountability and biodiversity monitoring.

6.2 Innovation and Commercial Value of Biodiversity

The use of biodiversity and associated knowledge has significant value in human well-being. Traditional innovations involved developments that helped sustained livelihoods and the environment. This knowledge remained private within the community. Exploitation of the knowledge associated with bioresources has been going on since long years. Advances in technology have enabled not only the collection of the species but also in quick conversion of the embedded knowledge into products of value. Investments into biodiversity have increased manifold and commercialization of biodiversity based innovations has grown to be an important area. Bioprospecting is no longer an individual activity but an enterprise. There are concerns that arise that bioprospecting can lead to disturbances in biodiversity and also lead to overexploitation of resources. It is widely known that the biodiverse rich countries have greater product manufacturing capabilities unlike the biodiverse poor countries. Industries that utilize biodiversity may be involved in raw material collection or development of intermediates or in final product formation. Several industries are involved in the entire stages including development of end user products. National bioeconomy strategies have been announced which largely focus on how sustainable bioeconomy approaches can be developed for food security, natural resource management, greater use of renewable resources, climate change positive approaches and identification of opportunities for human resource development.

One of the important contexts in sustainability is circular economy. This approach takes into consideration the utilization of waste products as further raw material for generation of new products. Circular bioeconomy utilizes biomass by increasing its use and creating more sustainable bioresource use and lowering carbon footprint. A survey of patents during the period of 2000–2020 indicates predominant contributions from North America and Europe that form 1/3rd of the patents with respect to physical processes alone. Another component includes a combination of physical, chemical and/biotech processes (Brandao et al. 2021). In a survey that involved scientists from five different continents, a platform was established to examine the context of biopiracy and patenting in relation to the bioresources in PubMed literature. Since 2011, more than 1500 publications per year have been found. While this is so, the number of papers related to biopiracy is very low (from 1990 to 2015). The study reveals that bioprospecting and biopiracy are the center of research on medicinal plants. Individual scientists who work on isolation and characterization studies must be aware of the potential violation of indigenous community rights as their work could downstream lead to development of products without no benefits accruing to the communities. Phytomedicine platforms not only help in being repositories of knowledge on the use of plant species but can effective to prevent biopiracy (Efferth et al. 2019).

Pharmaceutical prospecting involves several aspects such as the development of information about species and the chemical constituents. A large number of activities are involved beginning with harvesting biodiversity to the potential marketable drug. The activities include collection and identification of biological samples, screening, extraction, isolation and characterization, preclinical research, data generation and animal testing and clinical research. Advancements in technology have facilitated an increase in large number of species that can be screened. One of the prime concerns on pharmaceutical prospecting has been whether it can finance biodiversity conservation. Three major types of investments are envisaged in relation to pharmaceutical prospecting; investment in protection of biodiversity, investment in collection and identification of samples and investment in pharmaceutical R &D. A model for biodiversity investment choice has been proposed with the aspects of investment in biodiversity protection and species information to be complementary activities (Barbier and Aylward 1996).

There has been a growing interface of biotechnology in the development of value added products derived out of biodiversity. Along with this came two divergent views; less control of state due to the need to encourage research in biodiversity and increasing private rights in form of IP protection. Changes to patent legislations to achieve compliance in relation to TRIPS agreement have led to the introduction of product patent regime in member countries. Impetus to biodiversity prospecting and the development of biomedical innovations has been observed in some countries. In case of a study on the development of new medicines in Brazil, public private partnerships have led to effective marketing of the medicines. Reforms to the 1996 Patent Law of Brazil are purported to be the main driver for promoting innovation (Ryan 2010). The growth in collaboration with international agencies to promote conservation efforts on biodiversity while engaging in bioprospecting has been observed. In 1991, the Biodiversity Development and Conservation Programme (BDCP) formed in Nigeria involved preparation of an inventory of ethno botanical resources of the rain forests, assessment of the commercial value of the plant species and establishment of long term nature plots. Several collaborative projects were undertaken with companies and governmental agencies; the development of pharmaceuticals based on ethno medicinal plants with Shaman Pharmaceuticals Ltd, new drug development for treatment of malaria, leishmaniasis and trypanosomiasis with the US Army. Further, expansions of the international collaborations were done keeping in view sustainable development of bioresource economy, distribution of benefits to individuals, rural communities and local institutions (Iwu 1996).

In Brazil, a three year molecular ecology project called Bioamazonia involved a memorandum of understanding (MOU) with Novartis company. The agreement involved access to 30,000 biological samples from the Amazon rainforest to be exported to Switzerland for creation of new products for which Novartis would hold IP rights. Bioamazonia would in return receive stipulated upfront monetary payment for each clinically tested sample, any product commercialization and 1% royalty for a period of 10 years. While this agreement was cancelled in the year 2000, the need for balancing the interests of researchers and scientists some aspects of the biodiversity access were considered for exemption under the approval system under the Provisional Measure no 2186–16. Post Nagoya Protocol, the submission of a certificate of origin and prior informed consent was introduced in relation to research involving access of bioresources from Brazil (Filoche 2012).

In 2018, the Global Biodiversity Summit discussed the relevance of bioeconomy and its positive contribution to sustainability. A review on the number of studies that study the relationships revealed different approaches. Resource vision, biotechnology vision and agroecology visions have been identified as the three different strategies that support the relation (D'Amato et al. 2020).

The access to crop genetic resources for agriculture and food production has been emphasized incessantly in view of IPR. Granting of private property rights on plant varieties and inventions based on agricultural biodiversity has been opposed by those advocating human rights and national sovereignty over natural and genetic resources. Most IPR laws do not recognize benefit sharing and hence this has been dealt under the national laws or specific laws such as biodiversity legislations. Increased commoditization of bioresources has made governance of biodiversity very complex (Tsioumani 2018).

Crop biodiversity has been another area where the need for a comprehensive relook at the 2001 International Treaty on Plant Genetic Resources for Food and Agriculture (ITPGRA) has been realized. Annex 1 of the Treaty list the crop species that are shared with the wider aim of food security and common modalities for sharing are for agriculture purposes. Any non- agricultural use of Annex 1 would be subject to ABS under CBD. The need for an institutional design for incentivisation of the providers of genetic resources has been proposed with five aspects. The factors such as the narrow scope of application, higher transaction costs of the users, ambiguity on IP and benefit sharing, differentiated obligations on patent or plant variety protection and provider side implementation have been identified as the aspects that disincentivise benefit sharing (Rabit 2017).

While countries began to adjust existing legislations or enact new legislations to achieve compliance in relation to TRIPS and CBD, several concerns on biopiracy continued to be raised at the national as well as the international levels. Understanding country responses in relation to implementation of IP laws amidst prevalent biodiversity legislations or post enactment of legislations indicate to several interesting scenarios. In case of Australia, at the Commonwealth level, laws such as the Environment Protection and Biodiversity Conservation Act 1999 and the Environment Protection and Biodiversity Conservation Regulation 2000 are applicable. Queensland and Northern territory have the Biodiscovery Act 2004 and Biological Resources Act 2006 respectively. Model contracts for ABS agreements that take into considerations accruing of benefits for use of indigenous knowledge. The Commonwealth of Australia implemented the ABS agreement with the Craig Venter Institute to enable the latter to bioprospect the Australian waters for collection, DNA sequencing and analysis of microorganisms. In this case the IP vested with the Commonwealth government. During 1993–2007, the Griffith University and AstraZeneca established a partnership for natural product discovery under the Queensland biodiscovery collaboration. These examples demonstrate the need for addressing IP ownership issues vis a vis conservation of resources (Lawson 2010). A case study of the patents in relation to three plants and their fruits, the Camu Camu, Kakadu Plum and Acai indicate the issues in relation to patenting and ABS.

A search for names, components and uses of six medicinal plants of Mexico in the patent databases of the USPTO, EPO, JPO and Mexican Institute of Industrial Property for the period of 1976 to 2012 revealed that US patents recorded the highest followed by EU patents and Japan patents respectively in terms of the claims related to the species. Although further validation is required, the study identified that these patents could be potential sources of evidence of possible biopiracy (Soria-Lopez and Feuntes-Paramo 2016).

The Quassia case on biopiracy demonstrates the research groups and scientist challenges related to natural products research and how a misconstrued understanding of ABS could severely hamper the development of therapeutics. The study provides a chronological understanding of the identification of *Quassi amara* and its indigenous use for treatment of fevers and specific use in malaria (Bourdy et al. 2017).

6.3 Intellectual Property Rights and Biodiversity Conservation-TRIPS and CBD

Intellectual Property Rights have often been considered to be in conflict with biodiversity conservation. This is predominantly due to exploitation of bioresources at the cost of their conservation, lack of prior informed consent, violation of the rights of indigenous communities to their environment and no compensation to the communities. The unabated use of biological resources for commercial purposes resulted in resource exploitation. Traditional Knowledge consists of trade secrets as well as information in the public domain. A significant part of pharmaceutical industry is dependent on the knowledge associated with the use of medicinal plants. Such knowledge has been the main source for the isolation of active principles and development of drugs. While the knowledge and practices of the indigenous communities, was converted into products and marketed by industries, the profits earned by the commercialization of such finished end products were either not shared with or very little benefits were given to the communities. Prior to the year 1992, genetic resources were regarded as the "common heritage of mankind" and as such were treated as an unregulated and freely accessible good. There was no requirement that benefits need to be shared with the country of their origin for the use of the bio-resources.

TK started to gain greater prominence as it began to be raised at multiple fora. In response to resource exploitation, the Ad Hoc Working Group of Experts on Biological Diversity was formed by the United Nations Environment Programme (UNEP) in November 1988 to find the need for an international convention on biological diversity. In 1989, it established the Ad Hoc Working Group of Technical and Legal Experts to prepare an International Legal instrument for the conservation and sustainable use of biological diversity taking into account “the need to share costs and benefits between developed and developing countries” as well as “ways and means to support innovation by local people”. By February 1991, the Ad Hoc Working Group had become known as the Intergovernmental Negotiating Committee. Its work cumulated on 22nd May 1992 with the Nairobi Conference for the Adoption of the Agreed Text of the Convention on Biological Diversity (CBD) which entered into force on 29th December 1993. The three important principles of CBD are conservation, sustainable use and fair and equitable benefit sharing.

At the TRIPS Council under the WTO, developing countries made several attempts to evolve an effective framework so that the WTO members can meet their obligations under TRIPS and the CBD. As acknowledged in the Doha Ministerial Declaration, the protection of TK was among the important IPR issues to be resolved in the TRIPS Council. Paragraph 19 of the Doha WTO Ministerial Declaration called for the TRIPS Council to examine the issue (Gervais 2005). This included the study of the relationship between the TRIPS Agreement and the UN Convention on Biological Diversity in relation to protection of TK and folklore (Musungu 2008). The need to raise traditional knowledge (TK) to the status of one of the Intellectual Properties (IPs) during the TRIPS agreement is justified from the stand point that this represents the intellectual creation of the human mind which has the maximum public good character among the forms of IP. This also represents the IP that generations have not only accepted but also most effectively used and disseminated. In the review of Article 27(3) patentability of certain biotechnological inventions was discussed. During review of Article 27.3 in light of protection of plant varieties by the TRIPS council, many developing countries (including India) proposed arguments against the motion (TRIPS Council Secretariat Note IP/C/W/369/Rev,2006). The countries suggested that Article 27.3 (b) should be amended to prohibit the patenting of all life forms. In April 2011, WTO delegations of Brazil, China, Columbia, Ecuador, India, Indonesia, Kenya (on behalf of the African Group), Mauritius (on behalf of the African-Caribbean-Pacific Group), Peru and Thailand circulated a communication entitled: “Draft Decision to Enhance Mutual Supportiveness between the TRIPS Agreement and the Convention on Biological Diversity” to the Trade Negotiations Committee as part of the ongoing Doha talks. It proposed to include a new Article 29bis on “Disclosure of Origin of Genetic Resources and/or Associated Traditional Knowledge” in the TRIPS Agreement (**TN/C/W/59**). It was emphasized that the TRIPS agreement be amended such that Members shall require that an applicant for a patent relating to biological resources or Traditional knowledge, shall provide, as a condition to acquire a patent right:

- Country of origin details, disclosure of the source of the biological resource and/or the traditional knowledge used in the invention;
- Applicants to provide a copy of the Internationally Recognized Certificate of Compliance (IRCC);

- Evidence of prior informed consent through approval of authorities under relevant national regimes, where applicable; and
- Evidence of fair and equitable benefit sharing under the national regime of the country of origin, where applicable.

In the early 80s', WIPO worked on the issue of folklore together with the United Nations Educational, Scientific and Cultural Organization (UNESCO) leading to the announcement of the WIPO/UNESCO Model Provisions for National Laws on the Protection of Folklore against illicit exploitation and other Prejudicial Action. It was adopted in 1982 and used as a model by many Member States interested in *sui generis* protection of folklore (Antons 2010). The Standing Committee on Patents (SCP) was created in 1998 to serve as a forum to discuss issues, facilitate coordination and provide guidance concerning the progressive international development of patent law. In 2001, developing countries and many other members argued for the need for amendments to the TRIPS agreement under the purview of Article 71.1. In 2002, the WIPO recognized that the IP system is in direct conflict with traditional practices and lifestyles. TK holders face a dichotomy of the need to continue their own customary regime on one hand and the IP system administered by governments in another. In 2006, in the 8th Conference on Parties to the CBD held in Brazil highlighted the necessity for reform of domestic and global IP regimes to ensure proper protection to traditional knowledge.

In the field of TK, WIPO had initiated the formation of the Intergovernmental Committee on Intellectual Property and Genetic Resource, Traditional Knowledge and Folklore (IGC). Through this committee it has been addressing the substantive aspects of traditional knowledge and related areas. Beginning with its work in 2001 on the basis of the results of fact-finding missions conducted in 1998 and 1999.The 3rd Session of the IGC discussed the themes of sui generis protection of TK, definitions and terminology relating to TK and expressions of folklore. By 2005, the committee established the WIPO Voluntary Fund for Accredited Indigenous and Local Communities. In the 16th session held in 2010, the IGC requested the WIPO Secretariat to make available a list of brief technical explanation of various forms in which TK may be found. This was the first attempt towards an internationally agreed definition of different categories of TK. As there is an inherent complexity, variedness and dynamic aspect of TK, the different subsets of TK such as unfixed and fixed TK, document and non-documented TK, codified and non-codified TK and disclosed and undisclosed TK. Further, secular TK and sacred TK, relationship between indigenous knowledge and traditional knowledge, individual TK and collective TK and commercialized TK and non-commercialized TK are the other categories identified (as part of the gap analysis under WIPO/GRTKF/IC/13/5(b)Rev). The 25th session of the works of the IGC, held in Geneva from 15 to 24 July 2013, has discussed, among the others, two draft articles, one on the protection of TK and another on the protection of TCEs. In June 2019, Draft articles for the protection of TK were announced. This contains the purpose of protection of TK, usage of terms, the objectives of the instrument, scope and criteria for protection. Under Article 6 it recognizes appropriate legal and administrative measures for addressing violation of TK. The hallmark of the IGC's work has been the announcement of the three draft provisions; protection of TK, protection of TCEs and IP and Genetic Resources. In the 41st session of the IGC, the committee made recommendations to the WIPO General Assembly to renew the mandate of the committee for the 2022–2023 period.

TRIPS mandated the recognition of IP over bioresources. On the contrary keeping in view the principle objective of conservation, the CBD insisted on member countries to protect life forms by conserving and sustainably utilizing biodiversity (Article 1 CBD). The CBD encouraged the equitable sharing of benefits arising from the utilization of knowledge, innovations and practices associated with use of biological resources (under Article 8j). Sustainable use is yet another important aspect that is vital for conservation as well as utilization of biological resource. Sustainable use as per CBD 'means the use of components of biological diversity in a way and at a rate that does not lead to the long-term decline of biological diversity, thereby maintaining its potential to meet the needs and aspirations of present and future generations'. The initial COP meetings discussed on the setting up of various working groups to initiate the secretariat functions and also categorizing areas in the world into different zones (such as semi-arid, marine etc.,) for implementation of CBD. From the later COPs and particularly during COP1 the discussions related to Access and Benefit sharing (ABS) took concrete shape. The provisions related to ABS are Article 8j and Article 15 which set the minimum standards for member countries. Article 8j reads as "Each Contracting Party shall, as far as possible and as appropriate subject to its national legislation, respect, preserve and maintain knowledge, innovations and practices of indigenous and local communities embodying traditional lifestyles relevant for the conservation and sustainable use of biological diversity and promote their wider application with the approval and involvement of the holders of such knowledge, innovations and practices and encourage the equitable sharing of the benefits arising from the utilization of such knowledge, innovations and practices (emphasis added)". This article is by far the most important one in relation to ABS. The Working group on ABS has been reviewing this in relation with implementation in various countries.

Ownership and control of genetic resources became an intensely debated issue that led to the concept of Access and benefit sharing under the CBD. The developing countries wanted share in the profit derived from the commercialization of a product derived from their bio-resource. On the other hand, the developed nations needed to ensure that their development practices and interest of developing countries are not affected. Therefore, the agreement by contracting parties to facilitate access to genetic resources in return for a fair and equitable share of the benefits derived from their use represents a negotiated resolution between the technology rich countries and biodiversity rich countries. At the fourth COP meeting in 1998 a Panel of Experts on Access and Benefit-sharing was established to clarify principles and concepts related to ABS. This panel discussed issues such as prior-informed consent, mutually agreed terms, benefit-sharing, capacity-building and stakeholder involvement in ABS processes. At its fifth meeting, in 2000, the COP the Ad Hoc Open-ended Working Group on Access and Benefit-sharing, a subsidiary body of the Conference of the Parties, with the mandate to develop guidelines and other approaches to assist Parties with the implementation of the access and benefit sharing provisions of the

Convention. The Bonn Guidelines on Access to Genetic Resources and Fair and Equitable Sharing of the Benefits Arising out of their Utilization were adopted in 2002 to assist Parties when establishing administrative, legislative or;policy measures on access and benefit-sharing and/or when negotiating contractual arrangements for access to genetic resources and benefit-sharing. The application of Article 8j in conjunction with Article 15 of the CBD is necessary in bringing in ABS mechanisms in a country. Article 15 (1) of CBD reads as 'Recognizing the sovereign rights of States over their natural resources, the authority to determine access to genetic resources rests with the national governments and is subject to national legislation'. Further, Article 15(4) and Article 15(5) relate to access on mutually agreed terms and need for prior informed consent respectively.

It is pertinent to note that Article 16(5) of the CBD stipulates that intellectual property rights need to be supportive of the CBD's objectives. Article 16 (5) read as—The Contracting Parties, recognizing that patents and other intellectual property rights may have an influence on the implementation of this Convention, shall cooperate in this regard subject to national legislation and international law in order to ensure that such rights are supportive of and do not run counter to its objectives (emphasis added).

By inculcating access and benefit-sharing system, sui-generis system for legal protection, legal protection under IP law and public domain, efforts have been made at the international level (Phillips 2013). Another important component of the intellectual property aspect is digital sequence information (DSI). The CBD Decision XIII/16 and the Nagoya Protocol Decision 2/14 deal with the aspect of DSI. At the 13th meeting of the COP the decision was taken to establish an Ad-Hoc technical Expertise Group (AHTEG). A scoping and fact finding study was done by the AHTEG on the terminology and different types of DSI on genetic resources and potential implication for use of DSI in relation to the objectives of the Convention. The outcome of the AHTEG work identified the scope of applicability of DSI (SBSTTA/22/2 meeting 2018). There are more a thousand sequence databases of different organisms. As more whole genome sequences of organisms become available, the utilizing of digital sequence information will continue to increase. Big data analysis in relation to digital sequence has delinked the need to work with the physical organisms. Mining of DSI and utilizing information has opened greater opportunities for genetic manipulation. Governance norms in relation to digital sequences are not yet clear. While some countries have already adopted measures in this regard, many others are yet to do so (Bagley et al. 2019). With plant biotechnology companies holding a significant market in relation to improved seed varieties, there have been several concerns on the appropriation of DSI in the form of IP rights. This in relation to the crop species recognized under the ITPGRFA would affect access to germplasm and farmer rights. In view of the DSI discourse, one of the approaches that has been suggested is to either have an enhanced multilateral system under the ITPGFRA or have an expansive approach of recognizing not only sharing of germplasm but encouraging innovation by building an open source license that will preserve the right to use the resource for breeding and protect the rights of farmers (Wynberg et al. 2021).

6.4 Post Nagoya—ABS Implementation in Relation IP

The Nagoya Protocol on Access to Genetic Resources and the Fair and Equitable Sharing of Benefits Arising from their Utilization (ABS) to the CBD is a supplementary agreement to the CBD that was announced in 2010 after years of negotiations. It came into force from 12 October 2014. It provides details on how to establish the institutional mechanism and process of ABS. It is the first international instrument which included indigenous rights as a cross-cutting issue on international negotiations. The protocol also contains provisions catering to TK associated with genetic resources. Obligation of prior informed consent of the local communities and benefit sharing modes are an important part of the protocol. Post the Nagoya Protocol the implementation of user and provider measures has become an important aspect with respect to domestic legislation.

User measures and provider measures are key aspects that need to be implemented at the national level in member countries post Nagoya Protocol. Article 14.1 of the Protocol mandates the need for national check points. Article 14.1 states that 'Parties to the Nagoya Protocol are obliged to designate checkpoints which "collect or receive [...] relevant information related to prior informed consent, to the source of the genetic resource, to the establishment of mutually agreed terms, and/or to the utilization of genetic resources". As per Articles 15 -18 parties need to create measures involving principles of PIC and MAT for the utilization of genetic resources within their jurisdiction and benefit sharing. Article 17 identifies the need for internationally recognized certificates of compliance. ABS legislation of provider countries varies. Article 6 of the Protocol mandates the development of international access standards by providing legislative, administrative or policy measures as appropriate to bring in certainty and transparency in the application. Mandatory procedures are to be implemented by the users of genetic resources and checkpoints need to monitor access.

Botanical gardens need to acquire new plant genetic resources legally complying with ABS terms. Hence, awareness of national ABS measures, adherence to laws related to traditional communities, developing institutional ABS policy, PIC and MAT as required, use and transfer material as per ABS terms and share benefits are the aspects that need to be considered. In view of the IPEN (International Plant Exchange Network) code, principles on access to genetic resources and benefit sharing model of twenty-eight botanical gardens from twenty one countries and the CETAF (Consortium of European Taxonomic Facilitation) Code of Conduct and Best practices and the GGBN (Global Genome Biodiversity Network), a general framework of ABS application has been proposed as relevant (Davis 2019).

In the last meeting of the COP i.e., COP14 in Egypt, the parties decided to establish a science and policy based process on digital sequence information and the formation of an Ad Hoc Technical Expert Group, identify terminologies, scope, domestic measures, identify ABS framework in view of use of DSI, traceability, database access etc., Annex 1 of the ITPGRFA does not have restrictions that are imposed under the Nagoya Protocol for the purposes of ABS. Examining digital sequence information as big data is challenging in relation to agricultural IP. Several crop genomes have been sequenced. Some of the issues of ownership of big data, access and usage of digital data, governance norms are not yet fully addressed. In the context of DSI, there is a need for analyzing the context of ABS from a user and provider perspective. The analysis of the three components of science and technology development, alignment of geographical/political contours and governance mechanism led to the development of a model to identify to what extent governance mechanisms have developed to balance IP and ABS (Smyth et al. 2020).

There is a varied implementation of the Nagoya Protocol in the mega biodiverse countries. In the year 2020, the process of Brazil's ratification of the Nagoya Protocol started with the notification of the Legislative Decree No 136/2020. There is a need for implementation of the key rules of the Protocol such as designation of national focal point, PIC and access procedures, monitoring among others. From a practical standpoint it is expected that ratification should be helpful in a dual situation of Brazil not only being a provider but also a user of biological resources (Carls 2020).

Countries differ in the way they interpret 'access' and 'benefit sharing'. This leads to key differences in the interpretation of policy as well as law. Addressing intrastate implementation, elaboration if user measures, strengthening the administrative mechanism, addressing transboundary concerns continue to be some of the important aspects in the Nagoya Protocol implementation.

6.5 ABS Implementation and IP in India Under the Biological Diversity Act 2002

India is one of the mega-biodiverse countries in the world and is a party to key multilateral agreements on environmental issues. It has also played a lead role in many of these international conventions. The Biological Diversity Act, 2002 was enacted to meet the compliance under the CBD. Unlike the prevailing environmental legislations, the Biological Diversity Act is unique as biodiversity conservation is considered from the entire ecosystem and livelihood perspective. The three principal objectives of the Act are conservation of biological diversity, sustainable use of its components and fair and equitable sharing of the benefits arising out of utilization of genetic resources. The implementation of these three objectives is at three levels; the National Biodiversity Authority (NBA) at the national level, State Biodiversity Board (SBB) at state level and Biodiversity Management Committees (BMCs) at local levels.

The regulation of the activities related to the access of biological resources occurs within the respective authorities/jurisdictions. The objectives of the Act are interlinked. The implementation of the ABS mechanism in India is relevant from this perspective. The Act identifies access and use of biological resources and knowledge associated thereto for certain specific purposes such as commercial utilization, research, bio-survey and bio-utilization. Commercial utilization is defined under section 2 (f) of the Act and this covers uses of biological resources that involve certain end uses and in certain processes. The NBA has multiple roles. It facilitates conservation besides ensuring sustainable use and fair and equitable sharing of benefits arising from such use. It has a regulatory role with respect to access and use of biological resources occurring within India (under section 3, 4 and 6 of the Act). It has an advisory role with respect to Central Government and the State Governments in the matters relating to biological diversity. The NBA further acts on behalf of the Central Government in opposing grant of IPR in any country outside India based on any bioresource, or knowledge associated thereto, from India. It may also perform such other functions as may be necessary for the overall implementation of the provisions of the Act.

ABS is one of the important areas of biodiversity governance internationally and nationally. It secures fair and equitable share in the benefits arising out of access and utilization of biological resources. The benefits are given to the benefit claimers and for the conservation of biological diversity. The Act makes it mandatory for such persons to apply for prior approval from NBA in cases where they fall within the regulatory jurisdiction of the NBA. The transfer of research results relating to any biological resource occurring in or obtained from India and associated knowledge and application to obtain any Intellectual Property Right for any invention based on research or information relating to any biological resource obtained from India are regulated under the ABS mechanism. The sections 3, 4, 6, 7 and 21 of the Act are key for ABS regulation in India (Fig. 6.1).

It is worthwhile to understand the scope of the term 'biological resources'. As per section 2(c) of the Act "biological resources" are defined as "*plants, animals and micro-organisms or parts thereof, their genetic material and by-products (**excluding value added products**) with actual or potential use or value, but does not include*

Fig. 6.1 Regulation of ABS in India

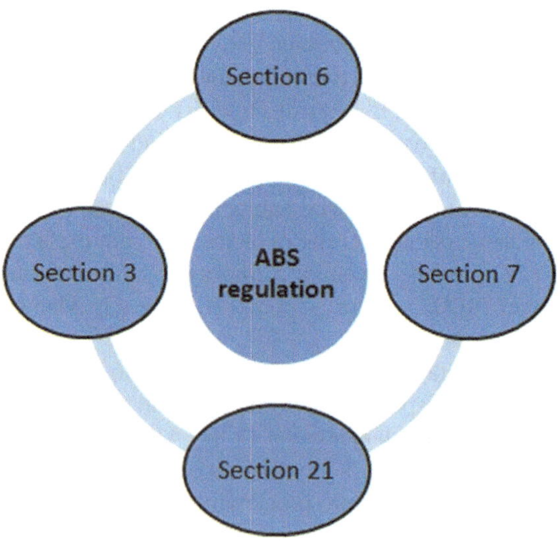

human genetic material" (emphasis added). This definition includes flora, fauna and microbes, in whole or in parts, along with their genetic materials and the by-products. But, it does not include the value added products of such plants and animals. The term "value added product" is defined under section 2(p) as "*products which may contain portions or extracts of plants and animals in unrecognizable and physically inseparable form.*" So, any such products which may possess "potions of extracts of" plants and animals, and not microorganisms present in unrecognizable and physically inseparable form will come under value added products.

As per section 3 of the Act, any person or entity having any foreign participation seeking access to any biological resource and/or associated knowledge or to obtain any biological resource occurring in India, for commercial utilization, research or bio-survey and bio-utilisation shall have to seek prior approval from the NBA before such access. As per section 4 of the Act, any person who seeks to transfer to any person or entity with foreign participation, the results of research relating to biological resource occurring in, or obtained from India, has to apply and obtain prior approval of NBA before such transfer. Exemptions to section 4 are in relation to collaborative projects between Government of India (or its agencies) and other countries and the publication of research papers or dissemination of knowledge in any seminar or workshop provided such publication is in conformity with Central Government Guidelines. In relation to section 6 of the Act, any person who seeks to apply for any Intellectual Property Right, anywhere, within or outside India, for any invention based on any research or information on a biological resource obtained from India has to apply and obtain prior approval of National Biodiversity Authority, before applying for such IPR. In cases where the IPR sought is a patent, the approval from National Biodiversity Authority may be applied and obtained prior to sealing of the patent. A clear distinction between the biological source and the plant varieties has been made and the Act won't apply to an applicant "making an application for any right under any law with respect to protection of plant varieties". Other exemptions are related to normally tradable commodities and collaborative research projects conducted between India and foreign institutions under international cooperation.

Under the functions of National Biodiversity Authority, as prescribed under section 18, the Authority may, on the behalf of the Central Government, take any measures necessary to oppose the grant of intellectual property rights in India or abroad on any biological resource that has been obtained from India. Further, the Authority may also call for joint ownership with regard to the intellectual property rights in order to secure benefits.

According to section 7, any Indian person or entity not covered under the earlier category, seeking to access and obtain biological resources for commercial utilization or bio-survey and bio-utilization for commercial utilization has to give prior intimation to the relevant State Biodiversity Board, from whose State the biological resources are sought to be accessed or obtained. Local people and communities practicing indigenous medicine including *vaids and hakims,* and growers and cultivators of biological diversity of the local area are exempted from this provision. At the time of writing of this chapter amendments have been proposed to section 3 and section 7 among other provisions and the Biodiversity amendment bill is under the discussion stage.

Beside the provisions mentioned above, section 21 of the Act outlines the criteria for equitable benefit sharing by elaborating the terms and conditions under which approval is granted when an application is made to the NBA. Sharing of the benefits that would arise from the utilization of biological resources and associated knowledge is to be in accordance with mutually agreed terms between the applicant, local body concerned and the benefit. Further, the Biological Diversity Rules 2004, provide the scope of application of ABS in setting out conditional access of bioresources where conditions may affect conservation.

Effecting an ABS agreement involves multiple processes. BMCs may be consulted on the applications filed. BMCs role is outlined in the relevant State Rules of respective states. This may include public announcements, meeting of the local body to discuss the complete information on the ABS application. After this, the BMC advises the SBB and/or the NBA on the request for access and accordingly NBA either approves or rejects and ABS application. It is only after the whole process that NBA signs the ABS agreement and where applicable, the BMC shall also monitor the flow of equitable share in the benefits that was agreed upon, back to the local communities by engaging with the SBB. In relation to section 21(4), the NBA introduced the ABS guidelines 2014 (known as 'Guidelines on Access to Biological Resources and Associated Knowledge and Benefits Sharing Regulations 2014'). The Regulations 8, 9 and 10 are relevant for the purposes of IP Rights. The Regulation 8 outlines the procedural formalities that need to be met by the applicant or the purposes of obtaining IPR in relation to bioresources. Under Regulation 9, the mode of benefit sharing in commercialization of IPR are identified under the two conditions of whether the applicant is involved in direct manufacturing or has licensed the use of IPR. Further, Regulation 10 deals with the obligation of the applicant with respect to the concerned State Biodiversity Board and the NBA as the case may be. The first sets of approvals for access of bioresources were given by the NBA in the year 2007.

Rule 18 states the "*procedure for the purpose of obtaining prior approval before applying for intellectual property protection*". For this purpose, any person seeking a patent grant or any other intellectual property on the basis of research conducted on biological material obtained from India shall make an application in Form III. For this reason, an applicant for the grant of the patent will do so under Form III with an application fee of INR five hundred rupees. The onus then shifts on the NBA where the expert committee on ABS within the specified period of three months will decide whether or not approval for access can be given. The Authority may also impose any further terms and conditions in this matter. As on date 2332 applications for IPR under Form III have been disposed by the NBA. The NBA is empowered to reject the request of the applicant after providing a reasonable hearing opportunity for the reasons to be duly recorded in writing. These rejections could be due to improper applications or due to conditions that do not favor access of a bioresource. There may be a restriction or prohibition of access to a biological resource under any of the conditions such as endangered taxa, endemic and rare species, if there is an adverse effect on the livelihoods of local people or on the environment, lead to genetic erosion, affect ecosystem and that which will be contrary to national interests and international agreements. Further, Rule 15 stipulates that if prevalent conditions threaten the conservation objectives, the approval given may be revoked.

Any contravention of provisions of sections 3, 4, 6, 7 and 24(2) are all cognizable and non-bailable. For violations of sections 3, 4 and 6, the penalty is fine which may extend upto ten lakh rupees and/or imprisonment which may extend upto five years. For violations of sections 7 and 24(2), the penalty is fine which may extend upto five lakh rupees and/or imprisonment which may extend upto three years. For violations of any order passed under the Act, the penalty is fine which may extend upto one lakh rupees for the first instance. For any second or subsequent violation, the penalty is fine which may extend upto two lakh rupees.

From the perspective of the Patents Act1970, as per Form-1 of the Patent Rules, the applicant needs to provide a declaration that "the invention as disclosed in the specification uses the biological material from India and the necessary permission from the competent authority shall be submitted by me/us before the grant of patent to me/us". If the applicant has utilized biological resources obtained from India and has failed to secure permission of NBA in this regard, the Examiner/ Controller, at the examination stage, or at any later stage, raises an objection with regard to the requirement of permission from NBA. Section 10 (4) of the Act mandates the need to identify country of source and geographical origin of the biological resource. The guidelines in relation to TK and biological material of the Patent office provide useful information to applicants in this regard.

In the year 2004, a patent held by Monsanto titled 'Plants' (EP 044599B1) was challenged at the EPO. This patent related to the engineered Galatea wheat plant which was developed based on the introduction of genes from the Indian landrace, Nap Hal. The Indian wheat variety is known for its low viscoelastic properties that make it suitable for chapathi and biscuit preparations. The filing of a writ petition in the Supreme Court of India by Research Foundation for Science and Technology and Ecology (in the case of Research Foundation for Science and Technology and Ecology and Anr *v* Union of India- WP (Civil) No. 64 of 2004) eventually led to the filing of the revocation of the patent at the EPO.

A European patent EP 1,962,578 titled 'Closter virus resistant melon plants' was revoked at the EPO in 2012. The patent claimed a new variety of melon resistant to Cucurbit yellow stunting disorder virus. This variety wad bred based using closterovirus resistant trait found in melons of Indian origin (which is registered in the international seed banks). Opponents of this application at EPO raised the grounds of non-patentability in view of public order and further contended that Monsanto, which developed the new variety, should have obtained permission from NBA under section 6 of the Biological Diversity Act.

There are several orders of the Controller of Patents that provide a joint application of the provisions of the Act and the Patents Act 1970. In the recent times, several cases related to the ABS have been taken by the Indian Courts. A summary of some of the case law are discussed below.

In a case of twin patent applications of ITC company, section 83 of the Patents Act 1970 was invoked. In one of the patent applications 1079/KOL/2009 titled 'Synergistic carotenoid compositions', applicants mentioned in a paragraph of the patent specification that the raw material for making natural lycopene and betacarotene to make the product are not obtained from India but from China and Spain and hence do not require a no objection from the NBA. Another application 852/KOL/2010 titled 'Formulation comprising 2- mercaptobenzamidazole derivatives and ZPTO' by the same company claimed in the complete specification that the ginseng extract, papain and adrandine extract (from pomegranate) are obtained from foreign sources. In relation to both these cases, the Controller in a speaking order indicated that when the company is set to commercialise products in India, the said extracts/compounds will be taken from India which has the same sources. Further, the Controller opined that the intention of section 83 sub-clauses (a) and (g) are not fulfilled keeping in view the need for working of patents in India keeping in view public requirements. A non-compliance under section 3 (b) of the Patents Act 1970 was invoked in relation to another patent application filed by ITC in 2010. Non-compliance due to lack of no objection from the NBA was also raised in relation to this patent application as it involved the use of tobacco leaves (Application number 1098/KOL/2010).

On the contrary to the decisions above, the Controller decided in favor of the application on no requirement of the NBA approval in case of the application 2937/DELNP/2009 titled 'Garlic processing'. The applicant Ecospray Limited filed patent application in relation to methods of processing extracts from Garlic in order to manipulate the polysulphides contained in it. The Controller in this case held that the extracts were obtained from UK and as such are 'value added products' and as such public at large would not be affected by an ensuing patent grant.

One case wherein NBA has opposed a patent post grant is that of Sunev Pharmaceuticals. In 2019, a post grant opposition was filed by the NBA objecting to the patents obtained by this company in several jurisdictions based on many bioresources from India such as neem, turmeric, capsicum, sesame, castor etc., Prior approval was not sought from the NBA before filing the patent in foreign jurisdictions.

While draft ABS guidelines 2019 are in the discussion rounds and yet to be announced, there is certainly consensus on the need for more definitive basis for ABS terms so that entities accessing biodiversity can be more certain and achieve compliance under the law.

6.6 Primacy of Conservation in ABS Process—Achieving Universal Accountability

It is pertinent to reiterate that ABS goals are linked to the other two objectives of the Convention on Biological Diversity. Any meaningful and long term success of ABS will depend on the inclusion of this aspect. ABS is not merely procedural. It is a process involving several stakeholders. Data and resource are two aspects that are reigning high in the current international discussions. While biosurvey and bioprospecting promise value to humans, identifying limits to access has been a key problem. Biodiversity monitoring as a necessary condition for the ABS process could help the administrative authorities to identify evidence of misappropriation and misuse of biodiversity. It is becoming evident that environment and human sustenance is significantly dependent on biodiversity conservation.

In the preamble of CBD, the value of traditional innovations has been recognized for conservation and sustainable use of biodiversity. With wider objectives, the CBD through its Article 8(j) has broadened the scope and mandate of protection. The discussion on TK was taken up by several inter-governmental bodies that include those dealing with human rights, IP, environment, trade and development, health. It is to be noted that Article 8j outlines the need to respect, preserve and maintain knowledge and innovations of indigenous and local communities. One of the ways in which misappropriation of TK has been addressed in many countries is by including TK as ineligible subject matter under the patent legislation. Several countries including India provide restrictive conditions under patent law to exclude traditional knowledge per se from patent protection.

Conservation and use are linked concepts. This is especially so in relation to agricultural biodiversity. Unless traditional crop varieties are conserved and also used in a sustainable manner, the continuation of crops and cropping systems may be affected. Further, in early 2019 the UN adopted a resolution, the UN declaration on the current decade as the Decade of Ecosystem Restoration (2021–2030) in the 73rd Session of the General Assembly. It urged countries to achieve international commitments and national priorities through restoration of ecosystems through process and policy implementation goals. This resolution coincides with the 2030 timeline of the Sustainable Development Goals. The developing countries are the centres for domestication agricultural crops. Due to the increase in use of genetic resources and misappropriation of genetic resources, equity and justice related concerns arise. Growing interest in commercial value of biodiversity leads to a greater propensity of biobased products. New technologies have posed a challenge to the ABS mechanism. Digital sequences and the need for a comprehensive ABS process has been a key point for discussion in the past COP meeting under the CBD.

In order to develop an accountability framework, it is important to know to whom ABS applies. ABS involves—providers and users. Providers include those from government, public research institutes, farmer communities and international centers for agriculture. Users include companies and public research institutions. ABS mandates the need for fair and equitable sharing of the benefits on mutually agreed terms. Benefits sharing can vary in different regions, it includes either monetary or non-monetary components or both. Many ABS regimes have monitoring and compliance requirements for the users. The development of the IRCC (Internationally Recognized Certificate of Compliance) compliance is one of the important achievements post the Nagoya Protocol. Under Article 17.3 of the Nagoya Protocol which states "An internationally recognized certificate of compliance shall serve as evidence that the genetic resource which has been accessed in accordance with prior informed consent and that mutually agreed terms have been established, as

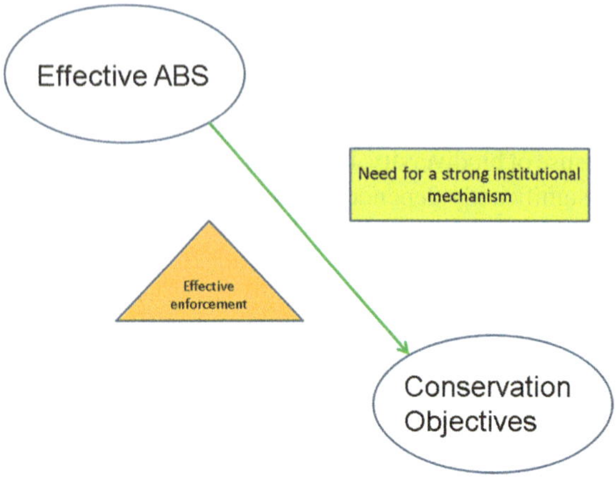

Fig. 6.2 Linking ABS process with conservation objectives

required by the domestic access and benefit sharing legislation or regulatory requirements of the party providing prior informed consent". The patent applicants in the member countries must furnish a copy of the IRCC. The ABS Clearing House is an important platform for sharing experiences. Sustainable utilization of bioresources is integrated with conservation activities. Article 10 (e) mandates member countries to '*Encourage cooperation between its governmental authorities and its private sector in developing methods for sustainable use of biological resources.* This has a great relevance in developing ABS governance frameworks that would help in building commitment in companies to contribute to the conservation efforts (Shova Devi and Padmavati 2016) (Fig. 6.2).

Analysis of the research in relation to ABS agreements and IP revealed several areas that need attention. Timely disposal, development of a portal for easy access of information by applicants, assistance to communities to prepare an inventory for bioresource and associated knowledge, lack of monitoring mechanism were identified as some of the key aspects. Analysis of the implementation of ABS at the various SBBs in India reveals that certain boards have utilized the Biodiversity Rules well, development of local measures, effective transfer of ABS funds. Traditional knowledge is integral to use of bioresources. It is found that largely there are no definite validation and protection mechanism identified for TK (unpublished results). To this end, national level accountability and reporting systems may help in creating better accountability in the ABS process. The development of pre-ABS and post ABS norms and ensuring monitoring of bioresearches through the ABS process will help in creating common principles.

Conflict of Interest The views and opinions expressed in this article are those of the author in her personal capacity. No conflict of interest.

References

Antons C (2010) Sui generis protection for plant varieties and traditional knowledge in biodiversity and agriculture: the international framework and national approaches in the Philippines and India. SSRN Electron J. https://doi.org/10.2139/ssrn.2617565

Bagley M, Karger E, Muller MR, Perron-Welch F, Thambisetty S (2019) Fact-finding study on how domestic measures address benefit-sharing arising from commercial and noncommercial use of digital sequence information on genetic resources and address the use of digital sequence information on genetic resources for research and development. Accessed from https://www.cbd.int/abs/DSI-peer/Study4_domestic_measures.pdf

Barbier EB, Aylward BA (1996) Capturing the pharmaceutical value of biodiversity in a developing country. Environ Resource Econ 8(2):157–181. https://doi.org/10.1007/bf00357362

Bourdy G, Aubertin C, Jullian V, Deharo E (2017) Quassia "biopiracy" case and the Nagoya Protocol: a researcher's perspective. J Ethnopharmacol 206:290–297. https://doi.org/10.1016/j.jep.2017.05.030

Brandão AS, Gonçalves A, Santos JM (2021) Circular bioeconomy strategies: from scientific research to commercially viable products. J Cleaner Prod 295:126407. https://doi.org/10.1016/j.jclepro.2021.126407

Carls S (2020) Brazil is one step closer to implementation of the Nagoya Protocol. J Intell Property Law Pract 16(1):19–20. https://doi.org/10.1093/jiplp/jpaa181

D'Amato D, Bartkowski B, Droste N (2020) Reviewing the interface of bioeconomy and ecosystem service research. Ambio 49(12):1878–1896. https://doi.org/10.1007/s13280-020-01374-0

Davis K (2019) Access and benefit sharing- What is it and how are botanic gardens responding J Botanic Gardens Conserv. Int 16(1):14–17

Efferth T (2019) Biopiracy of medicinal plants: finding fair solutions for the use of natural resources. Phytomedicine 53:294–295. https://doi.org/10.1016/j.phymed.2018.06.047

Filoche G (2012) Biodiversity fetishism and biotechnology promises in Brazil: from policy contradictions to legal adjustments. J World Intell Property https://doi.org/10.1111/j.1747-1796.2011.00434.x

Gervais D (2005) Trips, Doha and traditional knowledge. J World Intell Property 6(3):403–419. https://doi.org/10.1111/j.1747-1796.2003.tb00222.x

Guidelines for processing of patent applications relating to traditional knowledge and biological material. https://www.ipindia.gov.in/writereaddata/Portal/IPOGuidelinesManuals/1_39_1_5-tk-guidelines.pdf

Guidelines on Access to Biological Resources and Associated Knowledge and Benefits Sharing Regulations (2014). http://nbaindia.org/uploaded/pdf/Gazette_Notification_of_ABS_Guidlines.pdf

Hautier Y, Tilman D, Isbell F, Seabloom EW, Borer ET, Reich PB (2015) Anthropogenic environmental changes affect ecosystem stability via biodiversity. Science 348(6232):336–340. https://doi.org/10.1126/science.aaa1788

Iwu MM (1996) Biodiversity prospecting in Nigeria: seeking equity and reciprocity in intellectual property rights through partnership arrangements and capacity building. J Ethnopharmacol 51(1–3):209–219. https://doi.org/10.1016/0378-8741(95)01364-4

Lawson C (2010) Biodiversity conservation access and benefit-sharing contracts and patents. SSRN Electron J.https://doi.org/10.2139/ssrn.1612552

Meeting of the SBSTTA—Digital Sequence Information on genetic resources—Page 5. Accessed from https://www.cbd.int/doc/c/704c/70ac/010ad8a5e69380925c38b1a4/sbstta-22-02-en.pdf

Mori AS, Dee LE, Gonzalez A, Ohashi H, Cowles J, Wright AJ, Loreau M, Hautier Y, Newbold T, Reich PB, Matsui T, Takeuchi W, Okada KI, Seidl R, Isbell F (2021) Biodiversity–productivity relationships are key to nature-based climate solutions. Nature Climate Change 11(6):543–550. https://doi.org/10.1038/s41558-021-01062-1

Musungu SF (2008) The protection of Geographical Indications and the doha round: strategic and Policy considerations for Africa. Accessed from www.quno.org

Myers N, Mittermeier RA, Mittermeier CG, da Fonseca GAB, Kent J (2000) Biodiversity hotspots for conservation priorities. Nature 403(6772):853–858. https://doi.org/10.1038/35002501

Nielsen KS, Marteau TM, Bauer JM, Bradbury RB, Broad S, Burgess G, Burgman M, Byerly H, Clayton S, Espelosin D, Ferraro PJ, Fisher B, Garnett EE, Jones JPG, Otieno M, Polasky S, Ricketts TH, Trevelyan R, van der Linden S, Balmford A (2021) Biodiversity conservation as a promising frontier for behavioural science. Nature Human Behaviour 5(5):550–556. https://doi.org/10.1038/s41562-021-01109-5

Phillips PW (2013) Traditional knowledge. socio-economic considerations in biotechnology regulation, 159–175. https://doi.org/10.1007/978-1-4614-9440-9_11

Post grant opposition by NBA to the patent of Sunev Pharmaceuticals. http://nbaindia.org/uploaded/pdf/52nd%20Authority%20meeting%20final.pdf

Rabitz F (2017) Access without benefit-sharing: design, effectiveness and reform of the FAO seed treaty. Int J Commons 11(2):621–640. https://doi.org/10.18352/ijc.736

Ryan MP (2010) Patent incentives, technology markets, and public–private bio-medical innovation networks in Brazil. World Develop 38(8):1082–1093. https://doi.org/10.1016/j.worlddev.2009.12.013

Sala E, Mayorga J, Bradley D, Cabrai RB, Atwood TB, Auber A, Cheung W, Costello C, Ferretti F, Friedlander AM, Gaines SD, Garilao C, Goodell W, Halpern BS, Hinson A, Kaschner K, Kerner-Reyes K, Leprieur F, McGowan J, Morgan LE, Mouillot D, Palacios-Abrantes J, Possingham HP, Rechnerger KD, Worm B, Lubchenco J (2021) Prioritising the global ocean for biodiversity, food and climate. Nature 592:397–402

Seidl A, Mulungu K, Arlaud M, van den Heuvel O, Riva M (2021) The effectiveness of national biodiversity investments to protect the wealth of nature. Nature Ecology Evol 5(4):530–539. https://doi.org/10.1038/s41559-020-01372-1

Shova D, Padmavati M (2016) Biodiversity monitoring: a precondition to access and benefit sharing under the Indian Biological Diversity Act 2002. J IP Rights 21:288–294

Smyth SJ, Macall DM, Phillips PWB, Beer J (2020) Implications of biological information digitization: access and benefit sharing of plant genetic resources. J World Intell Property 23(3–4):267–287. https://doi.org/10.1111/jwip.12151

Soria-López M, Fuentes-Páramo I (2016) The identification of biopiracy in patents. World Patent Inf 47:67–74. https://doi.org/10.1016/j.wpi.2016.10.003

Tsioumani E (2018) Beyond access and benefit-sharing: Lessons from the law and governance of agricultural biodiversity. J World Intell Property 21(3–4):106–122. https://doi.org/10.1111/jwip.12094

UN Decade on Ecosystem Restoration. https://undocs.org/A/RES/73/284, www.decadeonrestoration.org

Wynberg R, Andersen R, Laird S, Kusena K, Prip C, Westengen OT (2021) Farmers' rights and digital sequence information: crisis or opportunity to reclaim stewardship over agrobiodiversity? Front Plant Sci 12.https://doi.org/10.3389/fpls.2021.686728

Part III
Access to Genetic Resources and to Associated Traditional Knowledge and Benefit Sharing

Chapter 7
Traditional Knowledge and Access and Benefit Sharing in the Context of Himalayan States

Ghanashyam Sharma and Bharat K. Pradhan

Abstract Access and benefit sharing (ABS) from genetic resources (GRs) or biological resources (BRs) is a major emerging opportunity in the entire Himalayan region. In India, the National Biodiversity Authority (NBA) and the State Biodiversity Boards (SBB) have been engaged in implementing the Biodiversity Act 2002 and facilitating the ABS arrangements in the region. In the Hindu-Kush Himalayan region, the International Centre for Integrated Mountain Development (ICIMOD) has been engaged in facilitating the ABS process for developing a common framework. Despite official support for the implementation of the Convention on Biological Diversity (CBD) by India, Nepal and Bhutan, the ground reality is still discouraging and the indigenous and local communities (ILCs) are yet to be benefited from the ABS agreements. Awareness about traditional knowledge (TK) protection has not received enough attention despite acts, rules and guidelines brought about by the governments. The belief of people in the region is that they want to protect their knowledge by law. They want to be benefited through ABS on GRs/BRs and associated TK, but paradoxically, they lack bio prospecting of the economically potential resources. People in the region fear of misappropriation, biopiracy and degradation of their knowledge and the resources. In Nepal, activities are supported to address the issues surrounding the use of TK by raising awareness and incorporating the concerns of local communities into national legislation, in cooperation with the Nepal Forum for Environmental Journalist (NEFEJ), Himalayan Grassroots Women's Natural Resources Management Association (HIMAWANTI), Centre for Law and Justice (CLJ), and the Government of Nepal. In Bhutan, in collaboration with the Ministry of Agriculture and National Biodiversity Centre, awareness is raised among local communities about TK, and local communities' concerns about TK are being included in the national draft Biodiversity Rules. Owing to enormously rich BRs/GRs and associated TK in the Himalayan region, increased market demand for GRs/BRs and associated TK could offer new opportunities for generating benefits and enhanced incomes for

G. Sharma (✉)
The Mountain Institute India, Tadong Daragaon, Gangtok, Sikkim Pin-737102, India
e-mail: banstolag@gmail.com

B. K. Pradhan
Sikkim Biodiversity Board, Forest and Environment Department, Deorali, Gangtok, Sikkim Pin-737102, India

O. V. Oommen et al. (eds.), *Biodiversity Conservation Through Access and Benefit Sharing (ABS)*, https://doi.org/10.1007/978-3-031-16186-5_7

ILCs. However, the current reality seems to be quite far from achieving this. The involvement and engagement of local, indigenous, and marginalised people in the ABS process is still unclear.

Keywords Traditional knowledge · Himalayan region · Access and benefit sharing · Genetic resources · Indigenous and local communities

7.1 Introduction

Traditional knowledge (TK), innovations and practices play an important role in practically all aspects of the lives and livelihoods of rural people in India: food and agriculture, human and animal health, clothing, shelter, architecture, art, culture, handicrafts, natural resource management, etc. (Sahai et al. 2005). TK is an inextricable part of the biocultural heritage of indigenous people and local communities (Singh and Pretty Pilgrim 2010; Pant 2015). Prior to the adoption of the Convention on Biological Diversity (CBD) in 1992, access to all GRs and associated TK was considered free for all mankind (Oli and Dhakal 2009). The biopiracy of GRs and TK from communities and countries by organisations and individuals were a common practice (Rao and Guru 2003; Singla 2020; Pant 2015), who monopolised the benefits accruing out of it and patented it disregarding the sovereign rights of individual nations (Javed et al. 2020). Many patents were held by companies on turmeric, neem, rice etc. but neither monetary nor social and cultural benefits flowed down to actual owners of the GRs and associated TK. Such unethical practices of large institutions enforced without any thought of respect and benefit sharing disempowered owner's negotiating power and violated the sovereign rights of countries over its own endemic resources (Pant 2015). Due to lack of policies and its strict enforcement, sovereign rights were violated by institutions, bioprospecting genetic materials owned by nations and the associated knowledge held by communities. The local communities and countries of origin of such resources were often not informed about the use of their genetic resources (GRs) and associated TK, which limited their bargaining power and prevented them from getting the benefits of their own resources (Oli and Dhakal 2009).

CBD in 1992 was organised by nations all over the world which then acknowledged and recognised the importance of the knowledge, practices, and innovations of Indigenous and Local Communities (ILCs). Nations worldwide came together to make provision for the enforcement of CBD in the form of national policies/laws in their respective countries ratified the CBD. The third objective of the CBD: ABS was to be implemented by nations who were signatory to the CBD by developing national laws to protect both diversity and its sustainable use along with providing procedures for bioprospecting so that access to GRs and associated TK are regulated and the accrued economic and social- cultural benefits are shared equitably.

The eastern Himalayan region is rich in biological resources and TK (Oli and Dhakal 2008; NBSAP Bhutan 2014; Sharma et al. 2016; O'Neill et al. 2017). From the ancient times until today, TK and practices have provided the basis for the wellbeing and livelihoods of the indigenous mountain communities, maintaining their health and replenishing the environment. In modern times, TK and BRs have been used by the food industry, the pharmaceutical industry, and the cosmetic industry, and there is a long history of TK in the evolution of modern food crops. Farmers in the Himalayan region domesticated and developed carrots, mustard, gooseberries, apples, pears, apricots, oranges, lemons, and cardamom (Oli and Dhakal 2008).

TK has no exact and universal definition because of its living nature, as acknowledged by WIPO (World Intellectual Property Organization). Nevertheless, CBD recognise it as knowledge, innovations and practices of ILCs that is developed, sustained and transmitted orally from generation to generation. TK is not held in private names of individuals of communities but is rather held as common pool of biological resources to be used by all and is transmitted within the communities orally and is guided by customary law for sustainable harvesting and use. TK is a living cultural knowledge and is broad based in nature, especially exists in almost all aspects of life and particularly held in the field of agriculture, fisheries, health, horticulture, forestry and environment management etc. and is transmitted orally in various forms such as stories, songs, folklores, proverbs, cultural values, beliefs, rituals, community laws, local language, and agricultural practices that includes animal and plant breeding as well. It also forms the basis of cultural identity, contributing to social cohesiveness and thereby reducing vulnerability and poverty. The TK has helped develop millions of farmers' food crop varieties in use today, as well as a wealth of traditional medicines and techniques for sustainable agriculture and resource use (Swiderska 2006; Oli and Dhakal 2008; Pant 2015; Javed et al. 2020).

The contribution of ILCs and their TK in biodiversity conservation and in sustainable development is globally recognized but the protection of TK of ILCs is one of the most contentious and complicated issue. The knowledge on the use of biological resources (BRs) are the efforts of individuals who contributed for the well-being of their community and society which gradually percolated under community's domain and became community knowledge with the demise of the creator (Oli 2009). TK has proved to be useful as an input in modern medicine (Gebru 2017). It provides leads/clues which forms the basis for new drug discoveries thus lessening the research efforts, time and wastage of financial resources in finding the compounds within millions of varieties of plants (Reid 2009).The World Health Organization (WHO) estimates that over 80% of the people in developing countries still depends on TK based herbal medicine for their primary healthcare and wellness (WHO 2002) and the number may increase in the coming decades as many health conscious people are now turning to natural remedies as a way to cure their diseases. The global market for herbal medicine found by following leads given by ILCs was estimated to be US$ 83 billion in 2019; the demand is growing at the rate of 15 to 25% annually and is expected to reach US$ 5 trillion by 2050.

There are growing concerns over the TKs of ILCs being exploited and patented by the third parties without their prior informed consent (PIC) and the accrued benefits are rarely shared with the communities to whom the knowledge originally belongs or where the knowledge originated and exists (Gupta 2011). Several claims have been made by the ILCs in the past, on the use of their TK on local flora and fauna for developing new drugs by the bio-prospector leading to lengthy legal battles (Oli et al. 2010).

From the past experiences, ILCs have been demanding for the protection of their TK through international legal framework, which has been a subject of debate in many international forums as it is very challenging to develop international framework for the protection of TK. Ouma (2017) suggested that, while crafting the international framework, the policyholders need to take into account the transboundary nature of the TK, which is often shared by the communities across the national boundaries. There are very few experiences globally of local communities or TK holders gaining substantially from the commercial use of their knowledge. On the contrary, cases of biopiracy and misappropriation of TK are becoming more apparent and have been on the rise in the last two decades (Bhattacharya 2014).

In this chapter, we briefly discuss about the international instruments on TK and ABS, policy instruments on protection of TK in India, Biocultural Protocols for TK protection of indigenous and local communities (ILCs), Traditional Knowledge Digital Library (TKDL) and *sui generis* system of protection, ABS programmes in the Indian Himalayan states and unsolved issues relating to the implementation of Biodiversity Act 2002 and ABS guidelines 2014 for protection of TK and associated GRs/BRs from mis-appropriators and bio-prospectors, and prospects of economic opportunity from ABS for the region.

7.2 International Instruments on TK and ABS

TK has been benefitting the entire mankind and the need for protecting it has been acknowledged in discussion and negotiations under the umbrella of a number of inter-governmental organizations that deal with biodiversity, the environment, indigenous people's rights, human rights, food and agriculture, etc. (Dagne 2014). CBD's third objective recognises the importance of the knowledge, practices, and innovations of ILCs, and makes provision for the prior informed consent (PIC) of the holders to be obtained by any public or private enterprise seeking access to GRs by way of putting bioprospecting procedures and benefit sharing mechanisms through enactment of National Biodiversity Law and Rules in each country that were party to CBD ratifications (Oli and Dhakal 2009).

A number of countries have developed national legislations to protect the TK of the ILCs but it is not able to bind other countries. Hence, the ILCs and the national government insisted on developing international legal instrument. Recognizing the biggest contribution of ILCs to the service of the human community at large which is inherited by them, the international community took a cue on the protection of TK for which, numerous policy documents are being prepared which recognizes the dependence of the ILCs on the bioresources as well as takes into account the protection of their TK, with legal implications. Not only its protection, but emphasis are also being given on the benefit sharing aspect, incase of commercial exploitation of

TK as benefit sharing mechanism help achieve number of Sustainable Development Goals (SDGs) such as, it could contribute to increasing the income of knowledge holders (SDG 1—no poverty) thus increasing their purchasing power (SDG 2—zero hunger) and leading a healthy lives (SDG 3 – Good health and well-being). It would also provide new opportunity for knowledge holders for economic and social development (SDG 8—decent work and economic growth), foster innovation (SDG 9—industry, innovation and infrastructure), provide incentives for maintaining biodiversity and climate change adaptation (SDG 13—climate action), promote benefit sharing, and sustainable trade with the ultimate objective of reducing biodiversity loss (SDG 15—life on land), opportunities to rearrange national laws and institutions (SGD 16—peace, justice and strong institutions), and partnership in conservation and sustainable utilization of GRs and associated TK (SDG 17—partnerships for the goals) (Upreti et al. 2020).

Amongst all the international agreements, CBD is the first and the most prominent multilateral agreement that recognized the close relationship of the ILCs with the biodiversity. It emphasizes on the need for the adoption policy, administrative and legislative measures to protect the rights of ILCs over their TK. It mandates the states to protect and encourage customary use of biological resources in accordance with traditional cultural practices that are compatible with conservation or sustainable use requirements (Kuei-Jung 2011). It is for this reason, TK of ILCs are considered and incorporated in the work programmes under the CBD. Further, their representatives are being invited to participate in the working group on TK and involved in the group's decision making. Additionally, a guideline has been prepared to involve the ILCs in the impact assessment of cultural, environmental and social impact of development projects proposed to take place on sacred sites, lands and water traditionally occupied or used by the ILCs. Later CBD adopted Bonn Guidelines during COP6 in 2002 to help countries, as providers and users of genetic resources, to implement ABS procedures effectively. It stipulates that the rights of indigenous and local communities should be respected and that their prior and informed consent (PIC) should be sought when genetic resources or associated traditional knowledge are being accessed (Kamau et al. 2010; Oli et al. 2010). Simultaneously, Governments are facilitating the participation of ILCs in developing policies for the conservation and sustainable use of resources, access to genetic resources and the sharing of benefits, and the designation and management of protected areas. Many governments are now in the process of incorporating conservation of TK in the national biodiversity strategy and action plan. A number of Governments have adopted specific laws, policies and administrative arrangements for protecting TK, emphasizing that the prior informed consent of knowledge-holders must be attained before their knowledge is used by others (https://www.cbd.int/traditional/intro.shtml).

The Nagoya Protocol on Access to GRs and Fair and Equitable Sharing of Benefits arising from their Utilization (Nagoya Protocol), which was adopted in 2010 and entered into force in 2014, expands upon the CBD provisions addresses TK associated with GRs with provisions on access, benefit sharing and compliance (UN 2014). It also addresses GRs where ILCs have the established right to grant access to them. Contracting Parties are to take measures to ensure these communities' prior informed consent, and fair and equitable benefit sharing, keeping in mind community laws and procedures as well as customary use and exchange (Ouma 2017). It also provides the list of possible monetary and non-monetary benefits.

Other relevant evolutions relating to TK which developed concurrently to progress in the CBD leading up to the Protocol include the establishment of: (1) the International Treaty on Plant GRs for Food and Agriculture (ITPGRFA) passed by the Food and Agriculture Organization Conference in 2001, and entering in to force on 29 June 2004, which provides for protections relating to 'farmers rights' including TK and traditional breeding practices, (2) the Intergovernmental Committee (IGC) on Intellectual Property and GRs, TK and Folklore established under the World Intellectual Property Organization (WIPO) in 2000, which provides a forum for negotiations on issues underlying development of a binding international instrument on TK, and (3) the 2007 United Nations Declaration on the Rights of Indigenous Peoples (UNDRIP) which vests rights relating to 'control, protection and development' of TK, as well as IPRs relating to TK, with ILCs (Swiderska 2006, Freedom-Kai 2016).

Concurrently, the World Trade Organization (WTO) has been negotiating the relationship between its intellectual property agreement, namely the Agreement on Trade-Related Aspects of Intellectual Property Rights (TRIPS), and the CBD. The CBD, WIPO, and the Declaration on the Rights of Indigenous Peoples, all sub organs of the UN, as well as the WTO, have become the major contributors towards global TK law making (Kuei-Jung 2011).

7.3 Policy Instruments on Protection of TK in India

7.3.1 Protection of Plant Varieties and Farmers' Right Act 2001

The PPVFRA seeks to protect the rights of the farmers and the breeders on plant varieties. The Act recognises the individual and community role played by farmers in improvement and conservation of varieties (Venkataraman and Latha 2008). The PPVFRA allows for the registration not only of new and essentially derived varieties, but also of "farmers' varieties" as well as of so-called "extant varieties." A "farmers' variety" is defined in Sect. 2(l) of the PPVFRA as a variety which: (i) has been traditionally cultivated and evolved by the farmers in their fields; or (ii) is a wild relative or land race of a variety about which the farmers possess the common knowledge. "Extant variety", according to Sect. 2(j) of the PPVFRA is a variety notified under Sect. 5 of the Seeds Act, a farmers' variety, a variety about which there is common knowledge, or any other variety which is in the public domain (Antons 2010).

PPVFRA does not have direct benefit sharing provisions between the users and the breeders but the farmers are entitled for recognition and reward from the "National Gene Fund" constituted by the Central Government. Credited to this National Gene

Fund are benefit-sharing payments, annual fees paid to the Protection of Plant Varieties and Farmers' Rights Authority, money received from compensation claims and contributions to the fund from national and international organisations and other sources. Nevertheless, the farmers have the right to claim for benefit sharing and compensation from the commercial breeder (Antons 2010). Ironically it is still unclear as to how this Act has helped ILCs of the Himalaya and farmers of the Northeast India.

7.3.2 *The Patent Act, 1970*

According to the Patent Act, 1970, traditional knowledge cannot qualify as an invention; hence cannot be patented. The Act also makes it clear that any existing TK or healing technique of well-established ethnic system of medicine such as Ayurveda, Siddha and Unani system and formulations therein cannot be patented. To avoid/ prevent patent grants to TK in India, an initiative has been taken to document and publish all the TK by an e-library, and such a library is called as Traditional Knowledge Digital Library (TKDL). TKDL provides details of scientific and TK arranged in a manner according to the classification of international patents. The objective of the library is to protect the ancient and traditional knowledge of the country from exploitation through biopiracy and unethical patents, by documenting it electronically and classifying it as per international patent classification systems.

7.3.3 *Biological Diversity Act 2002*

India has undergone many struggles in trying to safeguard her TK. In the wake of problems with patents related to neem, turmeric, basmati rice, etc. India enacted the Biological Diversity Act in 2002. The salient features of the Act are to:

1. regulate access to biological resources of the country.
2. conserve and sustainable use of biological diversity.
3. secure benefit sharing with local people as conservers and knowledge holders related to biodiversity.
4. respect and protect local communities' traditional knowledge related to biodiversity.
5. conserve and develop areas of importance from the stand point of biological diversity by declaring them as biological diversity heritage sites (BHS).
6. protect and rehabilitate threatened species.

The Act prohibits foreign nationals, institutions, companies, and non-residential Indians from accessing bioresources and associated TK for research or for commercial utilization or for bio-survey and bio-utilization without prior approval of the implementing agency at country level i.e., National Biodiversity Authority (NBA).

The Act forbids transfer of research result to foreign nationals; it also has the provision to deal with IPRs on biological resources and associated TK, both for Indian as well as the foreign nationals. Nevertheless, the transfer or exchange of biological resources and related information under collaborative research projects are exempted from prior approval of NBA provided such projects conform to the policy guidelines issued by the central government.

The Act is also binding on the Indian companies intending to commercialize bioresources and associated TK and, they come under the purview of the State Biodiversity Boards (SBBs). The Act mandates both NBA and the SBBs to consult Biodiversity Management Committees (BMCs), which are established at the local level, while taking any decision relating to the use of biological resources and associated TK within their territorial jurisdiction. The Act has also given power to the BMCs to levy fee and charges for biological resources collected within their areas.

The Biological Diversity Rules 2004 and Guidelines on ABS Regulation 2014 provide framework for ABS. The rules provides for the revocation of the approval granted to an applicant, in case violation of the provisions of the Act or conditions on which the approval was granted, non-compliance of the terms of the agreement, failure to comply with any of the conditions of access granted, and on account of overriding public interest or for protection of environment and conservation of biodiversity.

The rules also provide for restriction or prohibition to the request for access to the biological resources, if: the request for access is for any endangered, endemic and rare species, the request for access may likely to result in adverse effect on the livelihoods of the local people as well as on the environment, the request for access may cause genetic erosion or affecting the ecosystem function, use of resources for purpose contrary to national interest and other related international agreements entered into by India.

7.4 Geographical Indication (GI)

The Government of India enacted the Geographical Indications of Goods (Registration and Protection) Act, 1999 (GI Act 1999) which explains that "geographical indication", in relation to goods, means an indication which identifies such goods as agricultural goods, natural goods or manufactured goods as originating, or manufactured in the territory of a country, or a region or locality in that territory, where a given quality, reputation or other characteristic of such goods is essentially attributable to its geographical origin and in case where such goods are manufactured goods, one of the activities of either the production or of processing or preparation of the goods concerned takes place in such territory, region or locality, as the case may be".

Geographical indications (GIs) can help communities promote and protect markets for their biocultural products, and are gaining popularity globally as a tool for the protection and promotion of traditional knowledge. GIs specify the geographical origin of a product and link it with its essential qualities that are due to its place of origin.

Until June 2021, as many as 12crops, fruits and vegetables from the Northeastern states of India have been registered under GI Act 1999 (Table 7.1). It is evident from the list of registered GIs from other states of India, other states have quickly availed the opportunity and obtained GI tagging of their goods and produce unique to their region and thus they have ensured the protection of TK associated to their produce of goods. However, Northeast India, in spite of its rich horticultural diversity have not been able to take advantage of GI tagging (Fig. 7.1).

TK can be defined as the knowledge, innovations and practices of indigenous peoples and local communities. GI gives protection against misappropriation of

Table 7.1 Some horticultural crops registered under GI Act 2001 (Datta et al. 2021)

Sl. no	Crops and vegetables	Scientific names	GI registered from the state
1	Assam Orthodox Tea	*Camellia sinensis* var. *assamica*	Assam
2	Assam Karbianglong Ginger	*Zingiber officinale*	Assam
3	Tezpur Litchi	*Litchi chinensis*	Assam
4	Arunachal Orange	*Citrus reticulata*	Arunachal Pradesh
5	Large Cardamom	*Amomum subulatum*	Sikkim
6	Dalle Khorsani (Red Cherry Peeper)	*Capsicum annum* var *cerasiforme*	Sikkim
7	Mizo Chilli	*Capsicum frutescens*	Mizoram
8	Tripura Queen Pineapple	*Ananas comosus*	Tripura
9	Khasi Mandarin	*Citrus reticulata*	Meghalaya
10	Memong Narang	*Citrus indica*	Meghalaya
11	Kachai Lemon	*Citrus jambhiri*	Manipur
12	Naga Mircha	*Capsicum chinense*	Nagaland
13	Naga Tree Tomato	*Solanum betaceum*	Nagaland

Fig. 7.1 *Dalle-Khorsani* (Cherry peeper) and pickle dipped in salt water (GI registered 2021)

Table 7.2 Some horticultural crops which can be registered under GI Act 2019 (Datta et al. 2021)

Sl. no	Crops and vegetables	Scientific names	GI registry
1	Mizo banana	Attractive size, shape and taste	GI Registry 2019
2	Assam lemon	Cluster bearing habit, seedless	GI Registry 2019
3	Naga cucumber	Low in calories, high in potassium and contain high level of water	GI Registry 2019
4	Manipur black cherry	Natural edible colour, high moisture content	NERAMAC 2017
5	Assam elephant apple	Used in curries, traditional medicine	NERAMAC 2017
6	Leteku Assam Burmese grapes	High potassium content. Rich source of pectin	NERAMAC 2017
7	Siroi Lily Manipur	Bell shaped petals	GI Registry 2017
8	Mizoram Passion fruit	Small in size, its flavour is excellent and less acidic	GI Registry 2018
9	Manipur Temelong Orange	Sweet–sour taste and high juice content	GI Registry 2017
10	Arunachal Pradesh Rahre	Rich source of essential oil, adapted to biotic & abiotic stress	GI Registry 2017

TK, and its usage without any benefit-sharing. GI can create economic rewards for producers and enhance commercial value. There is still GI status not exploited post registration due to lack of background work for filing and awareness. Institutions and organizations in the Himalayan Region and Northeast region in particular have limited consumer awareness of GI, lack initiatives in building brands and visibility, lack formal organization structures and/or control mechanisms, lack support from R&Ds, lack forward and backward linkages, and defend against unfair competition. Therefore there is a need to spread GI literacy for developing, registering and enforcing a GI registration (Table 7.2) (Figs. 7.2, 7.3, 7.4 and 7.5).

Geographical indications (GIs) can help communities promote and protect markets for their biocultural products, and are gaining popularity globally as a tool for the protection and promotion of traditional knowledge. GIs specify the geographical origin of a product and link it with its essential qualities that are due to its place of origin (Pant 2015).

7.5 National Bureau of Plant Genetic Resources

Realizing the importance of collecting and conserving PGRFA, India has taken strategic steps for their ex situ conservation using appropriate approaches. Major part of this work has been carried out under Indian Council of Agricultural Research (ICAR) by the National Bureau of Plant Genetic Resources (NBPGR), New Delhi,

Fig. 7.2 A Large Cardamom farm at Hee-Gaon, West Sikkim, GI tag 2015

Fig. 7.3 Dzongu Mandarin Orange, a potential cash crop for GI registration

with its 10 regional stations/base centres/ quarantine centres' over different phytogeographic zones of the country and active collaboration and linkages with over 40 National Active Germplasm Sites (NAGS). The National Genebank of NBPGR has four types of storage facilities—seed gene-bank, cryogene-bank, in vitro gene-bank and field gene-bank and has 5,01,121 germplasm accessions currently, of nearly 2,791 species (www.genebank.nbpgr.ernet.in). Stock inventorization and monitoring of viability is performed regularly in most of the accessions.

The Indian Himalayan states being one of the gene-rich landscapes of the world, it faces a unique challenge of protecting its natural heritage while evolving mutually beneficial strategies for germplasm exchange with other countries. As of now the exploration, collection, exchange, characterization, evaluation, conservation and documentation of tens of thousands of plant germplasm is yet to be carried out by NBPGR. It also has the responsibility to carry out quarantine of all imported PGR including transgenics meant for research purposes.

Fig. 7.4 Women in Sumik in East Sikkim processing fruits of Mel (*Docynia indica*) for preparation of *chuk* (jam) which is used to cure blood dysentery, diarrhoea, and stomach disorders. This is a potential fruit for GI registration

Fig. 7.5 Temi tea, an orthodox organic tea grown in the pristine environment of mountains of Sikkim and is exported to many countries. This tea is potential for GI registration

The North-eastern region being the centre of origin/diversity of a number of horticultural crops, wild relatives and minor economic plants, for instance, beans (60–70 landraces), brinjal (50 landraces), yams (20 species and 103 landraces), chilli (200 landraces), bamboos (78 taxa), banana (18 species), citrus (17 species), aroids (15 species), orchids (700 species), medicinal plants (>2,000 species), wild relatives (132 species) and many other rare and endangered taxa (Hore 2007).

7.6 Biocultural Protocols for TK Protection of Indigenous People

The development of Bio-Cultural Community Protocols (BCPs) by indigenous people and local communities is another way in which communities can increase their capacity and legal empowerment to drive the local implementation of international and national laws on ABS. Article 12 of the Nagoya Protocol obliges states to "*take into consideration indigenous and local communities' customary laws, community protocols and procedures, as applicable, with respect to TK associated with GRs*". In 2009, in the lead-up to the Nagoya Protocol, Natural Justice, a nongovernment organization focusing on legal issues, started supporting communities to write down their claims. They were named "*biocultural community protocols*", or "*biocultural protocols*" for short. The first such protocol for livestock was created by the *Raika* pastoral community of western Rajasthan in India with the support of *Lokhit Pashu-Palak Sansthan*, a local NGO.

These communities have developed BCPs: *Raika* Pastoralist (Rajastan, India); *Samburu* Pastoralists (Kenya); *Vaidyas* of the Malayali Hills (Tamil Nadu, India); *Gunis* and Medicinal Plants Conservation Farmers of Mewar (Rajastan, India) and *Bushbuckridge* Traditional Healers (South Africa) (Bavikatte and Jonas 2009). These *biocultural protocols* have increased the visibility of livestock keepers as guardians of biological diversity and associated TK, empowered the communities by making them aware of their rights, contributed to the documentation of breeds and TK systems, inspired local communities to pursue conservation activities, and also contributed to a feeling of global solidarity between livestock keepers (Figs. 7.6 and 7.7).

Indigenous communities such as *Gaddi, Gujjar, Bakarwal, Kinnaura, Kauli* and *Kanet* of the north Indian Himalaya, *Bhotia* of Garhwal Himalaya, *Bhotia* and *Sherpas* of *Khumbu* valley of Nepal, *Kirat* of eastern Nepal, *Monpa* yak breeders of Arunachal Pradesh, *Bhutia, Lachenpa, Lachungpa, Dokpa, Gurung* and multi-ethnic

Fig. 7.6 Yak herding by *Dokpa* Community at Muguthang Valley (4500 m, North Sikkim)

Fig. 7.7 An agrobiodiversity landscape at Thangu (4200 m), North Sikkim

Nepali communities of Sikkim and *Changpa* of Changthang, Ladakh, and *Dropka* of Bhutan are some of the known pastoral communities of the Himalaya who have continued with the vastness of TK associated to BRs and GRs in the entire Himalaya. Therefore, there is an urgent need to develop *Biocultural Community Protocol* for these communities, document rich repository of TK associated to BRs and GRs for ABS arrangements. (Figs. 7.8 and 7.9).

Fig. 7.8 *Marcha,* a starter culture for fermented products (Microbes: *Pediococcus pentosaceus, Saccharomycopsis bayans, S. capularis, S. fibulijera, Candida glabrata,* etc.)

Fig. 7.9 *Kinema*, a soya fermented product (Microbes: *Bacillus subtilies, Enterococcus faecium, Candida parapsilosis, Geotrichum candidum*)

7.7 TKDL and *Sui Generis* System of Protection

Following patents on turmeric, brinjal, etc., in India, work began to prepare an easily navigable computerized database of documented TK relating to the use of medicinal and others plants (which is already under public domain) known as Traditional Knowledge Digital Library (TKDL). Such digital databases would enable Patent Offices all over the world to search and examine any prevalent use/prior art, and thereby prevent grant of such patents and biopiracy. A Traditional Knowledge Resource Classification (TKRC) is also in the process to enable the retrieval of information on traditional knowledge in a scientific and rational manner for patent examination.

In order to conserve and prevent misappropriation of country's TK, at international patent offices, India established Traditional Knowledge Digital Library (TKDL) in 2000 with an aim of identifying measures to facilitate the protection of patent rights and IPR on medicinal bioresources. In fact, India is the only country in the world to have set up an institutional mechanism to protect its TK and TKDL is a unique tool that plays a critical role in protecting the country's TK. TKDL has so far digitised about 330,044 traditional medical formulations (Javed 2016) and entered into agreements with 10 international patent offices—Europe, Malaysia, the USA, Japan, the UK, Canada, Australia, Germany, India and Chile, permitting them access to its archives to deter problematic patent petitions (GoI 2018).

Some TK is codified, that is, formalized in some way (eg. Ayurveda) however, a great part of it is non-codified, such as, "folk", "tribal" or "indigenous" medicine which is based on traditional beliefs, norms and practices accumulated during centuries old experiences of trial and error, and passed to successive generations through oral tradition. The TK associated to GRs/BRs particularly owned by the ILCs of northeast India is non-codified and thus is unprotected although Peoples Biodiversity Register is being developed in all the eight states. The protection under Intellectual Property Rights (IPRs) of traditional and indigenous knowledge has received growing attention since the adoption of the Convention on Biological Diversity (CBD) in 1992.

Some countries such as Thailand, Peru, Costa-Rica and Portugal have established laws relating to *sui generis* system of TK protection. *Sui generis* literally means "of its own kind" and consists of a set of nationally recognized laws and ways of extending plant variety protection (PVP) other than through patents (Dua 2014). Although TRIPS does not mention UPOV (International Union for the Protection of New Varieties of Plants), several countries believe that the UPOV convention meets the requirements for a *sui generis* system. However, countries do not have to join UPOV to implement a *sui generis* system to comply with TRIPS (http://www.southcentre.org/southletter/sl34/s134-10.htm). For codified systems and formulations of Traditional Medicine, the TKDL like databases are expected to play a major role in preventing bio-piracy but for the non-codified formulations of the Traditional Medicine (TM) such as Folklore practices, customary and rituals, etc., there is an urgent need of innovative regulations in the form of 'Petty Patents' and a '*sui generis*' system for protection and promotion of TK accompanied with some national level programs like the initiative taken by the National Innovation Foundation (NIF). This will enable potential of noncodified TM that is owned exclusively by the ILCs of the Himalayan region at large and Northeast states of India in particular can be identified, documented, standardized and optimally utilized to the benefit of the healer as well as the ailing mankind (Dua 2014). Some jurisdiction has succeeded in incorporating protective laws for their fading TK while some others are losing their valuable knowledge at a very fast rate due to non-protective or inadequate measures. However, a *sui generis* system to promote the TK has been proposed which greatly comply to the Nagoya Protocol, where India is a signatory (Narayanan 2018; Chakrabarty and Kaur 2021).

7.8 ABS Programmes in the Indian Himalayan States

India implemented two major programmes on Access and Benefit Sharing.

7.8.1 UNEP-GEF-MoEFCC ABS Project

The Project was commissioned by the United Nations Environment Programme under GEF funding. It was executed by National Biodiversity Authority in collaboration with the State Biodiversity Boards (Andhra Pradesh, Goa, Gujarat, Himachal

Pradesh, Karnataka, Odisha, Sikkim, Telangana, Tripura and West Bengal), Botanical Survey of India (BSI), Zoological Survey of India (ZSI), United Nations Development Programme (UNDP), United Nations Environment Programme—Division of Environmental Law and Conventions (UNEP/DELC), and United Nations University—Institute of Advanced Studies (UNU-IAS).

The objective of the project was to increase the institutional, individual and systemic capacities of stakeholders to effectively implement the Biological Diversity Act, 2002 and the Rules 2004 to achieve biodiversity conservation through implementing Access and Benefit Sharing Agreements in India. Facilitating the establishment of ownership of traditional knowledge associated with genetic resources by the holders of such knowledge is utterly urgent. If rights are not clear, there is no legal certainty and access cannot be facilitated.

7.8.2 Indo-German Biodiversity Programme

The ABS Partnership project was commissioned by German Federal Ministry for Economic Cooperation and Development (BMZ) in India. The project was implemented by Deutsche Gesellschaftfür Internationale Zusammenarbeit (GIZ) GmbH and executed by the National Biodiversity authority from November 2017 to April 2020.

The project aimed at strengthening the capacity of the National Biodiversity Authority, selected State Biodiversity Boards (Maharastra, Uttarakhand and Tamil Nadu), Biodiversity Management Committees of the selected states, as well as creating awareness amongst commercial users of biological resources and associated traditional knowledge for the effective implementation of ABS mechanisms under the Biological Diversity Act 2002, in keeping with India's commitments under Nagoya Protocol on ABS.

Apart from the above two programmes, Government of India in collaboration with the Norwegian Government has established a "Centre for Biodiversity Policy and Law (CEBPOL)" in the National Biodiversity Authority (NBA), Chennai, to develop professional expertise in biodiversity policies and laws and develop capacity building. This Centre is focusing on biodiversity policies and laws that cater to the needs of national and international rule-making and subsequent implementation on issues of biodiversity.

The CEPBOL has several themes and under the ABS theme, the centre is concentrating on effective implementation of Nagoya Protocol on ABS and linking its relevant articles with national provisions; identifying cross-sectoral issues such as ABS and REDD plus; link ABS and Protected Area management; supporting technology transfer; developing guidelines on ABS and governance; exploring ABS as an innovative financing mechanisms and developing capacities and raising awareness including human resource development both in India and Norway.

The Indian Himalayan states have established Biodiversity Management Committees at the local level who have been developing People's Biodiversity Register documenting all GRs and associated TK related to biomedicine, food, fodder and other relevant ecosystem services it provides. This has been done with assistance from State Biodiversity Boards.

7.9 TK and ABS in the Himalayan Region

The Himalayan region is home to a large number of different groups of ILCs who hold a great diversity of TK by virtue of remoteness of the area as well as late arrival of the development. These ILCs share common biological heritage; however, they have their own unique system or institution to govern them and they possess distinct knowledge and skills to use, manage and conserve the biological resources, traditionally. Besides, their skill of locating and identifying the plant, preparation of medicine, composition of medicine, doses, etc. for treating the particular ailment in diverse patients is beyond the understanding of present-day medical practitioners. Unfortunately, Himalayan region is no exception in respect to exploitation and misappropriation of long conserved TK by the modern pharmaceutical industries. The third objective of the CBD: ABS was to be implemented by nations who were signatory to the CBD by developing national laws to protect both diversity and its sustainable use along with providing procedures for bioprospecting so that access to genetic resources and associated TK are regulated and the accrued economic and social- cultural benefits are shared equitably. Since, all the Himalayan states including India, Nepal and Bhutan are the party to the CBD, they are obligated to develop state specific ABS policies and laws to give effect to the provisions of the convention and to fulfil its objectives of conservation and sustainable use of biodiversity and ensuring fair and equitable sharing of benefits under Article 6(a) of the CBD. Nevertheless, the development and implementation of ABS policy and legislation in the Himalayan countries have been slow as many complex issues on the ownership over resources and benefit sharing are unclear. So far, only India and Bhutan in the Himalayan region have developed ABS legislation while others are in the process of developing it (Oli 2009; Oli and Sharma 2010).

7.9.1 Bhutan and ABS

Bhutan was traditionally and historically referred to as Lhomenjong, meaning a valley of medicinal herbs due to its rich diversity of medicinal plants. The indigenous communities in Bhutan hold a strong tradition of the use of medicinal plants as a result of which, it is equally rich in associated TK. The government has mainstreamed traditional medicine called *gSo-ba-Rigpa* through the establishment of indigenous hospitals as well as a college and a pharmaceutical unit that manufactures and supplies medicine based on traditional knowledge (NBSAP Bhutan 2014). The indigenous communities in Bhutan have a long history of conservation practices because of their cultural and spiritual bond with nature; nevertheless, conservation is becoming an increasing challenge with the rapid socio-economic development of the country and the growing economic stress.

Bhutan became party to and ratified the UN CBD in 1995 and enacted Biodiversity Act of Bhutan in 2003 to regulate access to GRs and associated TK (BD Act 2003; NBSAP Bhutan 2014). The legislation provides that TK in Bhutan is a subject of protection and owners of the TK are the holders of the rights. The Act lays down procedures for securing access to associated TK, which involves obtaining prior and informed consent from the owners of TK in order to use TK for noncustomary uses (BD Act 2003).

The National Biodiversity Centre (NBC) of Bhutan was instituted in 1998 under the Ministry of Agriculture and Forests. It is the main implementing agency of the CBD responsible for coordinating biodiversity conservation and sustainable use and is a national focal agency to regulate access to BRs and ensure equitable sharing of the benefits. It is responsible for developing policies and legal frameworks for conservation and sustainable utilization of BRS and also serves as the national focal agency for bioprospecting and documentation of TK NBC (2018). The NBC includes a Traditional Documentation Unit, a Bio-Exploration Unit, and a Research Unit with the objective to tap GRs and associated TK within the country in order to contribute to the development of new and clean industries based on bioprospecting and genetic engineering.

Access to TK associated with biological resources (held by one community) will require an access and benefit-sharing agreement between the community based on their community protocol, with guidance from the national focal point and the user. Where TK is held by more than one community, an access and benefit-sharing agreement will be negotiated between the communities in accordance with their community protocols under the guidance of the national focal point and the user. Where this is not feasible, the national focal point shall enter into an access and benefit-sharing agreement on their behalf. The draft policy provides that monetary benefits derived from research and commercialization of genetic resources and/or associated TK (including the processing fee and the cash deposit payable at the scoping phase) is to be channelled to an access and benefit-sharing fund. This fund is managed by the national focal point, which shall disburse the funds for conservation and sustainable use of Bhutan's biodiversity and for the enhancement of rural livelihoods (BD Act 2003; https://nbc.gov.bt/wp-content/uploads/2016-03/draft-Biodiversity-Bill-2016.pdf; Barpujari and Sarma 2017).

The rights exist whether it be in material or other form and continue to be enjoyed perpetuity and are inalienable. It is also stated that anybody aspiring to use TK in Bhutan has to obtain prior informed consent from the owner of the knowledge i.e., local community who would have right to accept or reject the application. Under the Act, it is obligatory to maintain the inventory of TK within the country. While the number of TK holders are declining the associated GRs/BRs in Bhutan are also disappearing due to the process of globalization. Greater knowledge and awareness of the CBD and ABS, as well as clear recognition and a shared understanding of resource ownership, are pre-conditions for protecting the rights of communities and farmers in Bhutan (Wang et al. 2019). A study carried out by Wang et al. (2019) indicated generally low levels of awareness about most components of the CBD, particularly among rural communities, although they feel that benefits derived from local BRs and TK should be shared, while there is still an uncertainty about who owns these resources.

It also adopted Nagoya Protocol in 2011 in order to further the implementation of the third objective of the CBD by providing a strong compliance mechanism that ensures the sharing of benefits arising from the research and commercial utilization of GRs and associated TK with their countries of origin and community custodians and to enhance rural livelihoods, incentivize and ensure continuity of biodiversity conservation. In order to secure the economic benefits arising from the utilization of its GRs and associated TK through the national implementation of the Nagoya Protocol, Bhutan developed ABS Policy 2015 which provides comprehensive guidelines covering ownership of GRs, value of GRs, ABS collaborations, regulating access to GRs and associated TK and fair and equitable sharing of benefits. Further, Bhutan is in the process of amending its existing Biodiversity Act of Bhutan 2003, the Biodiversity Bill 2021 has already been drafted which has a dedicated chapter dealing with access and protection of TK (Chap. 5) and benefit sharing (Chap. 6).

7.9.2 Nepal and ABS

Until the beginning of the 1990s, ABS was an unknown concept in Nepal. There was no recognition that an ABS regime could form the basis for the protection of the rights of ILCs and farming communities over GRs and associated TK. It was only after Nepal became a party to the CBD, in February 1993, that the government and several NGOs began to discuss the importance of integrating ABS issues into national policies. With the ratification of CBD, work on establishing farmer's representative institution, biodiversity conservation and development committee (BCDC) initiated with the mandate to conserve and sustainably use GRs and associated TK for protecting the rights of communities during implementation of an ABS regime. Simultaneously, the work on protection of TK initiated in Nepal with the documentation of biological resources and associated TK in community biodiversity register (CBR) (Paudel et al. 2012). This defensive protection is aimed at preventing misappropriation of TK by private sectors (Oli 2009). Numerous government and non-government organizations like Li-Bird, research institutions, Universities, international organizations like ICIMOD, IUCN, etc. were involved in preparing the community biodiversity registers (CBRs). However, documentation process is being halted for the time being due to increasing risk of biopiracy in absence of ensured legal protection.

Lately, Nepal has also identified a *sui generis* system of protection for TK as one of its goal under CBD implementation program (UNCTAD 2016). It also joined WTO and became party to Nagoya Protocol and International Treaty on Plant Genetic

Resources for Food and Agriculture (ITPGRFA). Besides, the ABS bill which was drafted in 2002 based on Nepal Biodiversity Strategy and Action Plan (NBSAP) is in the process of final adoption. The Bill clearly specifies that the documentation of and access to genetic resources and traditional knowledge should be done only after obtaining prior informed consent (PIC) from the local communities or knowledge holders (Paudel et al. 2012). Recently, it has drafted ABS Strategy and Implementation Plan, and the Government of Nepal endorsed the first National Intellectual Property Policy in March 2017, which duly takes into account the protection of TK.

7.9.3 India and ABS

India ratified CBD in 1994, two years after signing the convention and to meet its obligation under the CBD, it enacted Biological Diversity Act 2002 (BD Act 2002) with special emphasis on fair and equitable sharing of benefits from the use of the biological resources and knowledge associated with it. The scope of the Act includes use of biological resources and associated knowledge occurring in India for commercial or research purposes or for the purposes of bio-survey and bio-utilisation, or for obtaining intellectual property rights for any invention based on research or information on a biological resource obtained from India. It also framed Biological Diversity Rules 2004 and notified guidelines on Access to Biological Resources and Associated Knowledge and Benefit Sharing Regulations 2014 which provides a framework for access to biological resources and associated TK and sharing of the benefits arising out of such access and use of TK in the Indian Himalayan States. The ABS Regulations provide some important and much needed clarity on benefit sharing obligations under the BD Act 2002 and makes an attempt to streamline the procedures for seeking approval to access and exploit Indian BRs and associated TK.

The biodiversity law establishes a three tier institutional structure for biodiversity governance in the country. One of the primary functions of the third tier, the Biodiversity Management Committees (BMCs), is documentation of biological diversity including habitats, landraces, folk varieties and cultivars, domesticated stocks and breeds of animals and micro-organisms, and knowledge relating to biological diversity. The other two tiers, the National Biodiversity Authority (NBA) and the State Biodiversity Boards (SBBs) are mandated to consult these BMCs when taking any decision relating to the use of biological resources and associated TK within the territorial jurisdiction of the BMCs. The Biological Diversity Rules, 2004, specify that the main function of the BMC is to prepare a People's Biodiversity Register (PBR) in consultation with local people (Pant 2015).

7.10 TK and ABS: Case from the Indian Himalaya

India, one of the 17 mega-biodiversity countries, is a home to approx. 645 indigenous tribes with roughly 54 million population inhabiting different landscapes of which 225 tribal communities are concentrated in northeast India. They have distinct culture and tradition, and biodiversity conservation is ingrained in their socio-cultural practices. They possess rich TK in diverse fields such as agriculture, climate, soils, hydrology, plants, animals, forests and human health, etc. TK has also contributed in forest conservation, soil conservation, seed conservation and crop biodiversity. The development of indigenous knowledge systems, including management of natural environment, has been a matter of survival to these communities (Kumar 2010). The indigenous tribes have planned and conserved natural resources by adopting indigenous knowledge for centuries. They have been using their traditional ecological knowledge for making their agriculture production better yielding through accumulation of experiences and intimate understanding of environment. The traditional agro-farming and husbandry practices adopted in India was first appreciated by Dr. John Augustus Voelcker way back in 1889, who was sent to India by the British Government to study Indian agriculture. In 1977, farmers managed to get rid from massive attack of crop pest i.e., brown plant hopper, which destroyed 40% of paddy throughout UP, by simply applying kerosene oil based on their past experiences (Prakash 2002).

The indigenous communities in India have traditionally been using plants for treating various ailments afflicting human and their domestic animals which has been documented long back in ancient literature such as *Susruta Samhita* and *Charaka Samhita*. In India, around 20,000 medicinal plants have been recorded; however, traditional practitioners use only 7,000–7,500 plants for curing different diseases. The proportion of use of plants in the different Indian systems of medicine is Ayurveda 2000, Siddha 1300, Unani 1000, Homeopathy 800, Tibetan 500, Modern 200, and folk 4500. In India, around 25,000 effective plant-based formulations are used in traditional and folk medicine (Pandey et al. 2013; Samal 2016). More than 1.5 million practitioners are using the traditional medicinal system for health care in India. It is estimated that more than 7800 manufacturing units are involved in the production of natural health products and traditional plant-based formulations in India and more than 1500 herbals are sold as dietary supplements or ethnic traditional medicines (Pandey et al. 2013).

The use of herbal medicine dates back as far as the Indus Valley civilisation in 2600 BC; Chinese medicine and Ayurvedic medicine developed by local people are still used in the Himalayan region. Such deep-rooted traditional biotechnological knowledge is very important to modern pharmaceutical research in that it works as an initial screen and can help isolate the medically significant properties of plants and animals. Several modern drugs, including anti-cancer drugs, antibiotics, malaria drugs, and analgesics, were developed from plant and animal resources based on traditional knowledge (Oli and Dhakal 2008). Realising the importance of the traditional medicine in modern allopathy, its documentation process has been given due priority in India and many research institutions and universities are taking up the task of TK documentation targeting different tribes. For example, Pradhan and Badola (2008) and Badola and Pradhan (2013) documented 118 and 124 species of plants being used in traditional medicine by the Lepcha and Limboo tribe for treating various human ailments in Sikkim. Similarly, Bharati and Sharma (2012) and Lepcha (2019) documented 27 and 37 plant species used for treating different diseases in livestock by the different ethnic communities including Lepcha tribe in Sikkim.

Sharma (2017) reported 486 species of common medicinal and aromatic plants including their indigenous use pattern, 25 edible fern species including their indigenous use pattern, 17 plant species used for veterinary ailments for domestic animals, 192 wild edibles and their parts used, 40 species of edible mushrooms used by indigenous communities, 11 species of oil yielding plants, 6 species used for fish poisoning, 7 fibre yielding plant species, and 19 plant species used as dye yielding plants for carpets and various local weaving, 35 different fermented products used as food/beverages of commonly used by the indigenous people, and 28 traditional meat products and their associated microbial biodiversity.

Debbarma et al. (2017) recorded 51 plant species of medicinal importance used by 19 different tribal communities in Tripura. Kumar et al. (2019) documented 200 plant species being used in herbal preparations by the ethnic communities in South India. Panghal et al. (2010) documented 57 medicinal plant species used by *Sapera* communities in Haryana for the treatment of various diseases. Kala et al. (2005) compiled 135 herbal drugs being used by the *Vaidyas* (traditional healers) in Uttaranchal. Tali et al. (2019) developed comprehensive database of 1123 medicinal plants used in traditional medicine in the state of Jammu and Kashmir. In the Kashmir Himalaya, a total of 190 plant species belonging to 70 families, growing along the migratory route of the Gaddi Shepherds are used to treat more than 80 different ailments and disorders (Dutt et al. 2015). In a study carried out in Kathua district of Kashmir, a total of 197 plants from 87 families and 174 genera were used for the ethnomedicinal purposes (Raoa et al. 2015).

Mao and Roy (2016) reviewed available publications on plant based TK from Northeast India to find out the gap and scope in further ethnobotanical research in the region. They reviewed a total of 470 published works consisting of 402 research articles, 41 contributory chapters in Books, 20 Books and 7 papers in Proceedings of different seminars and workshops. Of the eight states in the region, the highest report is from Assam (156) followed by Manipur (53), Arunachal Pradesh (49), Meghalaya (47), Sikkim (42), Nagaland (32), Mizoram (25) and Tripura (22). There are a good number of publications (44) which are covered in general in the North-eastern states. In terms of the different publication categories, 243 were on ethnomedicine and represent maximum number of works followed by 89 publications on ethno-botany, 84 on wild food and NTFPs, 11 on religious and supernatural beliefs, 10 on traditional conservation practices, 9 publications on ethno-veterinary, 6 on fermented drinks and beverages, 6 publications on herbal dyes, 5 on pest management and control, 3 on seasonal indicators for crops and agriculture, 2 on stupefying, and one each for family planning and birth control, and traditional weaving implements. Different species of plants (approx. 114) are used as piscicides by various people inhabiting Northeast

India from a long period of time for fish harvesting purpose (Moirangthem and Datta 2019).

7.11 GRs and Associated TK in the Khangchendzonga Landscape (KL)

Khangchendzonga Landscape (KL) is one of the six transboundary landscapes identified by the International Centre for Integrated Mountain Development (ICIMOD) in the Hindu Kush Himalayan (HKH) region. It covers an area of 25,080.8 km^2 and spreads across part of eastern Nepal (21%), Sikkim and West Bengal of India (56%) and the western and south-western parts of Bhutan (23%). Khangchendzonga National Park (KNP), Sikkim has been inscribed as India's first "Mixed World Heritage Site" on UNESCO World Heritage List, by fulfilling the nomination criteria under both natural and cultural heritage.

The KL is endowed with Non-timber Forest Products (NTFPs), an important provisioning ecosystem services, which are recognized for their contribution in rural livelihoods and forest and biodiversity conservation. Effective management through sustainable harvesting and market driven commercialization are two contrasting aspects that are bringing challenges in development of NTFPs sector. There are a number of avenues for ABS arrangements in the region. There are still uncovered areas on identifying potential species having market value, conducting value chain analyses, and sustainable management of NTFPs which all need analysis of their use patterns by communities and trends at a regional scale leading to bioprospecting and ABS agreements. Yadav et al. (2016) analyzed use patterns, trends, and challenges in traditional use and management of NTFPs in the southern slope of KL, and discussed potential implications for conservation and livelihoods. They reported a total of 739 species of NTFPs used by the ILCs of KL. Of these, the highest number of NTFPs was documented from India (377 species), followed by Nepal (363) and Bhutan (245) (Fig. 7.10).

The collection, conservation and sustainable utilization of NTFPs in Bhutan is mostly guided by the National Strategy for the Development of Non-Wood Forest Products 2008–2018. Other sectoral policies are the Forest Act 1969, Plant Quarantine Act of Bhutan 1993, Forest and Nature Conservation Act of Bhutan 1995, Environmental Assessment Act 2000 and Biodiversity Act and Framework of Bhutan 2003, 2006 (Ugyen and Olsen 2008). The Indian National Forest Policy (1988) makes a special mention of NTFPs emphasizing on protection, improvement and their enhanced production for generation of employment and income. Likewise, in Nepal there are several sectoral as well as specific policy provisions for sustainable use and management of NTFPs (Steele et al. 2015; Ghimire et al. 2008). The most comprehensive policy is the Herbs and Non-Timber Forest Product Development Policy 2004 (MoFSC 2004) (Table 7.3).

Fig. 7.10 Medicinal and Aromatic plants sold in the local markets

Indigenous farmers in the KL India have, through generations of innovation and experimentation, established a variety of land use systems to nurture a great diversity of both wild and domesticated plants and animals. Local agrobiodiversity features more than 126 landraces of cereals, including rice (77), maize (26), and millet (7); 18 cultivars of oilseeds; 34 cultivars of pulses/beans; 132 species of vegetables; 38 species of spices/ condiments; 33 landraces of tubers/roots; and 64 species of fruit. Sikkim's traditional system of cultivation also supports more than 200 species of wild edibles, 119 species of multipurpose agroforestry trees, 52 crops with high social and cultural value, and 69 species of plants sacred to indigenous communities. It also has a diversity of land uses, with 15 to 20 field types, and specific land use categories. Similarly, there is a high diversity of domestic animals, with about 21 different local and indigenous breeds. Homesteads on marginal farms make up 40–70% of Sikkim's total landholdings and account for 50–80% of these households' requirements. Homesteads, centres of agrobiodiversity and associated TK, are traditional sources of food and nutrition, and are important contributors to food and livelihood security among farming communities. The ethnic communities in the region prepare a number of fermented and non-fermented foods including as many as 15 different local instant pickles depending upon the season (Fig. 7.11).

O'Neill et al. (2017) reported some commonly used plants as medicine to prevent or manage gastro-intestinal afflictions, dermatological conditions, and respiratory-tract infections: *Swertia chirayita* (Gentianaceae), *Bergenia ciliata* (Saxifragaceae), *Oroxylum indicum* (Bigoniaceae), *Urtica dioica* (Urticaceae), *Acorcuscalamus* (Acoraceae), *Nardostachysjatamansi* (Caprifoliaceae), *Rhododendron arboreum*

Table 7.3 Major use category of NTFPs and frequency of taxa reported from Khangchendzonga Landscape (KL)

Use category	Frequency of taxa reported			Total (KL)
	Bhutan	India	Nepal	
Basket	4	3	–	7
Broom	2	–	1	3
Cigarette wrapper	–	–	1	1
Detergent	–	1	–	1
Dye	12	6	6	23
Edible	62	46	130	191
Fencing	2	1	1	4
Fermentation	–	4	3	7
Fibre	10	4	4	18
Fish poisoning	–	1	1	2
Fodder	3	4	12	19
Fuel	–	1	1	2
Gum	4	–	–	4
Incense	19	3	13	35
Insecticide	–	2	1	3
Medicinal	176	334	297	807
Oil	6	2	7	15
Ornamental	5	1	–	6
Preservator	–	1	–	1
Roofing	2	–	–	2
Ritual	–	3	5	8
Spiritual	–	1	1	2
Tea	1	4	4	9
Veterinary medicine	–	27	5	32

Yadav et al. (2016)

Fig. 7.11 Some NTFPs of the Eastern Himalaya

Fig. 7.12 Traditional bee keeping: harvesting of honey

Fig. 7.13 A herbal practitioner collecting *Artemisia vulgaris* plant

(Ericaceae), *Rumex nepalensis* (Polygonaceae), *Astilbe rivularis* (Saxifragaceae), and *Cheilocostus speciosus* (Costaceae). There is a lack of forward and backward linkage for bioprospecting these resources for ABS (Figs. 7.12 and 7.13).

7.12 Potential Threats to GRs/BRs for ABS

Despite expansive heritage of TK, the long conserved knowledge of ILCs on traditional medicine, which has evolved through thousands of years of constant interaction with nature, is facing multiple threats such as erosion, dilution, piracy, misappropriation and exploitation across the globe for not being in codified knowledge system. Apart from these, TK faces threats due to loss of pathways to TK transmission, change of traditional livelihood practices, change of traditional religion and beliefs, change of environment and natural resources, loss of traditional rights, change of traditional institutions, government policy and legislation, contact with other cultural groups, influence of outside market, colonisation, relocation, marginalisation by dominant societies, war, military occupation, population decline, migration, economic development pressure, etc. (Tang and Gavin 2016).

Swiderska (2006) identifies other drivers of TK loss such as 1. reduced landholdings, insecure land tenure and privatisation of communal land; 2. policies that restrict access to forests and sacred sites and participation in natural resource management; 3. erosion of traditional values and economies due to globalization; 4. weakening of traditional authorities/laws due to extension of government control; 5. spread of modern agriculture and technology; and 6. weakening of social cohesion and collective organisation caused by all of the aforementioned drivers.

There has been a rapid loss of traditional medical knowledge and practices due to peoples dependency on verbal transformation of knowledge, impacts of modern cultural transformation, and rapid land degradation. At the same time there is a depletion of resource base due to over exploitation and lack of management systems. However, several contradicting policies, plans, acts and laws interact to regulate and set the context in which medicinal resources are managed and utilised for subsistence and trade. In the Himalayan states of India including Bhutan, domesticated GRs and their wild relatives are threatened by various natural causes such as landslides and flash-floods, human activities such as conversion of forest or agriculture land into infrastructure development such as hydropower projects, pharmaceuticals, roads, airports, reservoirs and construction purposes. Further, extreme weather caused by global warming and climate change, such as heavy snow in high altitudes, heavy and untimely rainfall pattern or severe droughts, have resulted into phenological changes leading to decrease and even loss of many socioeconomically important species. The situation is again aggravated by various human activities such as land use change, habitat destruction, and over-harvesting.

Therefore, there is a clear threat of biopiracy especially in the Himalayan region. The traditional right claimed over the Hoodia plant (*Hoodia gordonii*) by the San people of South Africa; over Rooibos plant (*Aspalathus linearis*) by San and Khoi tribes, and over the *Arogyapacha* plant (*Trichopus zeylanicus*) by Kani tribe of Kerala are the most prominently cited examples of biopiracy / knowledge piracy. Biopiracy will damage the community knowledge, TK and national heritage of the Himalayan states of India, Nepal and Bhutan, and is also a potential threat to the economic interest of the region. Based on the history of biopiracy cases in India (Basmati rice, Darjeeling tea, Turmeric, Neem and so on), it has caused the problem of the intrusion to national sovereignty when an individual, corporation or a government from other countries utilize and benefit from the patenting of GR/BRs derived from native species and TK of another sovereign state. Furthermore, biopiracy violate the concept of common heritage of mankind, and also the state's sovereign rights to its own resources.

7.13 Role of Women in TK and ABS

Women in the Himalayan countries are primary custodians of TK and seed savers of traditional varieties of crops, fruits and vegetables. Recognition of the important role of economically deprived women in conservation, protection and preservation of BRs/GRs is largely lacking in policies and legal instruments; however, the vast TK held by them is yet to be documented. Gender inequalities persist in the mountain societies in decision making and local governance. Policies should also be based on contextual factors, considering through directives the recognition of women's subordinated situation in order to address its impact on their capability and freedom to participate in negotiations and decision making regarding benefit sharing as equals of men (Fig. 7.14).

Fig. 7.14 Women selling seeds of pulses and cucurbits

The women in this region are major stakeholders and custodians of knowledge, conserving food and medicinal plants in both *jhum* land and home gardens (Singh 2004; Mishra et al. 2009). Working with and studying tribal women can advance our understanding of community TK, including its typology and status among different age and social groups (Turner and Turner 2007; Mishra et al. 2009). It can also facilitate the development of sustainable technologies and products and could contribute to sustaining the biocultural resources and livelihood systems that persist (Ramakrishnan et al. 1996, 2002, Singh et al. 2009). Despite the rich knowledge of women, and their role in conserving biocultural diversity as experimenters, conservators and stabilisers of foods, medicines and other indigenous resources, their contribution is rarely recognised at policy level (Swaminathan 1998, Singh 2007, Singh 2007). Singh and Pretty Pilgrim (2010) identified a total of 1690 traditional practices carried out only by middle-aged and elderly women of the *Adi, Monpa* and *Khasi* tribes in Northeast India pertaining to ethnomedicine, cosmetics, plant based foods, aquatic resources and animal based foods, fermented foods, biopesticides and so on.

Women in the Northeast India have rich knowledge on food prepared from local crop species, seed saving; food prepared from ethnobotanicals accessed from agroforestry systems and community forest, *jhum* land and home garden; foods prepared from indigenous fishes; foods prepared from wild animals and insects, plants and animal based fermented foods, fish based fermented foods, traditional alcoholic beverages, medicinally important traditional foods, ethnoveterinary and ethnomedicine; traditional foods important for festivals and other cultural occasions; cosmetics; biopesticides and so on. This repository of knowledge is yet to be documented and recorded in the TKDL.

7.14 Way Forward

Kani Tribes's *Arogyapachya* bioprospecting was successful in India that followed all the legal process of bioprospecting enumerated in the law. Among the well-known cases are neem, turmeric, *Phyllanthus amarus*, etc., where patent offices in northern countries had wrongly granted patents on products or applications derived from the TK of local communities. In some cases, the Government of India took the lead and litigated against these cases and succeeded in getting the patents revoked (Pant 2015). However, bioprospecting from eastern Indian Himalayas have not been done despite of diverse biological pool of resources in the region.

Himalayas are transboundary in nature and use of resources is done by both Indian and Nepalese, thus this has led to illegal intrusion in the higher Himalayan region. Political boundaries do not limit the movement of biological resources and knowledge. Countries in the Himalayan region share common biological resources and traditional knowledge, and protection and sustainable use cannot be planned effectively by single countries in isolation. The unsustainable exploitation of wild flora and fauna is continuing through cross-border trade, and especially sensitive areas with high levels of biodiversity such as in the mountains need special protection and regulation. Regional cooperation needs to be strengthened to address the violation of national laws on genetic resources and associated traditional knowledge. A regional ABS framework will increase the bargaining power of countries sharing these common resources. By acting regionally, countries can express their priorities more strongly.

The Himalayan region is home to a large number of indigenous ethnic communities who share a common cultural and biological heritage with their immediate neighbouring countries. The region also contains two of the world's 10 mega-centres for biodiversity, along with all or part of four of the 34 global biodiversity hotspots (Conservation International: www.conservation.org; www.cepf.net). The Himalayas are also a 'mega language centre' with over 400 languages spoken by less than 100,000 speakers (Turin 2007). Despite these rich resources, the remoteness of many areas and the high-altitude environment means that many of the people are among the poorest in the world.

By acting regionally, countries can express their priorities in a stronger way. Operating independently, countries risk these priorities being played off against each other, both in international negotiations and by bio-prospectors seeking to access shared resources. Cooperation among all the countries in the region to develop a common regional ABS framework will help regulate access, promote a more equitable sharing of benefits with local communities, and promote the sustainable use of biodiversity. There has been growing concern over the ways in which genetic resources and TK are appropriated by corporations and research institutions without equitable sharing of the benefits with communities and countries of origin, and often even without their knowledge.

In the Himalayan region, the natural ecosystems existed long before political boundaries were established and do not respect them or adjust to them in any way. Naturally, these political boundaries do not limit the movement of communities of plants and animals, or confine TK systems within the same communities or geographic limitations. Thus the countries in the Himalayan region broadly share same or endemic genetic resources and TK systems within the ethnic groups and or among different ethnic groups. A regional ABS framework would help to govern and manage such an important common heritage.

7.15 Conclusion

The Himalayan region shares many common GRs and rich TK. Therefore, it is important to develop a common understanding about use and benefit sharing mechanisms applicable to all the Himalayan countries. This requires the development of a regional ABS framework or strategy incorporating the important issue of commercialization of GRs and associated TK. PBRs and GI registration can help communities promote and protect markets for their biocultural products, and are gaining popularity globally as a tool for the protection and promotion of TK. GIs specify the geographical origin of a product and link it with its essential qualities that are due to its place of origin (Pant 2015). The ABS regime can facilitate the transfer of GRs in return for economic and technological gains for poverty alleviation. ABS agreements regulate access and use rights between different parties and can, therefore, act as instruments for the redistribution of benefits generated from the use of GRs and associated TK. Industries using GRs are encouraged to pay higher research and development costs than in situations where there are no benefit sharing agreements. Similarly, local communities can receive direct benefits, which can be substantial (Oli and Dhakal 2009). While there is development in progressive ABS policies and laws in the region, illicit trade of biological resources and appropriation by users is ongoing despite defensive measures taken up by the national governments. Awareness among the ILCs of the value of their GRs and associated TK and their rights and duties, and the formation and legalization of their traditional institutions into a federation would help their negotiating and bargaining power and further benefit such communities. The lack of gene banks in the Himalayan states, as well as in community centres, is the greatest constraint to sustaining ex situ genetic resource collection and conservation. It is imperative to develop an adaptation strategy through policies that prioritize community-based biodiversity management programmes and registers maintained by local BMCs to take the advantage of ABS agreements (Sharma et al. 2016).

There are utterly serious unanswered questions that must be addressed with immediate effect as also raised by Chakrabarty and Kaur (2021). The inaccessible, marginal, fragile, heterogenous and diverse Indian Himalayan region is gradually encroached by modern scientific interventions like High Yielding Varieties (HYVs) and Genetically Modified Organisms (GMOs) which would soon result into environmental unsustainability including the immediate loss of BRs/GRs and associated TK. Would it be possible in India to protect the cultural, social and economic rights of the dying Indigenous communities? Or would we not reclaim our lost glory and help the vulnerable from their extinction?

Acknowledgements The authors are thankful to The Mountain Institute India and Sikkim Biodiversity Board, Forest and Environment Department for providing facilities to develop this chapter.

Conflict of Interest The views and opinions expressed in this article are those of the author in his personal capacity. None of the observations, opinions, or conclusions expressed herein are to be attributed to or construed as the views or positions of any organization, association, or company. No conflict of interest.

References

Antons C (2010) Sui Generis protection for plant varieties and traditional knowledge in biodiversity and agriculture: the international framework and national approaches in the Philippines and India. Indian J Law Technol 6:89–139

Badola HK, Pradhan BK (2013) Plant use in healthcare practices by Limboo tribe in south-west Khangchendzonga Biosphere Reserve, Sikkim, India. Indian J Tradit Knowl 12(3):355–369

Barpujari I, Sarma UK (2017) Traditional knowledge in the time of neo-liberalism: access and benefit-sharing regimes in India and Bhutan. Int Indigenous Policy J 8(1). http://ir.lib.uwo.ca/iipj/vol8/iss1/3, https://doi.org/10.18584/iipj.2017.8.1.3

Bavikatte K, Jonas H (2009) Bio-cultural community protocols–a community approach to ensure the integrity of environmental law and policy, p 21

Bhattacharya S (2014) Bioprospecting, biopiracy and food security in india: the emerging sides of neoliberalism. Int Lett Soc Humanistic Sci 23:49–56

BD Act (2003) The biodiversity act of Bhutan of 2003. Ministry of Agriculture, Royal Government of Bhutan. https://wipolex.wipo.int/en/text/490665

Bharati AK, Sharma BL (2012) Plants used as ethnoveterinary medicines in Sikkim Himalaya. Ethnobot Res Appl 10:339–356

Chakrabarty SP, Kaur R (2021) A primer to traditional knowledge protection in India: the road ahead. Liverpool Law Rev. https://doi.org/10.1007/s10991-021-09281-4

Dagne T (2014) Protecting traditional knowledge in international intellectual property law: imperatives for protection and choice of modalities. J Marshall Rev Intell Prop 14:25–49

Datta HS, Sharma G, Bora SS (2021) Geographical indications in horticulture: North East India perspective. Int J Current Microbiol Appl Sci 9(1):1207–1221. https://doi.org/10.20546/ijcmas.2020.901.134

Debbarma M, Pala NA, Kumar M, Bussman RW (2017) Traditional knowledge of medicinal plants in tribes of Tripura in northeast India. Afr J Tradit Compliment Alternat Med 14(4):156–168. https://doi.org/10.21010/ajtcam.v14i4.19. eCollection

Dua P (2014) Intellectual property rights: need for a sui generis regime for non-codified traditional medicine in India. Department of AYUSH, Ministry of Health & Family Welfare, Govt. of India. Am J Phytomed Clin Ther 2(1):116–125

Dutt HC, Bhagat N, Pandita S (2015) Oral traditional knowledge on medicinal plants in jeopardy among Gaddi shepherds in hills of northwestern Himalaya, J&K, India. J Ethnopharmacology 168(5):337–348

Freedom-Kai P (2016) Intellectual property rights in traditional knowledge: enabler or sustainable development. Utrecht J Int European Law 32(83). https://doi.org/10.5334/ujiel.283

Gebru A (2017) The Global protection of traditional knowledge: searching for the minimum consensus. J Marshall Rev Intell Prop 17:42–91

Ghimire SK, Sapkota IB, Oli BR, Parajuli-Rai R (2008) Non-timber forest products of Nepal Himalayas: database of some important species found in the mountain protected areas and surrounding regions. WWF Nepal, Kathmandu

Government of India (GoI) (2018) Traditional knowledge digital library. Ministry of AYUSH, Council of Scientific and Industrial Research, New Delhi

Gupta VK (2011) Protecting Indian traditional knowledge and biopiracy. https://www.wipo.int/export/sites/www/meetings/en/2011/wipo_tkdl_del_11/pdf/tkdl_gupta.pdf. Accessed 06 May 2021

HC, Bhagat N, Pandita S (2015) Oral traditional knowledge on medicinal plants in jeopardy among Gaddi shepherds in hills of northwestern Himalaya, J&K, India. J Ethnopharmacol 168:337–348. https://doi.org/10.1016/j.jep.2015.03.076.Ethnomedicinal plants of Kathua district, J&K, India

Hore DK (2007) Diversity in agri-horticultural plants: an experience with northeast India. Biodiversity and its significance. IK International Publishing House Pvt. Ltd., New Delhi, 167–191. https://doi.org/10.1007/s10991-021-09281-4.pdf

Javed G, Priya R, Deepa VK (2020) Protection of traditional health knowledge: international negotiations, national priorities and knowledge commons. Soc Cult South Asia 6(1):98–120. https://doi.org/10.1177/2393861719883069

Javed G (2016) Complementary measures and customary law for the protection of traditional knowledge: examples and lessons learned. Presented at Roundtable 3, World Intellectual Property Organisation, Geneva

Kala CP, Farooque NA, Majila BS (2005) Indigenous knowledge and medicinal plants used by Vaidyas in Uttranchal, India. Nat Product Radiance 4(3):195–206

Kamau EC, Fedder B, Winter G (2010) The nagoya protocol on access to genetic resources and benefit sharing: what is new and what are the implications for provider and user countries and the scientific community? Law Environ Dev J 6(3). http://www.lead-journal.org/content/10246
Kuei-Jung Ni (2011) Traditional knowledge and global lawmaking. Northwest J Int Hum Rights 85
Kumar JUS, Chaitanya MJK, Semotluk AJ, Krishna V (2019) Indigenous knowledge of medicinal plants used by ethnic communities of South India. Ethnobot Res Appl 18:4. https://doi.org/10.32859/era.18.4.1-112
Kumar KA (2010) Local knowledge and agricultural sustainability: a case of Pradhan tribe in Adilabad district. Working Paper No. 81. Centre for Economic and Social Studies, Hyderabad, India
Lepcha SR (2019) Ethno-veterinary practices by the people Lepcha primitive tribe living in Dzongu Tribal Reserve Area (DTRA) in North-Sikkim India. Pleione 14(1):71–82
Mao AA, Roy DK (2016) Ethnobotanical studies in North East India: A Review In: Ashok KJ (ed) Indian Ethnobotany: Emerging Trends, Scientific Publishers, pp 99–112
McGrath DG (1987) The role of biomass in shifting cultivation. Hum Ecol 15(2):221–242
Mishra S, Singh RK, Singh A (2009). Dynamics of Adi women's traditional foods in varying socio-ecological systems of Arunachal Pradesh: a source of learning and inspiration. In: Lindgreen A, Hingley MK (eds) The new cultures of food: marketing opportunities from ethnic, religious and cultural diversity. Farnham: Gower, pp 203–222
MoEF (Ministry of Environment and Forests) (1988) National forest policy. Government of India, New Delhi
MoFSC (Ministry of Forests and Soil Conservation) (2004) Herbs and non-timber forest products development policy 2004. Kathmandu: Ministry of Forests and Soil Conservation, Herbs and NTFPs Coordination Committee
Moirangthem K, Datta MP (2019) Piscicidal plants of Northeast India and its future prospect in aquaculture-a comprehensive review. Indian J Nat Prod Resour 10(3):165–174
Narayanan A (2018) Domestic biological diversity and ABS laws, policies and practice in India. In: Ramesh MK et al (ed) A primer on biological diversity laws, access and benefit sharing. UNDP & CELERA, NLSIU, Bangalore
NBC (2018) Biodiversity information and database. Natl Biodivers Centre, Sebithang, Thimpu, Bhutan
NBSAP Bhutan (2014) National biodiversity strategies and action plan of Bhutan, 2014. National Biodiversity Centre, Ministry of Agriculture and Forests, Royal Government of Bhutan. https://www.cbd.int/doc/world/bt/bt-nbsap-v4-en.pdf
O'Neill AR, Badola HK, Dhyani PP, Rana SK (2017) Integrating ethnobiological knowledge into biodiversity conservation in the Eastern Himalayas. J Ethnobiol Ethnomed 13:21. https://doi.org/10.1186/s13002-017-0148-9
Oli KP, Dhakal T (2008) Traditional knowledge in the himalayan region. regional programme on access and benefit sharing from genetic resources and associated traditional knowledge in the Eastern Himalayas, ICIMOD, GPO Box 3226, Khumaltar, Lalitpur, Kathmandu, Nepal. www.icimod.org/abs
Oli KP, Dhakal T (2009) Training of trainers and resource manual access and benefit sharing from genetic resources and associated traditional knowledge. International Centre for Integrated Mountain Development, GPO Box 3226, Kathmandu, Nepal
Oli KP, Sharma G (2010) Issues and challenges of access and benefit sharing mechanism in the Hindu Kush Himalayan countries. In: Warth A, Reyes-Knoche S (eds) Triggering synergies between intellectual property rights and biodiversity. Deutsche GesellschaftfürTechnischeZusammenarbeit (GTZ) GmbH, Postfach 5180, 65726 Eschborn, Germany, on behalf of Federal Ministry of Economic Cooperation and Development, Government of Germany, pp 219–233
Oli KP, Feyerabend GZ, Lassen B (2010) Towards an access and benefit sharing framework agreement for the genetic resources and traditional knowledge of Hindi-Kush Himalayan region: draft for discussion. ICIMOD, Kathmandu, Nepal

Oli KP (2009) Access and benefit sharing from biological resources and associated traditional knowledge in the HKH region-protecting community interests. Int J Biodiversity Conservat 1(5):105–118. http://www.academicjournals.org/ijbc

Ouma M (2017) Traditional knowledge: the challenges facing international lawmakers. https://www.wipo.int/wipo_magazine/en/2017/01/article_0003.html. Accessed 07 May 2021

Pandey MM, Rastogi S, Rawat AKS (2013) Indian traditional Ayurvedic system of medicine and nutritional supplementation. Evidences-based complementary and alternative medicine. Article ID: 376327. https://doi.org/10.1155/2013/376327

Panghal M, Arya V, Yadav S, Kumar S, Yadav JP (2010) Indigenous knowledge of medicinal plants used by Saperas community of Khetawas, Jhajjar district, Haryana, India. J Ehtnobiol Ethnomed 6(4)

Pant R (2015) Protecting and promoting traditional knowledge in India: what role for geographical indications? IIED Working Paper. IIED, London. http://pubs.iied.org/16576IIED. ISBN 978–1–78431–029–5

Paudel B, Adhikari K, Shrestha P, Tamang BB (2012) Nepal: innovative mechanisms for putting farmer's rights into practice. In: Ruiz M, Vernooy R (eds) The Custodians of Biodiversity, IDRC

Pradhan BK, Badola HK (2008) Ethnomedicial plant use by Lepcha tribe of Dzongu valley, bordering Khangchendzonga Biospehere Reserve in north Sikkim, India. J Ethnobio Ethnomed 4(22)

Prakash S (2002) Using indigenous knowledge to raise agricultural productivity: an example from India, IK Notes No. 45

Raoa PK, Hasanb SS, Bhellumc BL, Manhasa RK (2015) J Ethnopharmacol 171:12–27

RamakrishnanPS, Das AK, Saxena KG (1996) Conserving biodiversity for sustainable development. National Science Academy, New Delhi, Indian, p 246

Ramakrishnan PS et al (2002) Traditional ecological knowledge and managing biosphere reserves in South and Central Asia. UNESCO and Oxford IBH, New Delhi, India

Rao MB, Guru M (2003) Understanding TRIPS; managing knowledge in developing countries. Response Books, New Delhi. SAGE Publications Pvt. Ltd; First edition, pp 358

Reid J (2009) Bio-piracy: the struggle for traditional knowledge rights. Am Indian Law Rev 34(1):77–98. https://digitalcommons.law.ou.edu/ailr/vol34/iss1/2. Accessed 07 May 2021

Sahai S, Kumar U, Ahmed W (2005) Indigenous knowledge: issues for developing countries. Gene Campaign, Delhi

Samal J (2016) Medicinal plants and related developments in India: a peep into 5-year plans of India. Indian J Health Sci Biomed Res 9:14–9

Sharma G (2017) Tradable bioresources of Sikkim. Sikkim Biodiversity Board. Forest and Environment Department, Government of Sikkim

Sharma G, Partap U, Sharma E, Rasul G, Awasthe RK (2016) Agrobiodiversity in the Sikkim Himalaya: sociocultural significance, status, practices, and challenges. ICIMOD Working Paper 2016/5 Kathmandu: ICIMOD

Singh RK (2004) Conserving diversity and culture: Pem Dolma. Honey Bee 15(3):12–13

Singh RK, Pretty Pilgrim S (2010) Traditional knowledge and biocultural diversity: learning from tribal communities for sustainable development in northeast India. J Environ Planning Manage 53(4):511–533

Singh RK, Srivastava RC, Mukherjee TK (2009) Community based sustainable natural resources management and development in Northeast India. Curr Sci 96(1):19–21

Singh A (2007) Traditional foods and associated knowledge systems relating to health and nutrition among Adi women of East Siang district, Arunachal Pradesh. Thesis (MSc). Banaras Hindi University, Varanasi, UP, India

Singla A (2020) Protection of traditional knowledge in India with reference to Neem, Turmeric, Basmati Rice. file:///Users/ghanashyamsharma/Downloads/ProtectionofTraditionalKnowledgeinIndiawithreferencetoNeemTurmericBasmatiRice.pdf

Steele MZ, Shackleton CM, Shaanker RU, Ganeshaiah KN, Radloff S (2015) The influence of livelihood dependency, local ecological knowledge and market proximity on the ecological impacts of harvesting non-timber forest products. Forest Policy Econ 50:285–291

Swaminathan MS (1998) Gender dimensions in biodiversity management. Konark Publishers, New Delhi

Swiderska K (2006) Banishing the biopirtes: a new approach to protecting traditional knowledge. Gatekeeper Series 129, International Institute for Environment and Development, London

Tali BA, Khuroo AA, Ganie AH, Nawchoo IA (2019) Diversity, distribution and traditional uses of medicinal plants in Jammu and Kashmir (J&K) state of Indian Himalayas. J Herbal Med 17–18. https://doi.org/10.1016/j.hermed.2019.100280

Tang R, Gavin MC (2016) A classification of threats to traditional ecological knowledge and conservation responses. Conserv Soc 14(1):57–70

Tobin B, Swiderska K (2001) Speaking in tongues: indigenous participation in the development of a sui generis regime to protect traditional knowledge in Peru. IIED, London. pubs.iied.org

TRIPS, Plant Variety Protection and UPOV, The South Centre. http://www.southcentre.org/southletter/sl34/sl34–10.htm

Turin M (2007) Linguistic diversity and the preservation of endangered languages: a case study from Nepal. International Centre for Integrated Mountain Development (ICIMOD). 978–92–9115–055-7. ffhalshs-03083332f

Turner NJ, Turner KL (2007) Traditional food systems, erosion and renewal in Northwestern North America. Indian J Tradit Knowl 6(1):57–68

Ugyen PW, Olsen A (2008) Vulnerable medicinal plants and the risk factors for their sustainable use in Bhutan. J Bhutan Stud 19:66–90

UN (2014) The convention on biological diversity and the nagoya protocol: intellectual property implications, a handbook on the interface between global access and benefit sharing rules and intellectual property. United Nations Conference on Trade and Development (UNCTAD) Palais des Nations 8–14 avenue de la Paix, 1211 Geneva, Switzerland, p 202

UNCTAD (2016) Developing dimensions of intellectual property in Nepal: transfer of technology, access to medicines, genetic resources and traditional knowledge. United Nations, New York, Geneva

Upreti Y, Oli KP, Paudel KC, Pokhrel DN, Thapa P, Chaudhary RP (2020) Accessing genetic resources and sharing the benefits: the implications for research on biodiversity. In: Siwakoti M, Jha PK, Rajbhandary S, Rai SL (eds) Plant diversity in Nepal. Botanical Society of Nepal, pp 206–224

Venkataraman K, Latha SS (2008) Intellectual property rights, traditional knowledge and biodiversity of India. J Intellectual Property Rights 13:326–335

Wang SW, Lee WK, Brooks J, Dorji C (2019) Awareness of the convention on biological diversity and provisions regarding access and benefit sharing among multiple stakeholder groups in Bhutan. Oryx 54(5):735–742 © 2019 Fauna & Flora International. https://doi.org/10.1017/S0030605318000819.

WHO (2002) Traditional medicine strategy 2002-2005 (WHO/EDM/TRM/2002.1). Geneva, Switzerland

Yadav U, Paudel RC, Gurung J, Chettri N, Chaudhary RP (2016) Traditional use and management of NTFPs in Kangchenjunga Landscape: implications for conservation and livelihoods. J Ethnobiol Ethnomed 12:19.https://doi.org/10.1186/s13002-016-0089-8

Chapter 8
Access and Benefit Sharing in Indigenous Knowledge Stewardship and Sustainable Development

T. P. Ijinu, S. Rajasekharan, V. George, and P. Pushpangadan

Abstract The incredible ability of humankind in exploiting the natural resources surrounding them to their best advantage is perhaps the main reason that enabled humans to become the most successful organism on planet Earth. The growing concern over monopolization of benefits led genetic resource providing countries to restrict access to genetic resources and associated traditional knowledge. The ratification of Convention on Biological Diversity (CBD) in 1993, by the members of the United Nations brought forward the agenda of access and benefit-sharing (ABS) from the use of genetic resources. The World Trade Organization agreement on Trade-Related Aspects of Intellectual Property Rights (WTO-TRIPS, 1995), the Food and Agriculture Organization International Treaty on Plant Genetic Resources for Food and Agriculture (FAO-ITPGRFA, 2001), the Bonn guidelines (2002) and the Nagoya protocol (2014) provide a broad framework for ABS procedures. The Nagoya protocol exemplifies a trend toward using legal measures to clarify rights and drive sustainable development through market mechanisms such as contracts, intellectual property rights, and mutually supportive institutions at the national, regional, and community levels, with a substantive focus on traditional knowledge and Indigenous communities. Conversely, the present challenge is how the Parties implement the provisions of the Protocol so that legal, regulatory and administrative measures contribute to and fulfil the CBD objectives while facilitating the emergence and scale-up of BioTrade. This article provides a glance through historical aspects and recent developments in ABS regime and associated traditional knowledge.

Keywords Biodiversity · Traditional knowledge · Ethnobotany · Indigenous peoples · *Sui generis* protection · Nagoya protocol · Intellectual property right

T. P. Ijinu
Naturæ Scientific, Kerala University Business Innovation & Incubation Centre, Karyavattom Campus, Thiruvananthapuram 695581, Kerala, India

T. P. Ijinu · V. George · P. Pushpangadan (✉)
Amity Institute for Herbal and Biotech Products Development, Thiruvananthapuram 695005, Kerala, India
e-mail: palpuprakulam@yahoo.co.in

S. Rajasekharan
Kerala State Biodiversity Board, Thiruvananthapuram 695003, Kerala, India

O. V. Oommen et al. (eds.), *Biodiversity Conservation Through Access and Benefit Sharing (ABS)*, https://doi.org/10.1007/978-3-031-16186-5_8

8.1 Introduction

The history of human civilization and the development of economic systems are all inherently and inveterately interwoven with our biological resources. The economic activity of humankind continues to derive its sustenance directly or indirectly from biological resources (Ravi and Pushpangadan 1997). The unknown potentials of genetic diversity found in the biological organisms, particularly the plants, represent a never-ending natural frontier of inestimable value. Thus, biodiversity is our planet's natural capital, and it forms the foundation upon which human civilization is built (Pushpangadan 2002; Pushpangadan and Nair 2005a; Pushpangadan et al. 2012; Pushpangadan and Ijinu 2017). To understand this paradigm, the focus has to shift from the past to the current inequities. The core question is whether the North industrialized countries should be allowed to encash the biodiversity and knowledge of the South, which has been happening since the days of Vasco da Gama (Portuguese) and Christopher Columbus (Italian).

Historically, in the Indian context, the West was initially dependent on the Arab seafarers, intermediaries in trade for spices and other products from India and elsewhere to Europe, till the days of Vasco da Gama. Later, Europe gained access to the resources directly, and eventually, a series of clashes over the control of the seas and trade happened, first between the Arabs and the Western seafarers and later between competing Western nations themselves which ended in colonialism, a new phase in world history, leading to Western imperialism. Initially the Western traders had to bow down for products before the Southern people and later they started dictating terms. The daredevil Western seafarers started carrying the goods from the 'Indies' and elsewhere in the tropics, straight to Europe and made huge profits. Western scholars launched studies to categorize and document the precious natural resources with a view to identify economically potential natural products. Thus the native plants were collected, identified and classified by them, new nomenclatures were introduced, which laid the foundation for documenting the natural resources and classics of taxonomy. Having realized the immense industrial potential of the biodiversity, Western countries began to show interest in monopolizing the source countries. Internecine wars between the competing European nations later came to mutual agreements, and these are probably the initial trade agreements. This continued until the beginning of the twentieth century. Having realized the exploitation and humiliation by the colonial powers the people of the source countries revolted against them which lead to freedom struggle and finally the withdrawal of the colonial powers from the source countries in Asia, Africa and other continents.

The post-colonial world economic order continued to reel around the capital-rich Western countries. The ruling values, institutions, and jurisprudence remained Western. However, the thinkers and human rights movements helped to transform public opinion globally, and the Western world itself came to adopt more liberal values. It was the Brundtland Report submitted to the United Nations in 1987, chaired by Mrs. Gro Harlem Brundtland, then Prime Minister of Norway, which for the first time endorsed the rights of Third World Countries (TWCs) to get a share of economic profits from their resources. Norway, a past European colony, became a pioneer in the new mission. This was followed by the United Nations Earth Summit (Conference on Environment and Development) at Rio de Janeiro, where the Convention on Biological Diversity (CBD) was opened for signatures on 5 June 1992 until 4 June 1993 and finally it entered into force on 29 December 1993. To allow access to the South's genetic resources, the technology-rich nations had to give back a share of the rewards. Thus the sovereign rights for the Southern nations over their genetic resources were accepted for the first time. Today, the intricate imbalance in the core objectives and directives of CBD and WTO-TRIPS are a major concern for the Parties or members of these international laws. It is a matter of grave concern that certain well advanced countries like United States of America are still reluctant to become a Party to CBD, but continue to oppose the plea of the developing or the least developed countries for evolving an enabling and equitable legal mechanism for implementing the international trade and intellectual property laws concerning biodiversity, genetic resources and traditional knowledge systems.

8.2 Biodiversity

Biodiversity is the variety and variability of living organisms on the planet and it forms the bedrock for sustainable economic development. The total annual global markets for products derived from genetic resources are estimated to be worth between US$500 billion and US$800 billion (ten Kate and Laird 2021). Most of the countries located in these regions, interestingly the TWCs, are blessed with almost all known types of topographic and climatic conditions ranging from tropical to temperate and alpine zones (Pushpangadan and Nair 2005a). India is rated as one of the 17 megadiverse countries of the world, with about 7–8% of the total species of plants, animals, fungi and microorganisms. The floristic spectrum of India comprises approximately 45,500 species, of which over 5285 species are endemic (NBA 2018). The rich biodiversity of India is matched with an equally rich cultural diversity and a unique wealth of traditional knowledge system, developed, preserved and practiced by millions of ethnic and indigenous people living in the rural and forest areas (AICRPE 1998; Pushpangadan 1998). Local and Indigenous people used about 9,500 plant species for Indigenous medicinal preparations and more than 3,900 plant species as food, fibre, fodder, insecticides, pesticides, gum, resins, dyes, perfumes and timber (AICRPE 1998; Pushpangadan and Pradeep 2008; Pushpangadan et al. 2017, 2018; NBA 2018). Many of the plants are still unknown around the world, and the majority has never been studied for their therapeutic use. There are an estimated 500,000 known plant species worldwide (Corlett 2016), with a total number of angiosperm species at around 450,000, of which 10–20% are still unknown (Joppa et al. 2011; Pimm and Joppa 2015). Out of this, only about 1% of plants have been extensively studied in the laboratory (Newman and Cragg 2007; Newman et al. 2008).

8.3 Traditional Knowledge

From the very beginning of the human civilization, human communities living in different agro-climatic conditions have acquired unique knowledge about their ambient biodiversity by inherent instinct, or intuition, or by accidental discovery, or by error, or empirical reasoning or even by conscious trial and experimentation. The above said process enabled him to select a variety of natural resources, more particularly the biological resources, to meet his primary needs like food, nutrition, medicine, shelter, clothing, and to build his ever-growing material culture and civilization. Some human communities later established advanced material culture and began to build villages, towns and cities. However, majority of the people in the TWCs continue to live in and around the forests or in villages. The advanced societies developed written languages and documented their knowledge system. The valuable knowledge generated by innovative members was shared freely among the whole community. This knowledge is known as Traditional Knowledge (TK) or Indigenous Knowledge (IK) or Traditional Environmental Knowledge (TEK), or Local Knowledge (LK). The knowledge thus acquired through ages has always remained as part of their life, culture, traditions, beliefs, folklore, arts, music, dance etc. (George et al. 2014; Pushpangadan et al. 2014, 2015a, 2017; Pushpangadan and Ijinu 2017). TK, IK, TEK and LK are relative terms coined by the policymakers, researchers, or scholars belonging to different traditions from time to time according to its origin and utility of genetic resources, which is associated with biodiversity including plant, animal and microorganism diversity. Apart from the above, TK associated with non-genetic resources are also there in different cultures, which include martial arts, magico-religious cures, yoga etc. Definitions of TK is sometimes specifically focused on the knowledge belonging to different ethnic communities of the world, mainly it denotes the tribal people. Since TK is restricted to location-specific knowledge of common people, including ethnic communities residing in a particular region/ county, the knowledge is confined to the genetic and non-genetic resources available within their surroundings (Pushpangadan et al. 2016a).

TK is a general term, which refers to the collective knowledge, wisdom, beliefs and practices of Indigenous and Local People on sustainable use and management of their ambient resources. Through years of observations and analysis, trial, error or experimentations, the traditional communities have been able to identify useful as well as harmful elements of their ambient flora and fauna. TK thus can be of direct or indirect benefit to society as it is often developed, in part as an intellectual response to the necessities of their life. Protection and maintenance of TK of local and Indigenous communities is vital for their wellbeing, sustainable development, and for their intellectual and cultural vitality (Gupta 2004; Pushpangadan and Nair 2005b; Pushpangadan et al. 2017; Pushpangadan and Ijinu 2017). The World Intellectual Property Organization (WIPO) currently uses the term 'traditional knowledge' to refer tradition-based literary, artistic or scientific works; performances; inventions; scientific discoveries; designs; marks, name and symbols; undisclosed information; and all other tradition-based innovations and creations resulting from intellectual activity in the industrial, scientific, literary or artistic fields. 'Tradition-based' refers to knowledge systems, creations, innovations and cultural expressions which have generally been transmitted from generation to generation; are generally regarded as pertaining to a particular people or its territory; and are constantly evolving in response to changing environments (Gervais 2005; WIPO 2005).

In the light of global healthcare need, traditional medicine is gaining greater attention across the globe as its role in fulfilling the public health demands of TWCs. Traditional medicines have already grown into a multibillion-dollar global business, and the pharmaceutical sector is increasingly looking into the potential of genetic resources and associated TK. Documenting and safeguarding of these traditional medicine systems is becoming a greater priority. Many issues relating to the protection of traditional medical knowledge are unsolved. In attempting to protect TK, traditional medicine practitioners are confronted by a bewildering and diverse set of national and international regulations, regulatory structures designed primarily to accommodate pharmaceutical medicines, safety and efficacy concerns, and ownership issues in their efforts to safeguard traditional medicine (Abbott 2014). Traditional communities in India, as well as other TWCs have always considered the natural resources and the associated TK system developed by them as commonly owned properties that must be kept for and shared by everyone, and never to be commodified for sale or commercialization. It was with the coming of the Westerners that the process of commodification and trading of genetic resources started.

The study of the TK system of traditional societies about the plant world is the subject matter of the science 'ethnobotany'. Modern drug hunters consider ethnobotany as a cost-effective method of discovering novel and valuable molecules with high therapeutic value. It is widely believed that the chances of finding a bioactive molecule by random screening of plant samples are 1 in 10,000, and that of hitting a marketable drug is 1 in 4. In comparison, the success rate of identifying a bioactive molecule by selective screening based on ethnobotanical leads is 1 in 100, and discovering a marketable drug is 1 in 2 (Pushpangadan and Nair 2005b). Many modern drugs are derived from plants that were initially identified via ethnobotanical and ethnopharmcological research (George and Ijinu 2011; Pushpangadan et al. 2014, 2015a). Some examples are aspirin (acetylsalicylic acid, analgesic and anti-inflammatory) from *Salix* spp. (Willow tree, Salicaceae), reserpine (antihypertensive, tranquilizer) from *Rauvolfia* spp. (sarpagandha, Apocynaceae), quinine (anti-malarial) from *Cinchona* spp. (Rubiaceae), eserine (physostigmin, acetylcholinesterase inhibitor) from *Physostigma venenosum* (Fabaceae), vinblastine (anti-cancer) from *Catharanthus roseus* (Apocynaceae), diosgenin (cardio-tonic) from *Dioscorea* spp. (Dioscoreaceae), solasodine (anti-cancer, cholinergic) from *Solanum* spp. (Solanaceae), morphine (opioid analgesic) from *Papaver somniferum* (Papaveraceae), cocaine (anesthetic, stimulant) from *Erythroxylum coca* (Erythroxylaceae), atropine (anti-cholinergic) from *Atropa bella-donna* (Solanaceae), digoxin (cardiotonic) from *Digitalis lanata* (Plantaginaceae), taxol (paclitaxel, anti-cancer) from *Taxus brevifolia* (Taxaceae), etoposide (anti-cancer) from *Podophyllum peltatum* (Berberidaceae) etc. (Farnsworth et al. 1985; Balick and Cox 1996; Heinrich 2000; Duke et al. 2008; Kingston 2011; George and Ijinu 2011; Soejarto et al. 2012;

Newman and Cragg 2016, 2020; Che et al. 2017; Ijinu et al. 2020). Artemisinin (Qinghaosu) was first isolated by Youyou Tu in 1971 from *Artemisia annua* and recently in 2015 she was awarded the Nobel Prize in Physiology or Medicine for the discovery of its use as anti-malarial drug (Tu 2016) (Fig. 8.1).

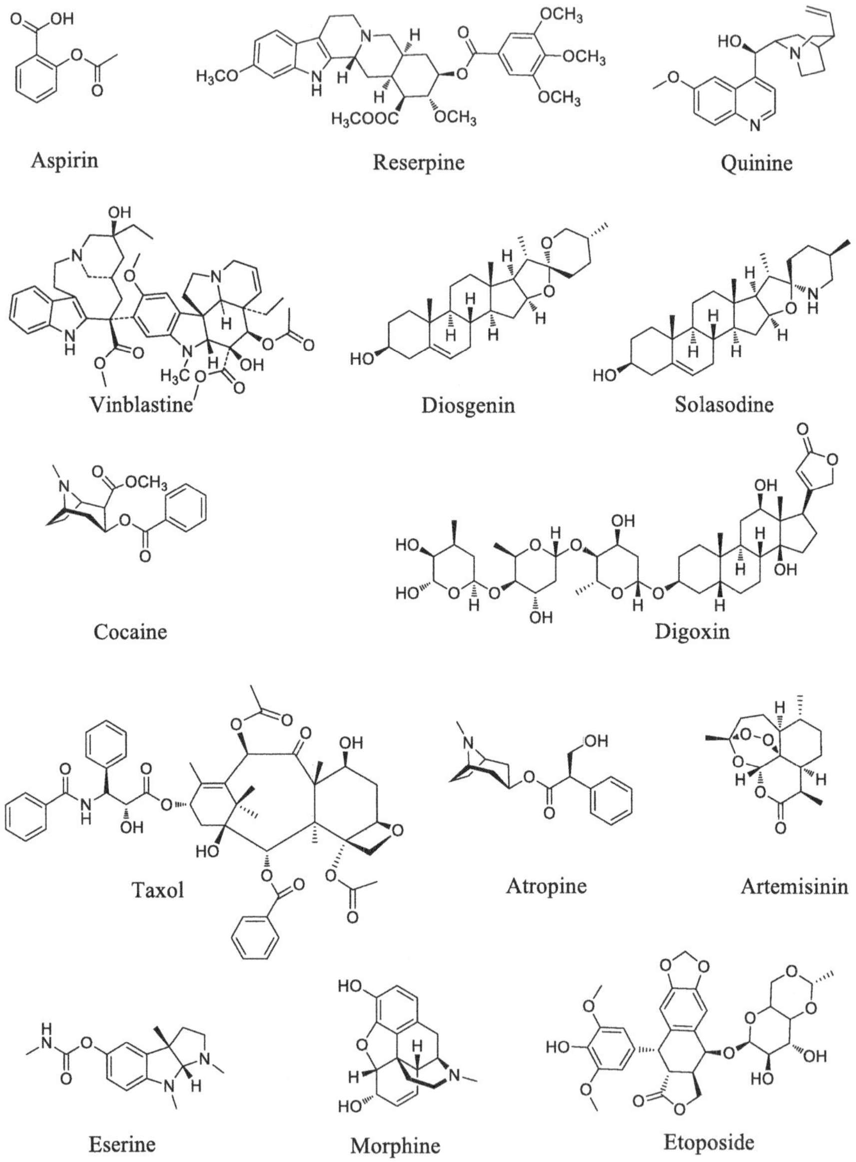

Fig. 8.1 Some important drug leads obtained through ethnopharmacological research

8.4 Indigenous Peoples

Indigenous peoples are unique social and cultural groups with shared ancestral links to the lands and natural resources they live on, inhabit, or have been displaced from. According to the World Bank, Indigenous peoples are found in over 90 nations, and there are between 370 and 500 million people worldwide, representing 5% of the population (World Bank n.d.). In this, about 15% are extremely poor. They have kept distinct social, cultural, economic, and political traits from the dominant societies in which they live. Indigenous peoples from all over the world, despite their cultural distinctions, face similar challenges in protecting their rights as distinct peoples (Cobo 2014). They are now probably one of the world's most underprivileged and vulnerable populations. The international community now recognizes that special steps are required to defend their rights and preserve their unique traditions and way of life. They hold invaluable ancestral knowledge and skills on how to adapt to, mitigate, and lessen the dangers of climate change and disasters.

Indigenous peoples are currently facing a legal situation that is a mixture of national and international law. Over the last 30 years, the rights of Indigenous peoples are set out in the UN Special Rapporteur on the Rights of Indigenous Peoples (UNSR, 1986) (Cobo 1986), International Labour Organization (ILO Convention 169) concerning Indigenous and Tribal Peoples in independent countries (1989), the UN Permanent Forum on Indigenous Issues (UNPFII, 2002), the UN Human Rights Council—Expert Mechanism on the Rights of Indigenous Peoples (EMRIP, 2007), the UN Declaration on the Rights of Indigenous Peoples (UNDRIP, 2007), and the American Declaration on the Rights of Indigenous Peoples (ADRIP, 2016). The UNDRIP is a non-binding declaration, but 150 states have adopted it, and is often held to be the most inclusive international instrument on the rights of Indigenous peoples (Medaglia and Perron-Welch 2019).

UNDRIP can serve as a model for the rights that Indigenous peoples can obtain under international law, as well as rights that should ideally be granted to them by States, even if they are not yet enshrined in binding customary or treaty law (Dörr 2019). The 46 Articles of UNDRIP primarily outline various rights pertaining to Indigenous individuals and groups, including membership of their institutions in accordance with their own procedures, political status of the community, land rights, and helping to live a culturally distinct life. Compared to all other rights, the 'right to self-determination' stated in Article 3, is the most important provision, which forms the foundation of the declaration. By virtue of that they have the right to freely decide their political status and pursue their economic, social, and cultural growth.

8.5 Post-CBD Developments

Biodiversity has been a commons of local communities throughout history, with resources and knowledge freely exchanged. Individual States can now assert their exclusive right to regulate and set conditions for access and use of their biodiversity, through access and benefit-sharing regulations, contrary to the widely accepted legal precept that biodiversity is a 'common heritage of humanity'. Traditional communities were almost unaccustomed to the concept of sovereign rights or property rights in genetic resources. The WTO was founded in 1994, less than a year after the CBD took effect, with a different agenda. The CBD is based on the idea that local communities rely on biodiversity and should continue to reap its benefits. But, the WTO oversees a global trading system heavily reliant on multi-national business and private monopoly rights over biodiversity and other goods. Thus, in the twenty-first century, we witness a paradigm shift in the global perception of biodiversity and its use.

CBD is built on the values of equity and ethics and, therefore, it has significantly greater flexibility in provisions concerning protection of the rights of traditional communities over their intellectual property and traditional resources. The Articles 8(j) and 15.7 of the CBD highlight the necessity of recognition and rewards for Indigenous peoples' contributions to biodiversity conservation and its sustainable use. It reads: 'respect, preserve and maintain knowledge, innovations, and practices of Indigenous and Local Communities embodying traditional lifestyles relevant for the conservation and sustainable use of biological diversity and promote their wider application with the approval and involvement of the holders of such knowledge, innovations and practices and encourage the equitable sharing of benefits arising from the utilization of such knowledge, innovations and practices'. Articles 10(c), 11, 12, 13 and 16 of the CBD contain other key clauses that demand for support and acknowledgment of Indigenous and traditional technologies. However, Article 16 is unambiguous about the protection of IPR of the traditional communities (Pushpangadan et al. 2017).

The term 'Access and Benefit-Sharing' is intricate to define and interpret. The ecosystems are not bounded by national or political boundaries; therefore, a piece of TK or a specific plant may not necessarily belong to a specific country or community. This further complicates the matter. The question of how to deal with it is one that global policymakers, primarily in Western industrialised countries, are grappling with. There are also opposing views from various groups that the patenting of life forms and the intellectual property rights (IPR) regime itself be scrapped. Here the common values, laws or procedures of the North clash with that of the South. Meanwhile, following the completion of the Uruguay Round (1986–1994), multilateral trade negotiations conducted within the framework of the General Agreement on Tariffs and Trade (GATT, signed on 30 October, 1947) gets institutionalized in WTO on 1 January, 1995 and is endorsed by majority of nation-states. Despite protests from marginal groups in the South, this got through. The WTO-TRIPS (signed on 15 April 1994) clauses do not accept the CBD principles of national sovereignty or Indigenous community rights on natural resources. Unlike conventions, the TRIPS agreement has more fangs and is now in conflict with the CBD principles, as the Western rule-makers claim. The TRIPS agreement establishes legally obligatory barriers to safeguard innovations through intellectual property rights (IPR), which must be followed by all the member countries. This means that each country must adopt laws and procedures that are imposed by Western institutions. Filing patents for TK is a time-consuming process that is impossible for those with limited resources. In this context, national sovereignty becomes a nebulous idea. A case-specific, *sui generis* legal system must arise at the global level if the interests of the Southern people are to be protected within the stipulates of WTO.

In the meantime, there were some paradigm shifts in the areas of biodiversity and trade. With scientific advancements in the West and the rapid growth of modern medicine, the interim period saw significant changes from trading of raw materials. The raw materials were no longer necessary in their whole since modern science and technology allowed for the isolation of active principles and more efficient use of them. Meanwhile, the Western legal system evolved, providing protection to new product innovators. Patents and the idea of having a monopoly on discoveries gained popularity. For successful products, the inventor may be entitled to get a share of profits for specified periods. After the CBD, every product made from a plant in the South, if known, had to pay a percentage of the profits to the country or community. However, in practice, this was nearly impossible. Molecules isolated from plant material are difficult to track down, and disclosing them is bad business. However, in the South, accessing such plant material was simple. Transnational companies had no trouble in obtaining such resources, and the impoverished countries and scientists were more than happy to assist Western institutions in exchange for some generosity. The composite results benefited both the takers and the providers. As a result, bilateral research projects have become the norm. Getting a scientific publication in Western peer-reviewed Journal became the pinnacle of a Southern scientist's professional ambitions. To support such collaborative research, large sums of money were sent to the South. Rich countries also amassed enormous tropical plant gene banks, and were mostly kept in ex situ conservation sites.

The mandatory requirements under TRIPS, for the signatories of WTO, to protect their inventions make Southern countries more vulnerable. Thus native governments are asked to safeguard the interests of transnational companies, which looks similar to the colonial period where native people were recruited as police to suppress legitimate freedom struggles for a payment. Here again, there are issues of TK and new inventions coming to clash. There are several well known cases of experimenting with the IPR and ABS models for TK and biodiversity across the world, and most of these have failed to take off (e.g., INBio–Merck collaboration; Suriname-Integrated Cooperative Biodiversity Groups project; San Hoodia case, Southern Africa; Argan oil commercialization, Morocco etc.). There are also innumerable cases of bio-piracy (e.g., the cases of neem, turmeric, basmati rice, maca, rosy periwinkle, hoodia, ayahuasca etc.), using the biodiversity and TK of the South for new products without their knowledge.

A combination of Articles 15 and 8(j) of CBD pursuant to the works of the two Ad Hoc Working Groups on ABS and Article 8(j) resulted in the Bonn guidelines. Unanimously, a total of 180 Parties to the CBD formally endorsed the Bonn guidelines at 6th Conference of Parties held at Hague in 2002. The Guidelines are premised on 8 key principles: voluntarism, ease of use, practicability, acceptability, complementarity, evolutionary, flexibility and transparency. The acceptance of the instruments amongst Parties to the CBD signalled a global commitment to tackle ABS seriously. There has been a profusion of national and regional regimes on ABS before and after the Bonn guidelines, with the majority of them focusing on business bio-prospectors and less directly on researchers. The conduct of researchers seem restricted and consigned to national or institutional ethics regimes for the conduct of research. Consequently, many countries in the South felt that, while the Bonn guidelines expounded on access, they were vague on benefit-sharing. This voluntary nature of the Bonn guidelines has been deemed insufficient for executing the ABS provisions of the CBD. In order to implement the third objective of the CBD and its ABS-related provisions, the World Summit on Sustainable Development, held at Johannesburg, urged for an action to negotiate within the framework of the CBD. That negotiation effectively commenced in 2004 and after six years, it resulted in the October 2010 Nagoya protocol, adopted at the 10th Conference of Parties of the CBD at Nagoya, Japan (Tsioumani 2015; Pushpangadan et al. 2017).

Today the rich countries are hundreds of times richer than many poor tropical countries, and this tells the end result of the game plan. This resulted through the past several centuries. Technology and IPR with the comforts on offer have made a strong lobby of politicians—academics—scientists—traders within the South, what some call the 'new North within the South'. From another angle, the Western cultures have valued intellectual calibre and the enthusiasm to pursue new interests. Like any other tangible physical property, intellectual property became a defined entity. This cannot be said about India and similar countries where intellectual merits are taken for granted. Perhaps traceable to a complex web of socio-religious values and history, the premium on intellectual attainments is less, or at times, absent in India. Thus the great traditions of folk wisdom with their invaluable repositories of knowledge are fast getting lost. The new generations are no more interested in these areas, and with unhealthy legalities in the post colonial era, they also face constraints in practicing their skills. This sees their TK being neglected and abused, where the need is to evolve systems that honour and economically reward such knowledge streams and conserve these for posterity.

8.6 Intellectual Property Rights

The IPR regime had its origin in the Paris Convention (20 March 1883) for the Protection of Industrial Property, and it developed predominantly in the Western economies of the USA, Europe and Japan. It was universalized by means of the WTO-TRIPS agreement of 1994 had as its purpose to provide a uniform international framework for the promulgation of IPRs within individual member states (Maister et al. 2012). The CBD acknowledges the IPR of the collective wisdom and common resources of the communities as a sovereign property, whereas TRIPS recognizes IPR as the monopolistic rights of individual or corporate innovators. This disparity provides developing countries like India with enormous challenges and opportunities

(Pushpangadan et al. 2015b, 2016a, b, 2017; Pushpangadan and Ijinu 2017; Ijinu et al. 2020).

There are two intellectual property (IP) components at stake-the property of the local communities that provide ethno-medico-botanical knowledge and the property of the researchers who isolate and develop single compound pharmaceuticals from the raw materials/ plant extracts or refined formulations of botanical drugs. Also, there are inherent general challenges when sharing IP. First, determination of who is the rightful owner of the ethno-medico-botanical knowledge can be difficult. Different communities may have similar ethno-medico-botanical knowledge on the use of plants. Determining who should be given benefits and a share in intellectual property is challenging and is best determined early on the process of bio-prospecting. Without protected IP, the economic incentive for developing pharmaceuticals could diminish for all parties involved. However, the IP used in the discovery process must be shared respectively among all parties involved in the development of the pharmaceutical, and contracts must be negotiated fairly and clearly across national and political boundaries. Naming individuals on patents is a more guaranteed way to ensure that benefits are distributed to locals, but such individual inventors must be defined (USPAT n.d.).

Benefit-sharing can take two forms: monetary or non-monetary. For monetary benefit-sharing, different payment schedules can be devised, including payments for the plants collected, research using the plants, and royalties from any product developed. Advance payments can also be used, as in the form of trust funds, which can help communities upfront during the long drug discovery process (Rosenthal 1997; Guerin-McManus et al. 2002). Patents may provide some monetary benefits but are generally not considered the best mechanism for benefit-sharing with local communities (Greaves 1994, 1995) and require that all parties to receive monetary benefit be individually named on the patent for effective distribution of monetary compensation (Rose et al. 2012; Cicka and Quave 2019). ABS does not solely imply monetary compensation for gains from a blockbuster drug. Monetary compensation may never be a reality as many potential pharmaceuticals fail clinical trials, and many products collected and tested never make it past the laboratory bench. ABS requires open communication with the host country and groups throughout the process.

Perhaps equally constructive for local communities are non-monetary benefits. Maintaining close connections with the host country and the local communities may help mitigate the uncertainty of loose ties between parties in the lengthy research and development process. These benefits may be in the form of training for the project, capacity building, commitment of research on local diseases, or provision for conservation needs. Firstly, for example, the US-National Cancer Institute established a partnership with a Panamanian research institute for access to plants, and to build local capacity and relations through sharing education and technological advances (Cragg et al. 1999; Kursar et al. 2006; Rose et al. 2012). Additionally, researchers can assist in the preservation of knowledge by reporting back collected data in a manner that is accessible to the local communities. For instance, books in the local language recording uses of local plants can be gifted to the community. Secondly, there is an untapped market for the pharmaceutical industry to contribute to combating diseases that primarily affect the source countries of their partnerships (Rose et al. 2012). Encouraging pharmaceutical companies to invest in those disease areas could increase the industry's involvement in the neglected disease sector. Lastly, outside organizations such as governmental or non-governmental may be necessary to ensure the conservation of natural resources long term. To further ensure adequate benefit-sharing, local organizations can empower local communities to become knowledgeable about the contracts they are signing (Cicka and Quave 2019).

Genetic resources themselves are not IP as they are not creations of the human mind, hence cannot be protected as IP (Nambisan 2017). The WIPO began work on TK in 1998 and at its twenty-sixth session held in Geneva in 2000 established the Intergovernmental Committee (IGC) on Intellectual Property and Genetic Resources, Traditional Knowledge and Folklore to primarily looking into IP issues that arise in the context of (WIPO 2001):

1. Access to genetic resources and benefit-sharing.
2. Protection of TK, whether or not associated with those resources.
3. The protection of folklore, including handicrafts.

The adhoc open-ended inter-sessional working group of Article 8 (j) and related provisions of the CBD has further discussed (5th meeting of Conference of Parties, Montreal, 15–19 October 2007) about the elements to be considered in the development of *sui generis* system for the protection of TK, innovations and practices of Indigenous and Local Communities. In this regard they framed a draft recommendation for future work on *sui generis* system. The major recommendations are:

1. Invites the parties and governments to consider that the development of effective *sui generis* systems for the purposes of convention, be local in nature and based on the relevant customary law of the appropriate Indigenous and Local Communities concerned, and recognised and/or created with the prior informed consent (PIC) and the full and effective participation of those communities for the purposes of respect, preservation, maintenance and promotions of the knowledge, innovations and practices of Indigenous and Local Communities.
2. Further invites parties and governments to recognize support and/ or develop through national guidelines and request the executive secretary to continue gathering and analysing information in consultation with parties, governments, international entities and Indigenous and Local Communities, to provide information regarding the development of guidelines for the consideration of the 6th meeting of working group on Article 8(j) and related provisions.
3. Requests the executive secretary to continue gathering and analysing information, in consultation with parties, governments, international entities and Indigenous and Local Communities on the possible development of proposals for an international frame work to set minimum standards and ensure protection of stake holders involved for the consideration for sixth meeting of the working group or Article 8(j), fully taking into account and collaborating with, the work of other

bodies concerning *sui generis* system and customary law and in particular WIPO and the United Nations Permanent Forum on Indigenous Issues (UNPFII, formed 28 July 2000).

To address the above issue WIPO-IGC was held at Geneva in the year 2008. The report of the committee stressed the need for international cooperation or practical measures, aimed at developing capacity and substantive materials for legal and policy processes, strengthening capacity of TK holders, building and guiding institutions, inter agency cooperation, coordination with in the UN system and awareness and capacity building for general public. At the regional level, the report of inter governmental committee listed certain options to address gaps in TK protection. Options to address gaps in TK protection includes:

1. Legal instrumentals concluded at regional, sub-regional or bilateral level including Sui generis instruments and conventional IP law.
2. Political or policy level declarations proclaimed at the regional, sub-regional or bilateral.
3. Model less and other forms of legislative guidelines adopted at regional levels.
4. Model protocols, guidelines and best practice recommendations adopted at regional or sub-regional levels.
5. Regional, sub-regional and bilateral initiatives and programmes to support community capacity building related to TK.

WIPO-IGC has also identified the elements that may be used to describe *sui generis* measures for TK protection. They are policy objective scope of protected subject matter, conditions of access to TK, conditions of protection of TK, scope of rights, right holders, acquisition of right, expiration and loss of rights, sanctions and enforcement, registration mechanisms and other procedures for the acquisition and maintenance of rights, access and benefit sharing elements (MAT and PIC), defensive protection, regional and international protection, including the problem of so called 'regional TK', institutional arrangements and recognition of customary laws and protocols. The elements mentioned here are to be deeply discussed and debated regional, sub-regional and bilaterally and find a solution under the *sui generis* system. The protective paradigms of TK are as under (WIPO n.d.):

- *Defensive protection*: Seeks to prevent others from using or securing IPR over the TK.
- *Positive protection*: Seeks to secure protective legal rights over the TK which can be achieved by either using the existing laws or using legislative means to enact new *sui generis* laws.

Some Indian initiatives based on international laws include:

- ***Defensive protection***
  - Founded National Biodiversity Authority (1 October 2003) under the Ministry of Environment, Forest and Climate Change (MoEF & CC), Govt. of India to implement the provisions under the Biological Diversity Act, 2002.

- Launched Traditional Knowledge Digital Library (TKDL, 2011) under the Council of Scientific and Industrial Research (CSIR), Govt. of India mainly to prevent misappropriation of country's traditional medicinal knowledge.

- ***Positive protection***
  - The Patents (Amendment) Act, 2005.
  - Biodiversity Act, 2002 and Rules 2004.
  - Scheduled Tribes and other Traditional Forest Dwellers (Forest Rights) Act, 2006.
  - Systematic documentation, scientific validation of TK, process/ product development, technology transfer, commercialization and benefit sharing.
  - Implementation of Article 8(j) of CBD.
  - To develop viable models under *sui generis* system.

Scheduled Tribes and other Traditional Forest Dwellers (Forest Rights) Act, 2006 (Chap. 2).

1. Right to protect, regenerate or conserve or manage any community forest resource, which they have been traditionally protecting and conserving for sustainable use.
2. Rights which are recognized under any State law or laws of any Autonomous District Council or Autonomous Regional Council or which are accepted as rights of tribals under any traditional or customary law of concerned rights of any States.
3. Right of access to biodiversity and community right to IP and TK related to biodiversity and cultural diversity;
4. Any other traditional right customarily enjoyed by forest dwelling Scheduled Tribes or other Traditional Forest Dwellers, as the case may be, which are not mentioned in clauses (a) to (k) but excluding the traditional right of hunting or trapping or extracting a part of the body of any species of wild animal.

8.7 Need for Systematic Documentation

TK associated with biodiversity includes both codified (documented) as well as non documented (orally transmitted information, expressions of folklore) information which are accepted as cultural components of biodiversity. If these are not preserved and documented, they are likely to be lost in the modern materialistic world. Hence, there is scope and urgent necessity to carry out systematic documentation of TK associated with biodiversity. It is essentially required for protecting the knowledge under *sui generis* system by establishing different viable models. Systematically documented TK will provide valuable information that is highly beneficial to the researchers for developing new processes, products, patenting, technology transfer, commercialization, benefit sharing etc. This will further lead to the integration of TK in to the planning process for the various developmental activities at grass root level. Implementation of effective legal frame work to check biopiracy is also highly essential while implementing sector wise programmes related to biodiversity and associated TK (Rajasekharan 2010).

Sector wise thrust areas identified for the documentation of Traditional knowledge include- traditional knowledge related to art and culture, agriculture, animal husbandry, architecture, biodiversity conservation and utilization, eco-friendly practices, fisheries, forest and wild life management, health care, medicinal and food plants, rural technology etc. Recently, Jawaharlal Nehru Tropical Botanic Garden and Research Institute (JNTBGRI), Thiruvananthapuram, India completed the systematic documentation of TK associated with plants used for food and medicine in 70 Gramapanchayats (village council) (5 each from all 14 Districts) of Kerala. JNTBGRI documented 11,000 (7667 single and 3333 combinations) information based on TK related to plants used for medicine. Out of this, 9467 (6584 single, 2884 combination) information were documented from non-tribal communities, and rest 1532 (1083 single, 449 combination) were gathered from tribal communities. The above information was recorded from 625 different plants exclusively used for medicinal purpose. Apart from this, they have also documented 650 information on food plants (Rajasekharan 2010). In addition to this, Amity Institute for Herbal and Biotech Products Development, Thiruvananthapuram, India documented 6008 information based on TK related to plants used for food and medicine from 34 Gramapanchayats of 12 districts of Kerala, India (Ijinu et al. 2011).

8.8 Indigenous Knowledge and Nagoya Protocol

It is widely recognised that Indigenous knowledge (IK) plays a vital role in the development of natural products based medicines, cosmeceuticals, nutraceuticals, functional foods, beverages, and fibres. To support livelihoods, to safeguard the environment, and to preserve culture and knowledge of Indigenous Peoples and Local Communities, IK must be suitably valued, protected and promoted. There should be a fair and equitable sharing of benefits when this knowledge of plants and animals is applied in the research and development of new products, and there are chances for Indigenous people to value-add and lead initiatives for the commercialization of these products. Nowadays some pharmaceuticals have been shifting their focus away from natural products drug discovery to synthetically producing novel molecules (Amirkia and Heinrich 2015); there are other industries that have come to the fore in ABS and TK contexts. Cosmetics and skincare, herbal supplements and nutraceuticals, novel foods, and related industries appear to be more likely to embrace and use marketing to emphasise their uses of TK relating to natural plant-based or even animal-based products. These can be important industries for Indigenous Peoples and Local Communities, can provide a culturally appropriate source of income and livelihoods. However, it is important to ensure that there are appropriate benefits and fair payments so that the Indigenous people get an opportunity to ‘value-add’ the raw materials, and also to protect their TK (Robinson and Raven 2019).

ABS regime remained a contentious issue when the CBD entered into effect in 1993. The TWCs, the major providers of genetic resources believed that the main users, Western countries, were not doing enough to ensure compliance with provider country access restrictions and demanded a legally obligatory ABS system to set user-country measures. Moreover, they have been unsuccessful in their attempts in the WTO-TRIPS council to obtain mutual support between the global ABS and IPR regimes through the imposition of requirements for disclosure of genetic resource origin in patent applications as a tool for tracking compliance with provider-country access regulations (Oberthür and Rosendal 2014). Later, the 10th meeting of the Conference of the Parties to the CBD held in Nagoya, Japan adopted the Nagoya protocol on Access to Genetic Resources and the Fair and Equitable Sharing of Benefits arising from their Utilization, and aimed at further developing a legal framework for ABS, provided by the CBD on the above and all other measures. The Protocol entered into force on 12 October, 2014. As of October 2020, it has been ratified by 128 parties, which includes 127 UN member states and the European Union.

The Nagoya protocol establishes a mandatory requirement that genetic resources be obtained on the basis of PIC and MAT to facilitate equitable sharing of benefits while taking into account the role of TK. Moreover, under the Nagoya protocol, users are required to observe any domestic laws of the country in which the subject matter was obtained (Articles 6(1) and 7). This could include, for example, legislation with respect to the environment or the collection of materials from national parks and other similar areas (Verkley et al. 2016). These compliance obligations are applicable to all parties to the Protocol, which specifies that each party 'shall take measures, as appropriate, to monitor and to enhance transparency about the utilisation of genetic resources' (Article 17). Moreover, Article 18 of the Protocol requires user countries to ensure that provider countries are able to obtain legal recourse within the jurisdiction of the provider country.

The Nagoya protocol on ABS regime cuts across several other international regimes with sectoral approaches. Despite relying on Article 8(j) of the CBD for its authority, the Nagoya protocol applies to a somewhat modified and broader conception of TK. This has led to turf battles but also the advancement of ABS approaches. The FAO International Treaty on Plant Genetic Resources for Food and Agriculture (ITPGRFA) is an ABS treaty for one specific type of genetic resources; the Nagoya protocol provides that other specialized ABS instruments may be developed and may prevail over the Protocol, if they are consistent and do not run counter to the Protocol's objectives. Among other regimes with close relations to the Nagoya protocol regime are the WHO's Pandemic Influenza Preparedness (PIP) framework for the sharing of viruses and access to vaccines and other benefits; the WTO-TRIPS; the Union for the Protection of Plant Varieties (UPOV); the WIPO; and the United Nations Convention on the Law of the Sea (UNCLOS) (Prip and Rosendal 2015).

Indigenous peoples from various global communities continue to raise their serious concerns to international organizations such as WIPO and before various UN committees on the lack of 'the fair and equitable sharing of benefits arising from the use of genetic resources in relation to the minimum benefits going to Indigenous communities; either as in-kind benefits or monetary payments by potential partners or contracting parties under the ABS principles of CBD. Thus, they have been granted the right to 'full and effective participation in all matters concerning them, as well as their right to remain distinct and pursue their vision on economic and social development' under the United Nations Declaration on the Rights of Indigenous Peoples (UNDRIP, adopted on 13 September, 2007). The Article 31 of UNDRIP states that 'Indigenous peoples have the right to maintain, control, protect and develop their cultural heritage, traditional knowledge and traditional cultural expressions, as well as the manifestations of their sciences, technologies and cultures, including human and genetic resources. They also have the right to maintain, control, protect and develop their intellectual property over such cultural heritage, traditional knowledge and traditional cultural expressions'. While non-binding, UNDRIP presents a 'comprehensive articulation of indigenous claims' and provides grounding for the final preambulatory clause of the Nagoya protocol, which concludes 'affirming that nothing in this Protocol shall be construed as diminishing or extinguishing the existing rights of Indigenous Peoples and Local Communities.

The preamble of the Nagoya protocol has seven paragraphs relevant to Indigenous Peoples and Local Communities and their TK. They refer to Article 8(j); the interrelation and the inseparable nature between genetic resources and TK; the owner or holder of TK, including countries; Indigenous Peoples and Local Communities' right to identify within their communities the rightful holders of TK; UNDRIP; and non-extinguishment of existing rights. The vital role of women in ABS, and their full and effective participation in biodiversity conservation, policymaking and implementation is also recognized in the preamble. However, the experience of Indigenous peoples who participated in the drafting and implementation of the Nagoya protocol demonstrates that there are still barriers to overcome in order to achieve this goal.

There is no doubt that the Nagoya protocol addresses some serious gaps and uncertainties in the CBD provisions on ABS. However, the Protocol has left a considerable number of aspects to the discretion of nation-states, with many of its provisions creating a lack of legal certainty and in some instances, even creating more serious loopholes. In addition, the Protocol has failed to address many concerns with respect to policing the use of genetic resources, and it lacks a meaningful enforcement mechanism to secure the interests of both provider states and user states. In general, providing such a wide degree of discretion to nation-states, as well as a lack of clarity and legal uncertainty, will most likely result in a low implementation rate, especially in relation to compliance and enforcement measures, thereby providing an incentive for user states to avoid these regulations. Moreover, these ambiguities and uncertainties have created serious loopholes that would suggest the interests of user states are better mirrored in the Protocol than the interests of provider states.

The rights of Indigenous peoples in the Nagoya protocol result from hard-fought battles by the International Indigenous Forum on Biodiversity over every comma and word. With solid support, especially from the African group and from Norway, Indigenous Peoples and Local Communities gained significant ground on their rights over TK, genetic resources, and the recognition of their customary laws by giving more content to Articles 8(j) and 10(c) of the CBD. Through these discussions to defend the conservation of biodiversity; ecosystems; Indigenous ways of life; land tenure and use; the right to culture, knowledge and practices; food; and economic security. Indigenous peoples generated the 'bio-cultural rights' that have an explicit link to conservation and sustainable use of biological diversity, which were advocated during international biodiversity negotiations as a defense against 'biopiracy' of knowledge and resources (Bavikatte and Robinson 2011).

Alongside the Nagoya protocol, the WIPO has its Intergovernmental Committee (IGC) on Intellectual Property and Genetic Resources, Traditional Knowledge and Folklore, which has sought also to develop legal instruments for the protection and promotion of TK. Moreover, in the WTO, some countries have sought to include checks and balances in the disclosures made by patent applicants, where they may have used genetic or biological resources and associated TK in their innovations. These negotiations in the WIPO-IGC and WTO are still ongoing. The WIPO voluntary fund has been established and utilised to support the attendance of representatives from Indigenous Peoples and Local Communities in the WIPO-IGC negotiations, and least developed countries are member states of WIPO and can attend and contribute to the negotiations shaping TK policy globally (Robinson and Raven 2019).

The new thinking centered on the concept of 'Knowledge Engineering' for building up future 'Knowledge Assistance' and 'Knowledge Industries' is now gaining attention and acceptance both nationally and internationally. Knowledge based development of value-added products from genetic resources and its commercialization has become one of the fastest developing economic activities in the world. The United Nations 2030 agenda for Sustainable Development with 17 Sustainable Development Goals (SDGs) and 169 associated targets seeks to facilitate a cohesive international response to address the 2.2 billion global citizens facing 'multidimensional poverty', the monumental threat of destabilisation to agricultural and economic markets posed by climate change, and mainstream ecosystem integrity and conservation of natural capital. Among the 230 indicators of SDGs, two directly refer to Indigenous peoples (2.3.2 and 4.5.1) and also several other indicators that are relevant for Indigenous peoples, particularly 1.4.2 and 5.a.1 on land rights. Among the 169 associated targets of SDGs, 73 have substantial links to the UNDRIP. Through recognition of Indigenous and Local Communities as holders of IPRs in TK relating to biodiversity and natural capital, the development of complementary legal infrastructure to support equitable benefit-sharing, facilitating community governance and self-determination, and protecting against misappropriation, TK as a technology of biodiversity conservation and innovation can be preserved for future generations and respectfully shared based on mutually equitable terms in support of sustainable development.

8.9 Conclusion

The issues of community rights/TK, ABS regime, and IPR protection related to genetic resources and TK are the most contentious ones that keep the developed and developing countries divided in their attitude and approaches. The prerequisite for developing an effective ABS regime is building up a comprehensive information system on all pertinent aspects of availability, diversity, distribution, economic uses and potentials, the conservation status of biogenetic resources and associated traditional knowledge, along with information on the existing science and technology infrastructure and capabilities, including human resource, national and international legal and policy frameworks, current achievements and plans and priorities for capacity-building for conservation, sustainable use, bio-prospecting and economic valuation of their great wealth of bio-resources and traditional knowledge. The new instrument like Nagoya protocol cannot reach these goals alone, and so much will rely on functional implementation by the Parties concerned for moving forward.

Acknowledgements The authors express their sincere thanks to Dr. Ashok K. Chauhan, Founder President, Ritnand Balved Education Foundation (RBEF) and Amity Group of Institutions, and Dr. Atul Chauhan, Chancellor, Amity University Uttar Pradesh (AUUP) for facilitating this work. The authors are also thankful to Dr. Sasikumar Parameswaran (Drug Testing Laboratory, Department of Rasasastra and Bhaishajyakalpana, Govt. Ayurveda College, Thiruvananthapuram, India), Ms. Parayangattil Subramanian Smitha (Department of Botany, Vimala College, Thrissur, India) and Ms. Sulochana Priji (Department of Botany, University of Kerala, India) for their technical support. Thadiyan Parambil Ijinu is receiving Young Scientist Fellowship from the Department of Science and Technology, Government of India (SP/YO/413/2018).

Conflict of Interest The views and opinions expressed in this article are those of the author in his personal capacity. None of the observations, opinions, or conclusions expressed herein are to be attributed to or construed as the views or positions of any organization, association, or company. No conflict of interest.

References

Abbott R (2014) Documenting traditional medical knowledge. Traditional knowledge documentation toolkit, pp 48. http://www.wipo.int/export/sites/www/tk/en/resources/pdf/tk_toolkit_draft.pdf. Accessed 25 June 2021

AICRPE (1998) All India Coordinated Project on Ethnobiology, Final technical report (1992–1998). Ministry of Environment and Forests, Government of India

Amirkia V, Heinrich M (2015) Natural products and drug discovery: a survey of stakeholders in industry and academia. Front Pharmacol (section Exp Pharmacol Drug Discov) 6:237. https://doi.org/10.3389/fphar.2015.00237

Balick MJ, Cox PA (1996) Plants, people, and culture: the science of ethnobotany. Scientific American Library, New York

Bavikatte K, Robinson DF (2011) Towards a people's history of the law: biocultural jurisprudence and the Nagoya protocol on access and benefit sharing. Law Environ Develop J 7(1):35–51

Che CT, George V, Ijinu TP, Pushpangadan P, Andrae-Marobela K (2017) Traditional medicine. In: Badal S, Delgoda R (eds) Pharmacognosy: fundamentals, applications and strategies. Academic Press, USA, pp 15–30

Cicka D, Quave C (2019) Bioprospecting for pharmaceuticals: an overview and vision for future access and benefit sharing. In: Joshee N, Dhekney S, Parajuli P (eds) Medicinal plants. Springer, Cham. https://doi.org/10.1007/978-3-030-31269-5_2

Cobo JRM (1986) Study of the problem of discrimination against indigenous populations, UN Document E/CN.4/Sub.2/1986/7/Add.4. United Nations, New York

Cobo M (2014) Study of the problem of discrimination against indigenous populations: final report submitted by the Special Rapporteur. https://www.un.org/development/desa/indigenouspeoples/publications/2014/09/martinez-cobo-study/. Accessed 25 June 2021

Corlett RT (2016) Plant diversity in a changing world: status, trends, and conservation needs. Plant Divers 38(1):10–16

Cragg GM, Boyd MR, Khanna R, Kneller R, Mays TD, Mazan KD, Newman DJ, Sausville EA (1999) International collaboration in drug discovery and development: the NCI experience. Pure Appl Chem 71(9):1619–1633

Dörr D (2019) Biopiracy and the right to self-determination of indigenous peoples. Phytomedicine 53:308–312

Duke JA, Duke PAK, du Cellier JL (2008) Duke's handbook of medicinal plants of the bible. CRC Press, Taylor and Francis Group, Boca Raton, USA, p 552

Farnsworth NR, Akerele O, Bingel AS, Soejarto DD, Guo Z (1985) Medicinal plants in therapy. Bull World Health Organ 63:965–981

George V, Rajasekharan S, Ijinu TP, Pushpangadan P (2014) Intellectual property rights (IPR), traditional medicine (TM) and traditional knowledge (TK). In: Pillai VNR (ed) Compendium on traditional knowledge, 26th Kerala Science Congress. Kerala State Council of Science, Technology and Environment, Thiruvananthapuram, India, pp 80–102

George V, Ijinu TP (2011) Scientific validation of traditional knowledge and development of herbal drugs, In: Rajasekharan S, Latha PG (eds) Traditional and folk practices-Contemporary relevance and future prospects. Jawaharlal Nehru Tropical Botanic Garden and Research Institute, Thiruvananthapuram, India, pp 56–63

Gervais DJ (2005) Traditional knowledge and intellectual property: a TRIPS-compatible approach. Mich St Law Rev 137–166

Greaves T (1994) IPR: a current survey. In: Greaves T (ed) Intellectual property rights for indigenous peoples: a sourcebook. Society for Applied Anthropology, Oklahoma, p 288

Greaves T (1995) The intellectual property of sovereign tribes. Sci Commun 17(2):201–213

Guerin-McManus M, Nnadozie KC, Laird SA (2002) The use of conservation trust funds for sharing financial benefits in bioprospecting projects. In: Iwu MM, Wootton J (eds) Advances in phytomedicine, Vol 1, ethnomedicine and drug discovery, vol 1. Elsevier Science. Amsterdam, Netherlands, pp 211–240

Gupta AK (2004) WIPO-UNEP study on the role of intellectual property rights in the sharing of benefits arising from the use of biological resources and associated traditional knowledge, WIPO-UNEP study no 4. www.wipo.int/edocs/pubdocs/en/tk/769/wipo_pub_769.pdf. Accessed 25 June 2021

Heinrich M (2000) Ethnobotany and its role in drug development. Phytother Res 14(7):479–488

Ijinu TP, Anish N, Shiju H, George V, Pushpangadan P (2011) Home gardens for nutritional and primary health security of rural poor of south Kerala. Indian J Tradit Knowl 10(3):413–428

Ijinu TP, George V, Pushpangadan P (2020) Bioprospecting of medicinal and aromatic plants. In: Peter KV (ed) Underutilized and underutilized horticulture crops. Brillion Publishing, New Delhi, India, pp 1–23

Joppa LN, Roberts DL, Pimm SL (2011) How many species of flowering plants are there? Proc R Soc B 278(1705):554–559

Kingston DG (2011) Modern natural products drug discovery and its relevance to biodiversity conservation. J Nat Prod 74(3):496–511

Kursar TA, Caballero-George CC, Capson TL, Cubilla-Rios L, Gerwick WH, Gupta MP, Ibañez A, Linington RG, Mcphail KL, Ortega-Barría E, Romero LI, Solis PN, Coley PD (2006) Securing economic benefits and promoting conservation through bioprospecting. Bioscience 56(12):1005–1012

Maister B, van Woensel C, de Jonge B (2012) Harnessing intellectual property rights for development countries: the double role of IRP's in the context of facilitating MDG's No 1 and 6. Wolf Legal Publishers. www.wolfpublishers.com/harnessingipr/pdf. Accessed 17 April 2021

Medaglia JC, Perron-Welch F (2019) The benefit-sharing principle in international law. J Intell Prop Law Prac 14(1):62–76

Nambisan P (2017) Protection of traditional knowledge associated with genetic resources. In: Nambisan P (ed) An introduction to ethical, safety and intellectual property rights issues in biotechnology. Academic Press, USA, pp 345–356

NBA (2018) National Biodiversity Authority, Annual report (2017–2018). Ministry of environment, forest and climate change, government of India, p 11

Newman DJ, Cragg GM (2007) Natural products as sources of new drugs over the last 25 years. J Nat Prod 70(3):461–477

Newman DJ, Cragg GM (2016) Natural products as sources of new drugs from 1981 to 2014. J Nat Prod 79(3):629–661

Newman DJ, Cragg GM (2020) Natural products as sources of new drugs over the nearly four decades from 01/1981 to 09/2019. J Nat Prod 83(3):770–803

Newman DJ, Kilama J, Bernstein A, Chivian E (2008) Medicines from nature. In: Chivian E, Bernstein A (eds) Sustaining life: how human health depends on biodiversity. Oxford University Press, New York, pp 117–161

Oberthür S, Rosendal GK (2014) Global governance of genetic resources access and benefit-sharing after the Nagoya protocol, 1st edn. Routledge, Taylor and Francis, London, p 288

Pimm SL, Joppa LN (2015) How many plant species are there, where are they, and at what rate are they going extinct? Ann Mo Bot Gard 100(3):170–176

Prip C, Rosendal K (2015) Access to genetic resources and benefit-sharing from their use (ABS)-state of implementation and research gaps. FNI Report 5/2015, PBL Netherlands Environmental Assessment Agency, Fridtj of Nansen Institute, Norway, pp 38

Pushpangadan P (2002) Biodiversity and emerging benefit sharing arrangements-challenges and opportunities for India. Proc Indian Natl Sci Acad B68(3):297–314

Pushpangadan P, Ijinu TP (2017) Intellectual property rights on traditional knowledge and their significance in sustainable societal development. In: Narasimharao BPR, Wright E, Prasad S, Joshi M (eds) Handbook of research on science education and university outreach as a tool for regional development. IGI Global, USA, pp 157–185

Pushpangadan P, Nair KN (2005b) Value addition and commercialisation of biodiversity and associated traditional knowledge in the context of the intellectual property regime. J Intellect Prop Rights 10(5):441–453

Pushpangadan P, Pradeep PRJ (2008) A glimpse at tribal India-An ethnobiological enquiry. Amity Institute for Herbal and Biotech Products Development, Thiruvananthapuram, India, p 167

Pushpangadan P, Ijinu TP, Dan VM, George V (2012) The role of science and technology in biodiversity based sustainable development. In: Sheng TK, Soljan D (eds) Science and technology policy for sustainable development. Daya Publishing House, New Delhi, pp 125–135

Pushpangadan P, Ijinu TP, George V (2014) Origin, growth and development of ethnobiology/ethnobotany, human culture and civilization. In: Kumar S (ed) Ethnobotanical studies in India. Deep Publications, New Delhi, pp 1–11

Pushpangadan P, George V, Ijinu TP (2015a) Traditional medicine and ethnopharmacology research in India. In: Chauhan AK, Pushpangadan P, George V (eds) Natural products: recent advances. Write and Print Publications, New Delhi, India, pp 120–146

Pushpangadan P, George V, Ijinu TP (2016a) Ethnobotany, ethnopharmacology, bioprospecting and patents. In: Pullaiah T, Krishnamurthy KV, Bahadur B (eds) Ethnobotany of India, vol 1. Eastern Ghats and Deccan. Apple Academic Press Inc., New Jersey, USA, pp 461–482

Pushpangadan P, George V, Ijinu TP, Rajasekharan S (2016b) Ethnobotany, ethnobiology, ethnopharmacology, bioprospecting of traditional knowledge and evolution of benefit sharing. In: Jain AK (ed) Indian ethnobotany: emerging trends. Scientific Publisher, Jodhpur, India, pp 1–23

Pushpangadan P, George V, Ijinu TP, Rajasekharan S (2017) Ethnopharmacology, traditional knowledge and intellectual property rights. In: Laladhas KP, Preetha N, Oommen OV (eds) Biodiversity for sustainable development. Springer International Publishing, Switzerland, pp 97–119

Pushpangadan P, Ijinu TP, Dan VM (2018) Evolution of ethnopharmacology research in India and their bioprospecting strategies. In: Das AP, Bera S (eds) Plant diversity in the Himalaya hotspot region, vol II. Bishen Singh Mahendra Pal Singh. Dehra Dun, India, pp 787–806

Pushpangadan P, Nair KN (2005a) Access to biological resources and associated traditional knowledge and benefit sharing (extended abstract). In: The group of like-minded megadiverse countries (LMMC) realizing the potential, Expert meeting of the Ministry of Environment and Forests, Government of India, New Delhi, pp 10–13

Pushpangadan P, Ijinu TP, George V (2015b) Intellectual property rights (IPR) regime and the traditional knowledge/indigenous knowledge holders. In: Chaubey NP, Sambasiva Rao N. (eds) Proceedings of 38th Indian social science congress. Indian Academy of Social Sciences, India, pp 45–60

Pushpangadan P (1998) Ethnobiology in India, A status report on all India coordinated research project on ethnobiology. Ministry of Environment and Forests, Government of India

Rajasekharan S (2010) Protection of traditional knowledge and sui generis system. In: Yesodharan EP (ed) Compendium on Intellectual property rights and development: a national perspective. 22nd Kerala Science Congress, Kerala State Council for Science, Technology and Environment, India, pp 79–91

Ravi K, Pushpangadan P (1997) Application of environmental valuation techniques for economic valuation of biodiversity a critical investigation. In: Pushpangadan P, Ravi K, Santhosh V (eds) Conservation and economic evaluation of biodiversity. Oxford & IBH Publishing House, New Delhi, pp 335–369

Robinson DF, Raven M (2019) Ethical trade in natural products based on traditional knowledge. Trade for Development News, Enhanced Integrated Framework Executive Secretariat at the WTO, Geneva, Switzerland. https://trade4devnews.enhancedif.org/en/news/ethical-trade-natural-products-based-traditional-knowledge. Accessed 11 April 2021

Rose J, Quave CL, Islam G (2012) The four-sided triangle of ethics in bioprospecting: pharmaceutical business, international politics, socio-environmental responsibility and the importance of local stakeholders. Ethnobiology and Conservation, 1

Rosenthal JP (1997) Equitable sharing of biodiversity benefits: agreements on genetic resources. In Investing in biological diversity: proceedings of the cairns conference, vol 253274

Soejarto DD, Gyllenhaal C, Kadushin MR, Southavong B, Sydara K, Bouamanivong S, Xaiveu M, Zhang HJ, Rong L, Franzblau SG, Fong HHS, Riley MC, Elkington BG, Waller DP (2012) An ethnobotanical survey of medicinal plants of Laos toward the discovery of bioactive compounds as potential candidates for pharmaceutical development. Pharm Biol 50(1):42–60

ten Kate K, Laird SA (2021) The commercial use of biodiversity: access to genetic resources and benefit-sharing. Routledge library editions: ecology. Taylor & Francis Group, p 414. ISBN: 9780367357641

Tsioumani E (2015) Moving towards fair and equitable benefit-sharing in research and development: the Nagoya protocol on access and benefit-sharing to the convention on biological diversity. http://www.benelexblog.law.ed.ac.uk/2015/02/18/nagoya-protocol/. Accessed 22 June 2021

Tu Y (2016) Artemisinin—a gift from traditional Chinese medicine to the world (Nobel lecture). Angew Chem Int Ed 55(35):10210–10226

USPAT (no date) General information concerning patents. United States Patent and Trademark Office, Alexandria, Virginia. https://www.uspto.gov/patents/basics/general-information-patents. Accessed 28 June 2021

Verkley G, Martin D, Smith D (2016) Best practice manual on access and benefit sharing, Version 1.0. Microbial Resource Research Infrastruture (MIRRI), Germany, pp 25

WIPO (no date) Traditional knowledge and intellectual property-background brief. http://www.wipo.int/pressroom/en/briefs/tk_ip.html. Accessed 14 May 2021

WIPO (2001) Matters concerning intellectual property and genetic resources, traditional knowledge and folklore-an overview. WIPO/GRTKF/IC/1/3. Intergovernmental committee on intellectual property and genetic resources, traditional knowledge and folklore. https://www.wipo.int/meetings/en/doc_details.jsp?doc_id=1662. Accessed 14 May 2021

WIPO (2005) Traditional knowledge: key to a diverse and sustainable future. In: Intellectual property and traditional knowledge, Publication No. 920(E). World Intellectual Property Organization, Geneva, Switzerland

World Bank (no date) Indigenous peoples. https://www.worldbank.org/en/topic/indigenouspeoples. Accessed 13 April 2021

Chapter 9
Access and Benefit Sharing: Agro Biodiversity and Farmers' Rights on Seeds—The Case of Indigenous Rice Varieties of Wayanad, Kerala, India

P. Prajeesh and N. Anil Kumar

Abstract By virtue of its location in the confluence of three major bio-geographic realms, India is rich in biodiversity and considered as a cradle of agro biodiversity. India is a farming country with majority of farming population falling under the categories of small and marginal holders who prefer to use saved seeds. Over the years, these farming communities have been remarkably developing and conserving varieties of crops that suit different habitats and ecological conditions. India is one of the leading countries which had developed its own legal measures to protect plant varieties and farmers' rights under the International Treaty on Plant Genetic Resources for Food and Agriculture. All these accolades demand for an inquiry into the Access to the Agro biodiversity heritage of the country and Benefit Sharing-ABS, with regard to the efforts of the farming communities. Studies show that the very idea behind the Indian legislation was to point that, farmers and breeders are allies in the struggle for sustainable food security, and that their rights must be mutually reinforcing and not antagonistic. This chapter examines the extent of realization of the concept of Farmers' Rights and ABS, along the lines of the historical journey and the grass root reflections from Kerala.

Keywords Agro Biodiversity · Access and Benefit Sharing · Farmers' Rights · PPV&FR Act · Wayanad

9.1 Introduction

9.1.1 Agro Biodiversity–The Indian Hotspot

India is considered as a cradle of agricultural biodiversity (Nayar et al. 2009) and one among the 17 mega diversity centres of the world (Mittermeier et al. 1997). Nayar et al. (2009) has also observed that by virtue of its location in the confluence of three major bio-geographic realms, i.e., Indo-Malayan, Palaearctic and Ethiopian.

P. Prajeesh (✉) · N. Anil Kumar
M. S. Swaminathan Research Foundation (MSSRF), Wayanad, Kerala, India
e-mail: prajeesh@mssrf.res.in

O. V. Oommen et al. (eds.), *Biodiversity Conservation Through Access and Benefit Sharing (ABS)*, https://doi.org/10.1007/978-3-031-16186-5_9

The Indian subcontinent is a crucible of tropical, temperate and semi-arid floristic elements, which in time and space have established, adapted and evolved to form the present spectrum of rich Biodiversity. They further observe that through Millenia, India is known for its civilizations and cultures and is also for the confluence of human migrations attracted by the fertile Indo-Gangetic plains. The flowering plants of India comprise about 15,000 species and represent 6% of the world's known flowering plants (Nayar 1977). Mittermeier et al. (2004) has given three hotspots of biodiversity in Indian subcontinent i.e., the Western Ghats and Sri Lanka, The Himalayas, and Indo-Burma region. Nayar (1996) has also explained 25 micro hotspots of flora in the country and given that the endemic species represent 33% of Indian flora. The country is also considered as one of the eight primary centres of crop origin and domestication of species, especially of Rice, Millets and Legume (Vavilov 1935). The country has 21 agro ecological regions with varying crop growing periods from 90 to 270 days spanning across different ecosystems and having soil profiles from rich alluvium, black and red soils to shallow skeletal soils and saline soils. This admixture of soil profiles, crop growing periods, soil moisture and rainfall patterns with hyper humid to semi-humid, semi-arid to arid conditions offer different ecological niches for different crops (Sehgal et al. 1992). Each of these agro ecological zones comprises of myriad microhabitats. It is reported that at least 166 crop plants and about 320 species of wild relatives of cultivated plants originated in India (Khoshoo 1988; Rana 1993). Within the spectrum of crop species and wild relatives, there are thousands of varieties, cultivars, landraces, and ecotypes which occur in India. Over the millennia, the Indian farming communities have been developing and sustaining an amazing variety of crops to suit different habitats and ecological conditions (Nayar et al. 2009). According to Balaravi (2010), a majority of Indian population farms with approximately 80% of Indian farmers classified as either small or marginal and preferring to use saved seeds. All these accolades demand for an inquiry into the Access to the Agro biodiversity heritage of the country and Benefit Sharing with regard to the efforts of Indian farmers in conserving/developing and sustaining a wide variety of crops that suit to different habitats and ecological conditions.

9.1.2 Indigenous Varieties—Varietal Improvement as Farmers' Innovations

The locally adapted farmer varieties have greater genetic variability providing yield stability in even the most challenging climatic conditions while the formally improved high-yielding varieties have a very narrow genetic base, observes Lehmann (1981), and Hasan and Abdullah (2015). The practice of seed improvement has moved out of the fields to formal labs and has started valuing genetic uniformity (Kochupillai 2016). As Vanloqueren and Baret (2009) observe, the varieties being cultivated presently exist due to the selection done by generations of farmers; yet, many do not view the process of cultivating and developing traditional varieties within farmers' fields as innovative. These farmers' varieties require further selection to meet the uniform standard required for intellectual property protection through plant variety registration (Salazar et al. 2007).

9.1.3 *The Idea of ABS and International Negotiations*

Meenakumari and Rana (2017) have provided a detailed note on implementing the Biological Diversity Act of India, focusing on ABS and Agro biodiversity. While discussing the international negotiations, the primary legal tools are: (1) The 1992 Convention on Biological Diversity (CBD); (2) The International Treaty on Plant Genetic Resources for Food and Agriculture—ITPGRFA (2001); and (3) The Trade Related Intellectual Property Rights, 1995, (TRIPS). The international negotiations on crop genetic resources started way back in 1983 with FAO's International Undertaking on PGRs. Later it gave way to The International Treaty on Plant Genetic Resources for Food and Agriculture (adopted on 3rd November, 2001) which was an effort to harmonize the provisions of the Undertaking with that of CBD (Meenakumari and Rana 2017). CBD of 1992, was the very first international agreement that addressed ABS in its objectives and provisions. It recognizes the sovereign rights of states over their natural resources and affirms the authority governments have, subject to their national legislation, to determine access to genetic resources. The Nagoya Protocol on Access to Genetic Resources and the Fair and Equitable Sharing of Benefits Arising from their Utilization to the Convention on Biological Diversity (Nagoya Protocol) is a supplementary agreement to the CBD. It provides a legal framework for the effective implementation of the third objective of the CBD, the fair and equitable sharing of benefits arising out of the utilization of genetic resources, including by appropriate access to genetic resources, with a view to contributing to the conservation of biological diversity and the sustainable use of its components (FAO 2019). As recognized by the Conference of the Parties to the CBD at its tenth meeting, the International Regime of ABS is constituted by the CBD and the Nagoya Protocol, as well as complementary instruments, including the Treaty and the Bonn Guidelines on Access to Genetic Resources and Fair and Equitable Sharing of the Benefits Arising out of their Utilization.

9.1.3.1 The International Treaty on Plant Genetic Resources for Food and Agriculture (ITPGFRA)

This Treaty covers designated accessions of 64 Annex-1 food and forage crops (which are in the public domain and managed by the national governments) and upholds the Farmers' Rights subject to national legislation. Unlike the bilateral system of agreements under CBD for accessing PGRs, this Treaty provides for a Multilateral System (MLS) of ABS to facilitate exchange of PGRs which are in public domain, covering initially only 35 food crops and 29 genera of forages. Union Ministry of Agriculture and Farmers Welfare is the nodal Ministry for implementing the ITPGRFA, assisted by the Department of Agricultural Research and Education and National Bureau of Plant Genetic Resources. A notification has been issued exempting the exchange of designated accessions of genetic resources of crops listed in Annex-1 of the ITPGRFA from the provisions of Sections 3 and 4 of the BD Act for research, breeding (Meenakumari and Rana 2017). While analyzing the case of Nepal, Joshi et al. (2017) observes that the interdependence of countries on Agricultural Plant Genetic Resources is a key rationale for the creation of the multilateral system of access and benefit sharing under ITPGRFA. They also point out that most people or countries are unaware of the current and future importance of interdependence and how the flow of genetic materials between countries contributes to national food security.

9.1.3.2 Seed-Related Treaties and Negotiations—A Different Perspective

From the perspective of the Global Citizens Alliance for Seed Freedom, TRIPS agreement was the corporates' solution to the problem of 'farmers save the seeds'. Later, different kinds of IPRs made seed, the common good to a commodity. Vandana Shiva has noted that a reductionist, mechanistic science and a legal framework for privatizing seed and knowledge of the seed reinforce each other to destroy diversity, deny farmers innovation and breeding, enclose the biological and intellectual commons, create seed monopolies and she calls it as 'the assault on seeds'. Whether it be breeders' rights imposed through UPOV, or Patents on Seed, or Seed Laws that require compulsory registration and licensing, an arsenal of legal instruments is being invented and imposed undemocratically to criminalize farmers seed breeding, seed saving and seed sharing (Navadanya 2012). Berhan and Edwards (2012) even urged the declaration of Article 27.3 (b) of TRIPs as immoral, because the patenting of crop varieties sucks in crop genetic diversity from primarily subsistence farming communities but restricts the resulting varieties into circulating only among the rich, especially when natural cross-pollination passes patented genes from genetically modified crop varieties to non-modified varieties. The 'Seed Freedom communities' had committed themselves to defend seed freedom as the freedom of diverse species to evolve; as the freedom of human communities to reclaim open-source seed as a commons. Given that seeds are the source of life, the embodiment of bio-cultural diversity; Seed Freedom is the birthright of every form of life, Seed Freedom is the birth right of every farmer and food producer and Seed Freedom is the basis of Food Freedom.

9.1.3.3 Implementation of CBD and TRIPS—Indian Measures

India's efforts to ensure compliance with the CBD and Article 27.3(b) of the TRIPS Agreement primarily included the introduction of three legislations viz., (1) The Protection of Plant Varieties and Farmers' Rights (PPV&FR) Act, 2001; (2) The Biological Diversity Act 2002; and (3) The Patent (Amendment) Act 2002. Since this chapter examines the ABS related to Agrobiodiversity, with special emphasis to indigenous rice varieties of Kerala, only the provisions with the PPV&FR Act of 2001 in comparison with that of the Biological Diversity Act of 2002 are discussed further.

9.2 The History of Farmers' Rights and the Roadmap to the Protection of Plant Varieties and Farmer's Rights Act of India (PPV & FR Act 2001)

Since the start of farming, farmers (both men and women), have happily enjoyed the free use and vending of the seeds that they cultivated and were continuously contributing to evolving new varieties that perform well in the changing agronomic and climatic conditions. Farmers have been engaged in collective systems of conservation and innovation (openly sharing planting materials and conserving them through use) since the earliest crop domestications. Relatively open flows of plant germplasm attended the spread of agriculture and have subsequently followed (or been driven by) imperialism, colonization, emigration, trade, development assistance and climate change (adapted from Halewood et al. 2013). The advent of scientific breeding after the rediscovery of Mendel's laws in the early twentieth century (first developed in the 1880s) had helped to guide the search for new material on the basis of known genetic characteristics (Byerlee and Dubin 2010; Pistorius 1997). The year 1930 marked the beginning of the acceptance of the idea of individual ownership rights over knowledge of vegetatively propagated plant varieties in USA. Subsequently in 1961, the Paris Convention on the Protection of New Varieties of Plants (popularly known as UPOV convention) came out with the concept of Plant Breeders' Rights. Following the General Agreement on Tariff and Trade, the World Trade Organization (WTO) brought in the agreement on Trade Related Aspects of Intellectual Property Rights (TRIPS). TRIPS insisted that all the signatories of WTO to protect the Intellectual Property Rights on plant varieties. Much later the concept of Farmers' Rights came, at the instance of members from the Third World to recognize the efforts of farmers in conserving plant genetic resources. As a result, the international instruments like The Convention on Biological Diversity of 1992 and The International Treaty on Plant Genetic Resources for Food and Agriculture of 2001 also have recognized the contributions of rural and tribal communities in the conservation and enhancement of biodiversity.

As Farmers' Rights Project describes,[1] the first use of 'Farmers' Rights' as a political concept dates back to the early 1980s, when Pat Roy Mooney and Cary Fowler of the then Rural Advancement Foundation International (presently known as the ETC Group) in the US, coined the term to highlight the valuable but unrewarded contributions of individual farmers in conserving plant genetic resources for food and agriculture by cultivating the plants—in situ conservation. The idea came up as a countermove to the increased demand for plant breeders' rights, as voiced in international negotiations. The concept of Farmers' Rights was first brought up in international negotiations in the Council of Food and Agricultural Organization chaired by Prof. M. S. Swaminathan in 1983. FAO defines Farmers' Rights as "rights arising from the past, present and future contribution of farmers in conserving, improving and making available plant genetic resources, particularly those in the centres of origin/diversity. The purpose of these rights is stated to be ensuring full benefits to farmers and supporting the continuation of their contributions." (Resolution 5/89, FAO, 1989).

India, being a signatory to WTO, had to respond to the Article 27(3)(b) where all the signatories were required to protect the Intellectual Property Rights on plant varieties either by (1) a patent or (2) an effective '*sui-generis*' system or by (3) a combination of these two. Discussions on the protection of Plant Breeders Rights in India are evident from the year 1989 during the Second Seed Seminar organized by the Seed Association of India (Seed Association of India 1990). The Keystone Dialogues 1988, 1990 and 1991 (which took place in Colorado, Madras and Oslo respectively and chaired by Prof. M. S. Swaminathan) were instrumental in framing the concept of Farmers' Rights (Keystone Centre 1990, 1991). It was soon after its establishment in 1989, the M S Swaminathan Research Foundation (MSSRF) has associated with the Keystone Centre to organize the second in the series of the Keystone Dialogues on Plant Genetic Resources. This Dialogue, held at Madras from 29 January to 2 February 1990, had stressed the urgency of implementing Farmers' Rights (Swaminathan 1996a, b).

In between in 1993, another draft of the Act was submitted to the Parliament by the Ministry of Agriculture. Quoting Shiva and Hola-Bhar (1993), Demangue (2005) has noted that the reward mechanism proposed by the 'Madras Model' Act was diluted in the version submitted to the Parliament and contained only a vague clause on farmers' rights. Subsequently this draft was also heavily critiqued. The draft final text embodying the results of the Uruguay Round of Multilateral Trade Negotiations at Brussels, popularly known as the 'Dunkel Draft' had provoked a larger public debate. A series of seminars and workshops were conducted in the Country to provide an interface between scientists and farmers and had come up with consensus recommendations in this regard. One of the most important papers that emerged from

[1] The Farmers' Rights Project database has listed the countries which have the provisions of either Farmers' Rights or Community Rights related to Plant Genetic Resources in their national/regional legislations. There are only 8 countries other than India (viz., Bangladesh, Malaysia, Nepal, Pakistan, Philippines and Thailand in Asian region; Ethiopia in African region Costa Rica in the American region) and 2 Intergovernmental organisations (ASEAN with 10 member states and The African Union with 54 member states) which are enlisted (http://www.farmersrights.org).

these exercises was called Dunkel Draft: Minimum Must For Renegotiation, which in 1992 listed alternative negotiating positions in GATT/TRIPS defined in consultation with experts. Quoting Shiva and Hola-Bhar (1993), Menon (1993), Rane (1993) and Tellis (1993), Demangue (2005)[2] has observed that NGOs and Farmers' organizations had organized several protest marches against the potential impacts of the WTO obligations on the Indian Agricultural sector. Gene Campaign[3] explains that, relentless campaigns led to a decisive event in 1993, which contributed to compelling the government to abandon its pro-patent stand. This event was a farmers' rally organized by Gene Campaign together with three farmer organizations, the All India Bhartiya Kissan Union, the BhartiyaKissan Union of Punjab and the Karnataka Rajya Ryatha Sangha. The farmers' rally led by the imposing figure of Mahendra Singh Tikait had 3, 00,000 farmers from UP, Karnataka and Punjab congregated behind the Red Fort on March 3. Jointly with Gene Campaign they issued a single call, "No Patents on Seeds". The coming into force of the Global Convention on Biodiversity in December 1993 and of the World Trade Agreement in April 1994, imparted a sense of urgency to incorporating provisions for Farmers' Rights in *sui generis* systems of legislation relating to Plant Variety Protection.

Meanwhile, the concept of Farmers' Rights was discussed by Professor M. S. Swaminathan in different National and International consultations and also had given the drafts of the Indian Act (Swaminathan 1994a, b). Sensing the urgency of the legislation, MSSRF organized an International Dialogue on "Farmers' Rights and Plant Genetic Resources" at Madras from 29–31 of January 1994. At the Dialogue, a draft of the Indian Act was circulated to the participants to build on during the session "Methodologies for Recognizing and Rewarding Informal Innovation in the Conservation and utilization of Plant Genetic Resources". The proceedings of this dialogue were published in early 1995 [Swaminathan, M. S. 1995 (ed.); Swaminathan, M. S. (ed.) 1996b]. A Technical Consultation was also held at MSSRF from 15–18 January 1996 for developing and implementation framework for Farmers' Rights. Incorporating the inputs from the workshop, a revised model act was published [Swaminathan, M. S. (ed.) 1996b]. This draft, combining breeders' and farmers' rights in single legislation was taken up by the Joint Parliamentary Committee (JPC) headed by the late Shri. Sahib Singh Verma, who had paid a visit to MSSRF Chennai for a detailed discussion in this regard.[4]

In 1999, the draft with the modifications by the JPC was introduced in the LOK SABHA and was referred back to the committee for inviting public opinion. After considering a large number of suggestions received from various public or private agencies, associations and individual, "the Committee felt that the Bill has very inadequate provisions for protecting the interest of farmers". Later, the revised draft,

---

[2] Demangue (2005) has drawn the pathway to the PPV&FR Act after having personal discussions with the lead-players in this area and also quoting many government documents which she had collected personally.

[3] Delhi-based Gene Campaign is dedicated to food, livelihood security of rural communities and rights of farmers (http://genecampaign.org/farmers-rights).

[4] Personal communication by Prof. M. S. Swaminathan, e-mail dated 31st March, 2015.

introduced in the LOK SABHA in August 2001, contained two outstanding modifications. First, the possibility of farmers to register heterogeneous local varieties and enjoy the same exclusive rights as formal sector breeders; second, a separate chapter on farmers' rights in the largest sense, partly regrouping the elements already present in the 1993 Bill. In late August 2001, both Houses of the Parliament finally passed the Protection of Plant Varieties and Farmers' rights Bill, 2001. The Protection of Plant Varieties and Farmers' Rights Act of India (PPV&FR Act) received the assent of the President of India on 30th October, 2001. Draft rules of this Act was submitted to the Govt. of India after having been developed and discussed at an FAO—sponsored consultation in Chennai during January 21–23, 2002 (M. S. Swaminathan Research Foundation 2002; Swaminathan 2002a). Many of the elements proposed in this draft were later incorporated in the official Rules promulgated in 2003.

9.3 Protection of Plant Varieties and Farmers' Rights Act of India

The Act does not need any introduction at all; it has the provisions to recognize the rights of Farmers, Breeders' and Researchers in single legislation. A Farmer is not just one who cultivates crops but is also a breeder who experiments with the seed. The Act defines farmers in three roles, as cultivators, conservers and breeders. Specifically, nine Rights have been given to the farmers by the Act which include the right on seeds and right to receive reward and recognition for their conservation efforts. The provision of registration of Farmers' Varieties allows the farmers to register those varieties cultivated and evolved or bred by them, wild relatives or landraces about which the farmers possess common knowledge. Other categories of registration are New Variety, Extant Variety and Essential Derived Variety. Now the registration is offered to 161 crops as notified by the Authority. As of 2021 August, 4,065 varieties have been given registration certificates (Figs. 9.1 and 9.2). It has also the provision to recognize Farmers' efforet toward the conservation of agro biodiversity. The Plant Genome Savior Community Recognitions and Awards given by the PPV&FR Authority are for the rural and tribal farming communities, groups or individuals engaged in conservation, improvement and preservation of genetic resources of plants and their wild relatives (GoI 2001).

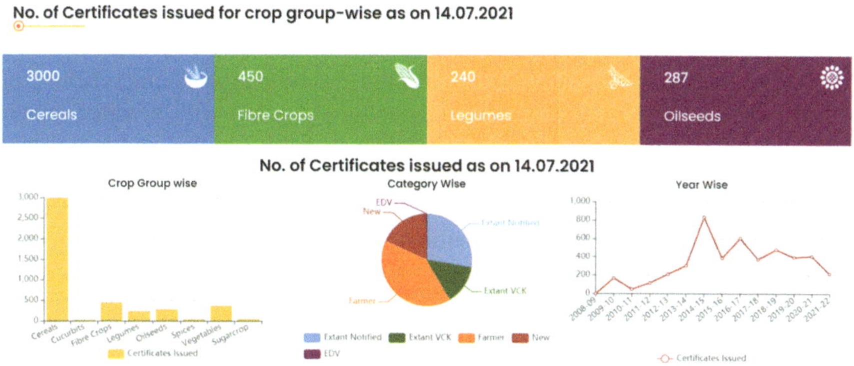

Fig. 9.1 Status of Registration of Varieties under PPV&FR Act, August, 2021 (*Source* http://plantauthority.gov.in)

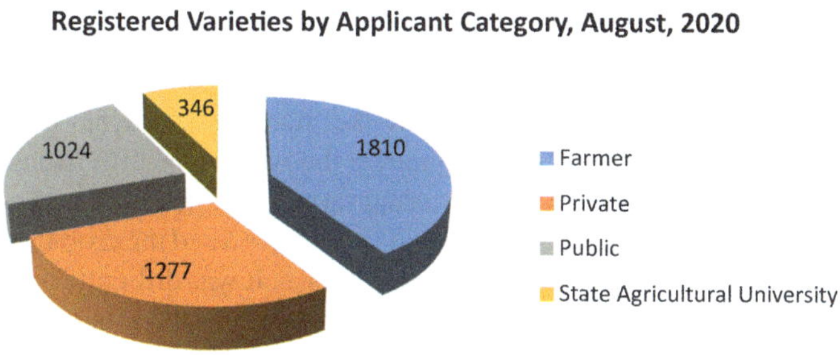

Fig. 9.2 Registered Varieties under PPV&FR Act by Applicant category, August, 2020 (*Source* http://plantauthority.gov.in/list-certificates)

9.3.1 PPV&FR Act-Reflections from Kerala, India with Special Emphasis to the Efforts of Farmers of Wayanad

Balaravi and Suchithra (2003) and Kochhar (2008) have rightly observed that the most gigantic and uphill task in the implementation of the PPV&FR Act is the realization of farmers' rights. It is appreciable that some of the farmers from Kerala State were proactive in learning the Act and accessing its provisions, with the help of regional facilitators. Since the inception of Plant Genome Savior Recognitions in 2007–08 and Awards in 2009–10 by the Government of India, Kerala has dominated with 26 entries, viz., 4 Plant Genome Saviour Community Awards, 11 Plant Genome Saviour farmer redemption/awards, and 11 Plant Genome Saviour farmer honor winners (PPV&FRA 2021). The farmer who had won the Plant Genome Saviour Recognition in 2008–09 had also served as a Farmers' representative member of PPV&FR Authority during the period 2010–12. Wayanad farming community headed by SEED CARE is the very first to hold the titles of Farmers' Varieties in Kerala State.

Chandrasekaran and Vasudev (2002) observed that the marginal farmers belonging to tribal regions rarely rely on modern agriculture and they will therefore remain unaffected by any growth in the plant breeding industry. On the contrary, if they continue to remain 'illiterate' and unaware of their rights, they stand to lose from the unauthorized use of their landraces and other germplasm. Conscious and selfless efforts to compensate for their contributions will have to be made by Government agencies and non-partisan NGOs if their role as preservers of biodiversity is to be recognized. In this context, the initiative taken by SEED CARE is very important in Wayanad. SEED CARE is a farmer–led Association of Traditional Crop Conservators of Malabar, including both men and women, established in 2009 in Wayanad, Kerala. It essentially has tried to bring the marginal farmers belonging to tribal regions, the small and medium farmers and the large progressive farmers under one roof for a cause, for what its name denotes. Unlike in the other States of India, there is not much pressure from the private seed industry in Kerala and the small and marginal farmers are still relying to a great extent on their indigenous varieties; they save and use farm-produced seeds, freely exchange or vend them. Some farmers largely depend upon the seed system maintained by the State Government. While starting the grassroot activities of MSSRF to popularize the Act and its provisions, the basic question raised was on the need of registering farmers' varieties. Farmers were not familiar with ownership on plant varieties; in fact, they owned the crop they produce, but did not own the seed or the variety. In such a case, it was difficult for the farmers to appreciate that registering farmers' varieties would help them in protecting their varieties and they might start getting benefits shared out of the commercial use of those varieties at a later point in time. MSSRF, through its genetic and legal literacy programme had organized the traditional farmers of the district under SEED CARE and they have forwarded 27 applications for Farmers' Varieties of Rice to the Govt of India of which 6 were registered in 2013 (*Veliyan, Thondi, Chennellu, Chomala, Gandhakasala and Jeerakasala;* Reg. nos: 56 to 61 of 2013); 10 in 2015 (*Mullankayama, Valichoori, Onavattan, Kurumottan, Kunjoottimatta, Marathondi, Chenthadi, Koduveliyan, Thuroodi* and *ThonnooranThondi*; *Reg.nos: 220 to 228 and 231 of 2015) and 4 in 2016 (Kottathondi, Kayama, Mannuveliyan, Adukkan; Reg.nos: 20 to 23 of 2016).* (PPV & FRA, 2021, http://plantauthority.gov.in/list-certificates).

9.4 PPV & FR Act-Challenges in the Implementation

Though through these years, farmers of Wayanad in Kerala have managed to get practical access to the important rights of farmers, further reflections show that the structure and the functioning of the government authority need to be extended to better implement the provisions of the Act. The reflections and suggestions are detailed in this section.

9.4.1 Nature and Identity of the Farmers and Gender Relations Are to Be Considered

The very first question in the application for registration of varieties (Form 1 of 2009) was about the identity of the applicant(s). The options were: (1) Individual Breeder, (2) Successor of Breeder, (3) Institutional Applicant, (4) Farmer, (5) Community of Farmers, (6) Group of Farmers, (7) Assignee of any of the above, (8) Convention Country, (9) Any Other. There was a single template for all categories of varieties and now the application template for Farmers' Variety is separate (farmer variety-2013, PPV&FRA 2021) and has only 3 options viz., (1) Farmer, (2) Community of Farmers and (3) Group of farmers.

The registration certificates issued in 2013, 2015 and 2016 to SEED CARE, a group of farmers state that:

*Whereas (Secretary, SEED CARE…..) has declared that **he** has developed (Name of the Variety) as Farmers' Variety and that **he** is the true breeder thereof (or the legal representative or assignee of the true breeder) and that **he** is entitled to a plant variety right on the said variety, having regard to the provisions of the Protection of Plant Varieties and Farmers' Rights Act, 2001 and that there is no objection to the registration of plant variety in favour of **him**. And whereas **he** has, by an application, requested….may be allowed to **him** for the said variety…….*

The certificate template provided in the PPV&FR Rules, 2003 and also the format (Plant Variety Registry, Form O-2) both show the same words. The application template asks about the nature or identity of the applicant (but not sex) and unfortunately, it doesn't get reflected in the certificate. Neither does the certificate accept the identity/nature, nor does it give the sex of the applicant(s).

9.4.1.1 Discussion Point: Farmers' Knowledge and Their Contribution in Conserving Genetic Resources—The Gender and Community Perspectives

The type of knowledge farmers possess varies by age, gender, roles and responsibilities and environment. Access to or control over resources as well as education, training, information, and control over the benefits of production also influence the type of knowledge rural women and men have. Experience-based local knowledge interweaves with cultural values and develops and adapts continuously to a gradually changing environment. Rural women's and men's local knowledge, skills, and innovations raise the issue of recognition and protection of farmers' rights. The different tasks and responsibilities of rural women and men in agriculture have enabled them to accumulate different types of local knowledge and skills. Women are typically involved in the selection, improvement, and adaptation of local plant varieties, as well as seed selection, seed storage, seed management, and exchange, management, and saving. They usually keep home gardens where they grow traditional varieties of vegetables, herbs, and spices selected for their nutritious, medicinal, and culinary advantages. It is widely studied that women play an important role in maintaining biodiversity at the farm level. But women's roles and knowledge are often overlooked or underestimated in natural resource management and related policies and programs. The new agreements under the World Trade Organization have an influence on biodiversity and also have gendered impacts. None of these trade negotiations consider women's and men's different knowledge, skills, and uses of agro biodiversity which is unfortunate and wrong. But in many situations, women work against the decreasing biodiversity, which is caused partly by men who favor cash-oriented monocultures (Howard 2003; Momsen 2007; Food and Agriculture Organization 2005; Randriamaro 2006; Alzina and Gabriela 2007; Gender in Agriculture Sourcebook 2009). A study from Wayanad district, where SEED CARE is facilitating the seed care movement has critically analyzed the gendered roles in agriculture, especially in paddy cultivation. It is observed that, in paddy cultivation, along with other responsibilities (sowing, weeding, harvesting) the winnowing activities are done by women and they separate the matured, healthy grains and the chaffy ones (Vedavalli and Anil Kumar 1998).

The question here is "who is a farmer?" As per the definition in the PPV&FR Act (Sec. 2 (k)),

"farmers" means any person who -

(i) *Cultivates crops by cultivating the land* ***himself****; or*
(ii) *Cultivates crops by directly supervising the cultivation or land through any other person; or*
(iii) *Conserves and preserves, severally or jointly, with any other person any wild species or traditional varieties or adds value to such wild species or traditional varieties through selection and identification of their useful properties.*

While greatly appreciating the efforts of rural and tribal genome saviors, the PPV&FR Act does not clearly mark the efforts of women farmers in biodiversity conservation. The Act ascertains the (male) farmers' provisions and the very immediate result is biased templates of applications and certificates under the Act. While giving a gendered critique of the Act, Parthasarathy (2003) mentioned that gender bias and gender blindness persist in the legislation. Farmers are generally perceived as 'male' by policymakers, development planners and agricultural service deliverers and for this reason; women find it more difficult than men to gain access to various valuable resources and other provisions. It is noted here that neither the applications (which the farmers need to submit to the Authority for registration) nor the certificates (which the authority distributes to the farmers on the confirmation of registration) indicate the sex of the applicant and also the latter do not indicate the nature or identity of the applicant.

MSSRF had organized a workshop in August 2003 to specifically discuss the gender and community perspectives of farmers' rights and biodiversity and had come up with concrete recommendations to include these perspectives in both the Indian legislations, the PPV&FR Act and the Biological Diversity Act. Regarding the PPV&FR Act, the recommendations were contextualized with reference to the numerical order of Draft Rules developed and submitted by MSSRF to the Government of India in February, 2002. Out of the 5 recommendations, the important 2 were regarding the definition of the term 'community' and incorporating gender justice in to the Rules (M. S. Swaminathan Research Foundation 2003). The workshop had identified that, since a farmer is defined as 'a person who cultivates or has the means to cultivate', the issue of women's access and control over property becomes relevant. It is in this respect that we need to consider women's rights keeping in mind her status as an individual, inheritor/natal, matrimonial/spousal and member of the community. Under the Indian law, women have no right even to the matrimonial home or to the farm as well. Therefore, both the Acts have to take into account the varied legislations that govern women's access and control over any property in different ethnic groups/minorities etc. A specific suggestion was that in the case of farming families, it shall be made essential that applications for registration of a plant variety arising from a farm family shall always include the names of both spouses (woman and man) as applicants. It was also observed that, considering the dynamic nature of any Indian community, there needs an operational definition of the term 'community' under the guidelines of the Act. It could be user groups; groups referring to geographical limitation, occupation; caste or class distinctions etc.

It is noted here that, beyond the recommendations given by the 2003 workshop, these concerns have not been contested fairly. As a result, the documents associated with the provisions of the act have been found stuck with the issue of 'nature and identity' of the farmers or farming communities whose rights are said to be protected.

9.4.2 Farmers and Conventional Breeders Have to Be Distinguished

The certificates of registration of Farmers' Varieties issued to SEED CARE read like:

> ... Now, these presents that the above said applicant (including his legal representatives and assignees or any of them) shall, ..., have the exclusive right to produce, sell, market, distribute, import or export the variety for a term of initial period of 6 years and renewable for the remaining years from the ....day of ...... and of authorizing any other person to do so ...

When the rights offered are to produce, sell, market, distribute, import or export the seeds, the Act does not provide the opportunity to brand the seeds [section 39 (iv)]. After issuing the registration certificates of 6 Farmers' Rice Varieties to SEEDCARE, the Authority has requested SEED CARE to bring the registered varieties under an informal seed chain by producing Truthfully Labeled (TL) Seeds and has kindly extended their necessary cooperation.[5]

It is argued that there needs to be much clarity on the provisions of producing, selling, marketing, distributing, importing and exporting and also bringing the seeds

[5] A letter to the Secretary, SEED CARE from the Registrar of Protection of Plant Varieties and Farmers' Rights Authority dated 13th August 2013.

into the seed chain. Even though the act mentions farmer as a breeder in the capacity of developing a variety [section 2 (c)], it distinguishes breeders' rights and farmers' rights with respect to the marketing of seeds under a brand name. But the certification of registration does distinguish farmers and breeders (who are in different levels of power struggle) neither in the nomenclature nor with respect to the extent of their rights. Moreover, the objective of the Act was to regulate the quality of seeds in course of business of the breeders and to safeguard the rights of the farmers in conserving and developing plant varieties. One of the reasons of emerging the very problem of imposing extraneous fees for farmers' variety is the ambiguity in the definitions of the farmer, as a farmer and a breeder.

Shalini (2015a) has observed that many seed saver groups in the country are convinced that this European-styled Intellectual Property Right is neither useful nor desirable in their context, since among other things, it attacks the very idea of sharing of seeds, standardizes seed quality as per the industry viewpoint and creates unnecessary inter-farmer competition in an already hostile market. While discussing the problems of defining Farmers' Rights as IPRs, Borowaik (2004) has pointed out the asymmetries which create the impression that there is parity among competing rights and in reality; this may promote a further shift from farmer-centered agriculture. Srinivasan (2003) also pointed out that the IPR-based Farmers' Rights approaches are also unlikely to provide significant economic returns to farmers and farming communities.

9.4.3 Communications of PPV&FR Authority Need to Reach to the Farmers

It is observed that neither the farmers nor the PPV&FR Authority have a chance to communicate in a better way, thus preventing the provisions of the Act to reach to the grassroots. Hanchinal et al. (2014) observes that the awareness and capacity-building programs of PPV&FR Authority has reached successfully to farmers through a number of facilitators like institutes under the Indian Council of Agricultural Research (ICAR), Krishi Vigyan Kendras (KVKs) under ICAR, State Agricultural Universities, and Non-Governmental Organizations. In Kerala, ICAR KVKs, Kerala Agricultural University and Non-Governmental Organizations like MSSRF are supporting and helping farmers to access their rights under the said Act. It is strongly argued that there needs to be a structure, like the one which is operational under the Biological Diversity Act, 2002 (BDA). BDA defines a three-tier system of implementation with the National Biodiversity Authority (NBA), State Biodiversity Boards (SBBs) and Biodiversity Management Committees (BMCs-at the level of Local Self Governments) which makes the implementation of the Act very smooth.[6]

---

[6] Referring to the case of issuing the very first Internationally Recognized Certificate of Compliance (IRCC) under the Nagoya Protocol on Access and Benefit Sharing, Parameswaran (2017) had observed that India through the NBA has taken a giant leap and left a mark in the history of

But in the case of the PPV&FR Act, the farmers are communicating directly to the PPV&FR Authority in Delhi (may or may not be with the help of a facilitator) and there is no way to fill up the application or make representations in languages other than English and Hindi. Importantly these two Acts which share (somehow) common objectives of conservation and protection of biodiversity need to co-exist, mutually reinforcing and could be implemented jointly. As noted by Swaminathan (2002b) there is a need for the effective and integrated implementation of the PPV&FR Act, Biological Diversity Act and the Seed Act (which was then under the consideration of the Parliament). He had also given the importance of awareness generation and 'information empowerment' through young lawyers and Panchayati Raj members.

Kochhar (2008) observed the need for a harmonious institutional mechanism/capacity for the application of the Indian IPR laws, with a focus on the implementation of the Protection of Plant Varieties and Farmers' Rights in the country. The allocation of business related to agriculture and biodiversity is attributed to two different central ministries. The BDA is administered by the National Biodiversity Authority and the Ministry of Environment, Forest and Climate Change is the nodal agency. The PPV&FR Act is administered by the Protection of Plant Varieties and Farmers' Rights Authority and the Department of Agriculture and Cooperation is the nodal agency. It was also suggested that a bi-focal nodal mechanism as followed by China or any other mutually complementing system would be of importance here. Lushington (2012) has noted that the multiplicity of legislation in India creates overlapping institutions and the divergent objectives of these laws tend to create confusion in terms of policy in India. Interestingly, the applications for different provisions under PPV&FR Act already have the terms to route them through the BMCs and the applicant needs to state if the varieties proposed for registration have already appeared or not in People's Biodiversity Registers (PBRs) maintained by BMCs. The minutes of the 14th meeting[7] of the Authority held on 20th October, 2011 has given that the National Biodiversity Authority and Indian Council of Agricultural Research have been notified in the Official Gazette vide S.0.1912(E) dated 18.08.2011 to which the PPV&FR Authority will forward the copies of registration certificates issued under the PPV&FR Act, 2001.

The Act and the Rules [(section 45(2)(c) of Act and Rule 70(2)(b)] envisage the expenditure from the gene fund, for supporting the conservation and sustainable use of Plant Genetic Resources (PGR) for *in-situ* and *ex-situ* conservation, for strengthening the capability of the Panchayats and local bodies and for capacity building on ex-situ conservation in agro biodiversity hotspots. The 16th meeting of the Authority held on 23rd May, 2012 has already made the decision to support BMCs recommended by the National Biodiversity Authority. Annexure-I of the 16th meeting minutes of the Authority has given the guidelines for supporting the conservation, sustainable use of Plant Genetic Resources and capacity building. The National

---

biodiversity governance demonstrating to the rest of the world, the manner in which access to biodiversity can be regulated with assured monitoring for equitable sharing of benefits.

[7] The minutes of the Authority meetings are available at: http://www.plantauthority.gov.in/minutes.htm.

Biodiversity Authority has to recommend the BMCs which satisfy the criteria and a fund ranging from 1.00 lakh to 3.00 lakhs can be allotted through State Biodiversity Boards to the concerned BMCs. It is also welcoming that the Project Appraisal Committee chaired by Dr B. S. Dhillon has recommended the capacity-building programmes for Panchayats and local bodies in agro biodiversity "hotspots" of the country and the funding also has been proposed.[8]

Balaravi and Suchithra (2003) discussed that no law by its mere enactment and enforcement benefits all stakeholders. Stakeholder capability in understanding the law and using it to his/her advantage may help in achieving the goals of the legislation. The present awareness and capacity-building programs have to be also redefined to go into the programmes of the BMCs instead of spending Crores of rupees through the existing channels. Regarding the reach of the Authority, it has 5 regional offices, one each in Ranchi (Jharkhand), Guwahati (Assam), Palampur (Himachal Pradesh), Pune (Maharashtra), and Shivamogga (Karnataka). The Authority had envisaged much larger activities during the 11th plan period and had proposed the opening of 5 Regional Offices in different parts of the Country in its 2nd meeting itself held on 20th December, 2006. The New Indian Express (2012) reported that the Authority has announced its South Indian regional office in Kerala. The report continues that, though initially the branch office was planned to be set up in Coimbatore, it was later shifted to Thrissur (Kerala) after consultations and the office would cater to the needs of Kerala, Karnataka, Tamil Nadu and Andhra Pradesh and the Union Territory of Puthucherry. Differing from this, in the 17th meeting of the Authority held on 19th October, 2012, the members of the Authority have approved the opening of branch offices Thiruvananthapuram (Kerala) and in Palampur (Himachal Pradesh).

It was also hoped that the Expenditure Finance Committee of the 12th Plan might include this proposal. Sri. Narayanan Unny, the then member of the Authority from Kerala has been also appreciated and acknowledged for his help in identifying space for the branch office at Thiruvananthapuram. Unfortunately, these decisions have not been materialized yet. It is argued here that the farmers' communications have to reach the authority (and vice versa) in better ways either through the regional offices or through an alternate mechanism arranged through the State Governments. The basic idea is to increase the opportunity for farmers to interact better with the PPV&FR Authority and help them to ease their access to the Authority and the provisions of the Act.

9.4.4 Is Deadline for Application is a Death-Line? 'NO', Says the Authority

The time allocated for registering Farmers' Varieties of important crops had passed in 2014 and many of such varieties were left behind. PPV&FR Rule 24 [Registration

[8] From the Newsletter for Intellectual Property Rights in Plant Varieties 2(1), April-June, 2014, PPV&FR Authority (http://plantauthority.gov.in/pdf/Newsletter05.pdf).

of extant plant varieties under sub-section (2) of section 15-(1)] provides that the extant varieties are to be registered within three years from the date of their notification under the Act. Later, "*keeping in view the less number of applications filed by farmers probably due to lack of awareness and understanding of the process of filing applications*",[9] duration of filing of farmers' varieties under extant category was increased from three years to five years vide Gazette notification No. 783(E) dated 27th October 2009″. The amendment which had come to extend the time for 10 more years from the original deadline is welcoming. For 17 crops including Rice, Farmers' Varieties can be registered until the date 01.07.2024.

9.4.5 The Extent of Realization of the Act

21st meeting of the Authority held on 31st October, 2014 has decided that the facilitators of applications may be suitably rewarded (with recognition and cash) for facilitating the filing of applications, encouraging them on their efforts and also motivating others to come forward for such activities. This kind of a move will only contribute to the flooding of applications at the registration desk but not to realizing the rights conferred by the Act. Out of the 4,457 registered varieties, only 1810 are farmers' varieties (as of 14.07.2021, PPV&FRA 2021).

Under the applications in the category of Farmers' Varieties, there may not be more than 50 applications from Kerala, of which 27 are by SEED CARE, Wayanad.[10] From these figures, it is assumed that the provisions of the Act have not fairly reached the farmers of Kerala. It is observed here that, if we go in a hurry to register farmers' varieties in a country like India, the primary centre of origin of different crops, thousands of farmers' varieties may keep out from the very needful protection. Section 28 (1) of the Act also needs to be read with this, which deems the Central or State Governments as the owners of seed rights on extant varieties, if the breeder or his successor fails to establish their rights. Prajeesh (2015) mentioned the concerns regarding the annual maintenance fee prescribed for the farmers' varieties registered which stands sorted out with effect from 2016. A study conducted with 401 farmers of Kerala by Blakeney et.al. (2020) has reported that most of the farmers were either unaware of the legislation, or unaware of its functions. The farmers were not much involved in the registration of farmers' varieties and have not made any benefit-sharing claims in relation to the varieties which got registered under PPV&FR Act. Farmers also have confusion regarding the functions of the registration of Geographical Indications, they observe. Though most of the varieties registered have been published for inviting claims of benefit sharing under subsection 1 of section 26 of the PPV&FR Act, none of them were opposed or no claims have been received so far, it seems,

---

[9] As given in the minutes of the 12th meeting of the Authority held on 18th February, 2011.

[10] Unfortunately, there is no option in the authority's website for tracing the total number of applications for registration unless having the exact information like the acknowledgment number, name of the applicant etc.

a search through the Plant Variety Journal reveals (search done from the Authority webpages in July 2021).

9.5 Access and Benefit Sharing in Case of Indigenous Crop Varieties: The Provisions and Prospects

Section 26 of the PPV&FR Act assures the breeders & farmers who provide genetic resources for developing new varieties, a fair share of benefit from the commercial gains of the registered varieties. Meenakumari and Rana (2017) has given the following exemptions provided under the Biological Diversity Act (BD Act, 2002) to promote the bonafide use of bio resources for research and non-commercial use:

- Provisions of section 6 of BD Act, 2002 shall not apply to any person making an application for any right under the Protection of Plant Varieties and Farmers' Rights Act, 2001.
- Accessing bio resources for conventional breeding or traditional practice in use in any agriculture, horticulture, poultry, dairy farming, animal husbandry, beekeeping, etc. in India is exempted from provisions of this Act. However, "End Uses" of biological resources for "Commercial Utilization" (such as drugs, industrial enzymes, food flavours, fragrance, cosmetics, emulsifiers, oleoresins, colours, extracts and genes used for improving crops and livestock through genetic interventions are not exempted.
- The exchange of designated accessions of genetic resources of crops listed in Annex-1 of the ITPGRFA has been exempted from provisions of Sections 3 and 4 of the BD Act.
- Unlike the bilateral system of agreements under CBD for accessing PGRs, the ITPGRFA provides for a Multilateral System (MLS) of ABS to facilitate the exchange of PGRs which are in the public domain, covering initially only 35 food crops and 29 genera of forages. A notification has been also issued exempting the exchange of designated accessions of genetic resources of crops listed in Annex-1 of the ITPGRFA from the provisions of Sections 3 and 4 of the BD Act for research, breeding and training purposes.

9.6 Conclusion

The Protection of Plant Varieties and Farmers' Rights Act of India is to protect the concerns of both farmers and breeders over the varieties. Being a 'progressive' one that recognizes Farmers' Rights extensively, the Indian legislation has to inspire the farmers to continue their efforts in conserving, cultivating and breeding the varieties. Shalini (2015b) has viewed that the word 'progressive' can be misleading since the idea of exclusive economic rights on seed is far from progressive and situating farmers' rights within an IP law is a step backward. The progressive law on IP on seed does not become any better if some farmer communities begin to endorse it, or until the farmer conservers getting benefitted of the provisions of the Act. She has also noted that the farmer movements have started discussing 'seed sovereignty' than mere 'seed rights' but the legal systems are yet to. Kochupillai (2011) observes that, as a pioneer and role model in the protection of farmers' rights, India is duly bound to closely monitor the effectiveness of this regime, not only for the benefit of its own large farming community, but also for the benefit of the farming communities in other developing countries. It should be the concern of the Government to ensure Farmers' Rights and the farmers should not be put in an obligation to go after the legislation which has administrative or procedural complexities. In a detailed case study done by Ramanna (2006), the main barriers to the realization of the Farmers' Rights in India have been discussed. The 'bottom-up' approach was suggested as the key strategy to improve the situation. It was also observed that the coordination between various legislations and various bodies are essential for this. Defining Farmers' Rights as development rights to securing economic and social gains for farmers is also need to think of.

It is hoped that the Farming communities/Groups and Individual farmers should be provided with separate certificates of varietal registration in which the nature, identity and sex of the right holders and other legal terms and conditions are clearly mentioned. The certification of registration should also distinguish farmers and breeders. Importantly, the infrastructure for implementing the PPV&FR Act needs to be strengthened with the cooperation of State governments, regional agencies or bodies such as Biodiversity Management Committees. Until these concerns are taken care of, imposing a deadline for registering farmers' varieties shall not be advisable. As Shalini (2015a) observes, the Act needs constant citizen vigilance to keep it 'pro-farmer' in its implementation because compliance with can itself become a burden on the small and marginal peasants, which is becoming increasingly evident and the big concern should be to bring forth a change in the mindset of the small farmers and seed savers' groups who view the current intellectual property regime with skepticism and continue to keep away from it. The next level challenge which needs to be discussed loudly with regard to the registered varieties will be the monitoring of the use of the varieties in breeding, benefit sharing thereof and other proceedings in the events of infringement of rights. As Lushington (2012) observes, we must continue to analyze the impact of the registration of farmers' varieties on farmers' livelihoods, and with respect to the social benefits that the right to registration was intended to generate.

Acknowledgements The authors gratefully acknowledge Ms. Mina Swaminathan for her unconditional support to sharpen the perspective; Prof. M. S. Swaminathan for all the encouragement and input towards sketching the history of Farmers' Rights; Drs S. Balaravi, V. Arivudai Nambi, and R. Vedavalli for the relevant helps during the preparation of the manuscript; members of SEED CARE for all the discussions; & the reviewers of this manuscript.

Conflict of Interest The views and opinions expressed in this article are those of the author in his personal capacity. None of the observations, opinions, or conclusions expressed herein are to be attributed to or construed as the views or positions of any organization, association, or company. No conflict of interest.

References

Alzina L, Gabriela D (2007) Gendered production spaces and crop varietal selection: case study in Yucatan, Mexico. J Tropical Geogr 28(1):21–38

Balaravi S (2010) Intellectual property rights in Indian agriculture. In: Yesodharan EP (ed), Compendium IPR and development: a national perspective Thiruvananthapuram: Kerala State council for science technology and environment, pp 11–22

Balaravi S, Padmanabhan S (2003) Protection of plant varieties and farmersrights act: what are farmers rights and why they are important. In: Proceedings of the workshop, Farmers' rights and biodiversity: a gender and community perspective, MSSRF/PR/03/49, M. S. Swaminathan Research Foundation, Madras, pp 49–54

Berhan Tewolde Gebre Egziabher, Sue Edwards (2012) The loss of crop genetic diversity in the changing world. In: Seed freedom, A global citizens 'Report, Vandana Shiva, Ruchi Shroff, Caroline Lockhart (Co-ord.), Navdanya, New Delhi

Blakeney M, Krishnankutty J, Raju RK, Siddique KHM (2020) Agricultural innovation and the protection of traditional rice varieties: Kerala a case study. Front Sustain Food Syst 31–11. https://doi.org/10.3389/fsufs.2019.00116/full. Accessed 5th July 2021

Borowaik C (2004) Farmersrights: intellectual property regimes and the struggle over seeds. Politics Soc 32(4):511–543

Byerlee D, Dubin HJ (2010) Crop improvement in the CGIAR as a global success story of open access and international collaboration. Int J Commons 4(1):1–19

Chandrashekaran S, Sujata V (2002) The Indian plant variety protection act beneficiaries: the indian farmer or the corporate seed company?. J Intell Property Rights 7(6):506–515

Demangue S (2005) Intellectual property protection for crop genetic resources, a suitable system for India, Herbert UtzVerlag GmbH, München, Germany

Food and Agriculture Organization FAO (1989) Extract of the twenty-fifth session of the FAO conference, Rome, 5–23 November 1983

Food and Agriculture Organization (FAO) (2005) Building on gender, agrobiodiversity and local knowledge, a training manual, FAO, Rome

FAO (2019) ABS Elements: Elements to facilitate domestic implementation of access and benefit-sharing for different subsectors of genetic resources for food and agriculture—with explanatory notes. FAO, Rome. 84 pp. Licence: CC BY-NC-SA 3.0 IGO. http://www.fao.org/3/ca5088en/ca5088en.pdf. Accessed 5th July 2021

Gender in agriculture Sourcebook (2009) The World Bank. Food and agriculture organisation and international fund for agricultural development, Washington, DC

GoI (2001) The Protection of plant varieties and FarmersRights Act (No.53 of 2001), The Gazette Of India extraordinary, Registered No. DL-33004/2001. http://plantauthority.gov.in. Accessed on 5th July 2021

Halewood M, Noriega I, Louafi S (2013)The global crop commons and access and benefit-sharing laws examining the limits of international policy support for the collective pooling and management of plant genetic resources. In Halewood M, Noriega I, Louafi S (eds) Crop genetic resources as a Global Commons, Routledge, Abingdon UK

Hanchinal RR, Agarwal RC, Prakash R, Stephen T, Jaiswal J (2014)Impact of awareness programmes and capcity building in farmers, plant variety registration under the PPV&FR Act. J Intell Property Rights 19:347–352

Hasan M, Abdullah HM (2015) Plant genetic resources and traditional knowledge: emerging needs for conservation. In: Salgotra RK, Gupta BB (eds) Plant genetic resources and traditional knowledge for food security. Springer, Singapore, pp 105–120
Howard, Patricia (2003) Women and plants, gender relations in biodiversity management and Conservation, ZED Books, London
Joshi BK, Vernooy R, Chaudhary P (2017) Crop interdependence, adaptation to climate change and the multilateral systems of access and benefit sharing: the case of Nepal. Indian J Plant Genetic Resour 30(3):210. https://doi.org/10.5958/0976-1926.2017.00026.2
Keystone Center (1990) Madras plenary session. Final consensus report of the keystone international dialogue series on plant genetic resources, SecondPlenary Session, 29 January–2 February, 1990 (Madras, India), pp 25–26
Keystone Center (1991) Oslo plenary session. Final consensus report: global initiative for the security and sustainable use of plant genetic resources. Third Plenary Session, 31 May-4 June 1991, Oslo, Norway (Colorado: Keystone Center). https://idl-bnc.idrc.ca/dspace/handle/10625/12710. Accessed on 5th July 2021
Khoshoo TN (1988) Conservation of biological diversity. In: Plant genetic resources—Indian perspective. National bureau of plant genetic resources, New Delhi, India, pp 76–90
Kochhar S (2008) Institutions and capacity building for the evolution of IPR regime in India: protection of plant varieties and farmers' rights. J Intell Property Rights 13(1):51–56
Kochupillai M (2011) The Indian PPV&FR Act, 2001: historical and implementation perspectives. J Intell Property Rights 16:88–101
Kochupillai M (2016) Promoting sustainable innovations in plant varieties, vol 5. Springer, Berlin Heidelberg
Lehmann CO (1981) Collecting European land-races and development of European gene banks-historical remarks. [J Article] Die Kulturpflanze 29(1):29–40
Lushington K (2012) The Registration of plant varieties by farmers in India a status report. Rev Agrarian Studies 2(1):112–128. http://www.ras.org.in/the_registration_of_plant_varieties_by_farmers_in_india. Accessed 5th July 2021
Meenakumari B, Rana RS (2017) Implementing the biological diversity act in India: focus on ABS and agro-biodiversity. Indian J Plant Genetic Resour 30(1):25. https://doi.org/10.5958/0976-1926.2017.00003.1
Menon P (1993) Seeds of protest—farmers take on multinationals. Frontline, pp 106–109
Mittermeier RA, Robles Gil P, Mittermeier CG (1997) Megadiversity. CEMEX, Mexico City (Mexico)
Mittermeier RA, Gil PR, Hoffmann M, Pilgrim J, Brooks T, Mittermeier CG, Lamoreux J, da Fonseca GAB (2004) Hotspots revisited: earth's biologically richest and most endangered ecoregions. CEMEX, Mexico City (Mexico)
Momsen J (2007) Gender and biodiversity: a new approach to linking environment and development. Geogr Compass 1(2):149–162
MS Swaminathan Research Foundation (2002) Implementing farmers' rights for conservation and Utilisation of plant genetic resources in the Asia-pacific region: from legislation to action, M. S. Swaminathan Research Foundation, Madras
MS Swaminathan Research Foundation (2003) Proceedings of the workshop, farmers' rights and biodiversity: a gender and community perspective, MSSRF/PR/03/49, M. S. Swaminathan Research Foundation, Madras
Navdanya (2012) Seed freedom, a global citizens report, Vandana Shiva, Ruchi Shroff, Caroline Lockhart (Co-ord.), Navdanya, New Delhi
Nayar MP (1977) Changing patterns of the Indian Flora. Bullet Botanical Surv India 19:145–154
Nayar MP (1996)Hot spots of endemic plants of India, Nepal and Bhutan. Tropical Botanic Garden and Research Institute, Palode, Thiruvananthapuram, 252 p
Nayar MP, Singh AK, Nair KN (2009) Identifying hotspots of agricultural biodiversity. In: Agrobiodiversity hotspots in india: conservation and benefit sharing. Vol 1, i-xii+ total p 217

Prajeesh P (2015) 'Farmers' rights to seeds, issues in the Indian law'. Comment Article Econ Political Weekly L(12):16–18

Parameswaran P (2017) India lays cornerstone of biodiversity access and benefit sharing system. Current Sci 112 (1):24–28

Parthasarathy M (2003) 'A gendered critique of the plant varieties protection and farmers' rights act, draft rules and related legislation'. In: Proceedings of the workshop, farmers' rights and biodiversity: a gender and community perspective, MSSRF/PR/03/49, M. S. Swaminathan Research Foundation, Madras, pp 39–48

Pistorius R (1997) Scientists, plants and politics: a history of plant genetic movement. International plant Genetic Research Institute, Rome

PPV&FRA (2021) Protection of plant varieties and farmers' right authority. http://plantauthority.gov.in/functions-awards. Accessed 5th July 2021

Ramanna A (2006) 'The farmers rights project—background study 4 farmers rights in India: a case study', FNI report 6/2006, Farmers' Rights Project, Fridtjof Nansen Institute, Lysaker, Norway

Rana RS (1993) India takes a lead in conservation of plant genetic resources. Indian Farming 43(7):37–44

Randriamaro Z (2006) Gender and trade (Brighton, U.K.: BRIDGE/Institute of Development Studies). www.bridge.ids.ac.uk/reports/CEP-Trade-OR.pdf. Accessed 5th July 2021

Rane W (1993) Farmers. Rally against GATT Proposals', Economic and Political Weekly 28(44):2391

Salazar R, Louwaars NP, Visser B (2007) 'Protecting Farmers' new varieties: new approaches to rights on collective innovations. In: Plant genetic resources. World Dev 35(9):1515–1528

Seed Association of India (1990) Plant variety protection pros and cons, proceedings of the second national seed seminar 13–14 March 1989. New Delhi, Seed Association of India

Sehgal J, Mandal DK, Mandal C, Vadivelu S (1992) Agro-ecological regions of India. Second Edition, Tech. Bull. No. 24, NBSS and LUP, 130 p

Shalini B (2015a). 'Pay up, or give up your IP!'http://indiatogether.org/plant-variety-protection-law-and-seed-ip-costs-for-farmers-agriculture. Accessed 5th July 2021

Shalini B (2015b) Seed Sovereignty. Lett Econ Political Weekly L(21):4

Shiva V, Radha HB (1993) Protection of plants, people and intellectual rights—proposed amendments to the draft plant varieties act, 1993, Research foundation for science, technology and natural resource policy, New Delhi

Srinivasan CS (2003) Exploring the feasibility of farmers' rights. Dev Policy Rev 21:419–447

Swaminathan MS (1994a) A plant variety protection system for India. In: Ramachandriah V (ed) GATT accord: India's Strategic Response, Commonwealth Publishers, New Delhi

Swaminathan MS (1994b) Draft plant varieties recognition and protection act: rationale and structure. In: Ramachandriah V (ed) GATT accord: India's Strategic Response, Commonwealth Publishers, New Delhi

Swaminathan MS (1995) Draft plant varieties recognition and protection act: rationale and structure. In: Swaminathan MS (ed) Reaching the unreached: farmers' rights and plant genetic resources—recognition and reward: a dialogue, McMillan India Ltd, Madras

Swaminathan MS (1996a) Agrobiodiversity and farmers' rights; proceedings of a technical consultation on an implementation framework for farmers' Rights, M. S. Swaminathan Research Foundation, Madras

Swaminathan MS (1996b) Agrobiodiversity and farmers' rights; proceedings of a technical consultation on an implementation framework for farmers' Rights, Konark Publishers, New Delhi

Swaminathan MS (2002a) The protection of plant varieties and farmers' rights act: from legislation to implementation. Current Sci 82(7):778–780

Swaminathan MS (2002) 'The protection of plant varieties and farmers' rights act: from legislation to implementation. J Intell Property Rights 7:324–329

The New Indian Express (2012) 'Farmers' rights authority office to be opened in Kerala soon, 25 June. http://www.newindianexpress.com/states/kerala/article550182.ece. Accessed 5th July 2021

Tellis O (1993) 'Indian farmers' movement: KissaKisanKa, The Sunday Observer, March 14–20, 1993

Vanloqueren G, Baret PV (2009) How agricultural research systems shape a technological regime that develops genetic engineering but locks out agroecological innovations. Res Policy 38(6):971–983

Vavilov NI (1935) Theoretical basis for plant breeding, moscow. origin and geography of cultivated plants. The Phytogeographical basis for plant breeding (D. Love, transl.). Cambridge Univ. Press, Cambridge, UK 1:316–366

Vedavalli L, Anil Kumar N (1998) Case Study; Wayanad, Kerala. In: Swaminathan MS (ed) Gender dimensions in biodiversity management. Konark Publishers, New Delhi, pp 96–106

Chapter 10
Insects and ABS

Chandish R. Ballal

Abstract Most people are oblivious to the innumerable and indispensable roles played by insects in pollination, decomposition, dung burial, pest control and supporting animal and human life. Thus, considering the importance of insects and the fact that insect diversity faces the same ecological threats as all other biodiversity, the need to conserve insect diversity can never be over emphasised. Under the Convention on Biological Diversity and Nagoya Protocol, the concept of Access and Benefit Sharing (ABS) was to ascertain a country's sovereign right over its biological resources, to ensure that the endangered and vulnerable species are protected from being exploited and in the case of insects specifically prevent the "beetle mania". Though insects constitute over 50% of all known species, less than 3% of all species listed in the IUCN red list as threatened or endangered are insect species. The lack of complete scientific data on insect species is considered to be a reason for this distortion. While insect taxonomy is undoubtedly one of the most important areas to support documentation and conservation of insect diversity, there are certain hurdles in the form of "taxonomic impediments", the major one being the restrictions imposed on the free exchange of dead specimens between taxonomists in different countries based on the ABS regime. Internationally the accepted taxonomic practice involves the exchange of dead specimens between taxonomists for comparison, which is imperative not only to arrive at the correct identity of the insect species, but also for precise documentation and recommendation of ideal management/conservation strategies. The ABS regime is also seriously hampering the future of classical biological control (CBC) initiatives to manage alien invasive insect pests. CBC is applied world-wide and is dependent on exchange of exotic natural enemies. The main beneficiaries of CBC are the global farmers, as their pest problems can be managed through reduced use of pesticides, resulting in lower pesticide residues in food. Thus, the implementation of ABS regulations, while safe-guarding our rich insect diversity from being illegally subjected to commercial exploitation, should ideally recognize and exempt non-commercial taxonomic research and biological control initiatives, where there are clear benefits for the society, environment and public health.

C. R. Ballal (✉)
Former Director of ICAR-National Bureau of Agricultural Insect Resources, Bengaluru, India
e-mail: ballalchandish@gmail.com

O. V. Oommen et al. (eds.), *Biodiversity Conservation Through Access and Benefit Sharing (ABS)*, https://doi.org/10.1007/978-3-031-16186-5_10

Keywords Access and benefit sharing · Augmentation · Biological control · Classical biological control · Conservation · Insect diversity · Insect taxonomy

10.1 Introduction

Insects form a very large group of organisms as more than 8 lakh species of insects are reported to occur in the world. In India, about 61,000 insect species have been recorded out of which about 20,000 species are endemic to the country, with new species being reported every year. Under the Convention on Biological Diversity (CBD) and Nagoya Protocol, it is clearly established that countries have sovereign right over their genetic resources, which includes insect genetic resources. If these resources have to be accessed and used, agreements have to be established between the country seeking access and the country providing access to share the benefits resulting from the usage of the shared resources (CBD 1993). The original idea behind the concept of ABS was to ensure that the endangered wildlife species are protected and to prevent illegal wildlife trade. Hence the rule generally says that to kill or collect any non-domestic animal will require Government permission. While it is important to keep the endangered species safe, the sharing of dead insect specimens for taxonomic studies and live insects for research and as biocontrol agents also fall under the ABS regime causing serious hurdles in taxonomic and biocontrol research.

It has emerged that insects are equally vulnerable to unplanned human activities and human interference with nature as are bigger animals. For the bigger species of animals, information and funds are available, helping researchers to document and engage in conservation efforts as per need. For the small, "insignificant" insects, if lack of protection and conservation is reported to be pushing them to oblivion, dearth of information is equally responsible. This clearly proves that as a first step, insects have to be identified and documented (with focus on the endangered species), only then can initiatives to protect and conserve them follow. For Insect taxonomic research, insect collections globally from different ecological zones and sharing insect specimens between global taxonomists are essential, thus bringing this area of research under the ABS regime. Similarly, biocontrol research and practice are also under the ABS regime as they involve collection of natural enemies from one country to be utilised in another for managing notorious invasive pests.

10.2 Insects Moving Towards Extinction and Illegal Trade vs. Formal Exchange of Insects

As of July 2017, the International Union for Conservation of Nature (IUCN) lists 343 insect species and 21 sub-species as endangered, while 195 are listed separately as critically endangered. It is a matter of concern that 5.7% of all evaluated insect species are listed as endangered and greater concern that 52%

of insects are in danger (https://www.endangeredspeciesinternational.org/insects7.html). Over-exploitation of many vulnerable species led some Governments of various countries to adopt the Convention on International Trade in Endangered Species of Wild Fauna and Flora (CITES) in 1973 to prevent unregulated international trade of vulnerable species. The treaty entered into force in 1975 and now has 183 Parties. Among the insect species of India, only few butterfly species are included in Appendices of CITES—one species included in Appendix I and 8 species are in Appendix-II (Hegde et al. 2013). Documentation of the species which are endangered and facing extinction is extremely important so that serious steps can be initiated to conserve them. An article entitled "Vanishing butterflies mean a vanishing planet" gives a clear overview on the threat of insect extinction in the Indian context as expressed by various biodiversity experts (http://indiasendangered.com/vanishing-butterflies-could-mean-a-vanishing-planet/). It is a matter of great concern that more than a hundred Indian butterfly species face extinction and as per a survey conducted in the Sunderbans by the Zoological Survey of India, there is also a drastic reduction in the number of honeybees. Considering the fact that insects play very important roles as ecosystem service providers, conservationists and funding agencies have to focus on these small sized species as much as they do on the mega fauna. There are several reasons for insect extinction. While it is a matter of concern that some of the insect species like butterflies and beetles are traded internationally for their aesthetic value, there are other factors which contribute significantly to the decline in populations of insect species. These factors include habitat destruction and over-exploitation, over use of chemical pesticides, environmental pollution and advertent or inadvertent introduction of exotic species and human interference is primarily responsible for these factors. Human activities have impacted the landscape globally through fragmentation, degradation and destruction of natural habitats and the creation of new anthropogenic ones (Cardoso et al. 2020; UNEP 2021).

To emphasise on the importance of insects and the need to conserve them, quantified data should be available on the ecosystem services they provide. For eg. FAO has documented that around 70% of all food crops are pollinated by insects and in Europe this proportion is recorded as 84%. It is important for India to generate such data to project the importance of conserving Indian insect diversity and to drive home the point that insect extinction can have catastrophic effects on human beings. For example, data has indicated that in the absence of animal pollination there is a reduction in agricultural production to the tune of 3% to 8%, indicating that agriculture is highly pollinator dependant. In India, due to lack of quantified data, we are not able to ascertain the direct threat on crops yields due to decrease in pollinator numbers. The fact that most of the wild pollinators are highly sensitive and susceptible to habitat degradation and fragmentation clearly indicates the need to focus on habitat conservation to conserve wild pollinators too.

Thousands of insect species are estimated to have vanished from the earth's surface, for instance in the USA, 160 species are recorded as extinct (https://www.endangeredspeciesinternational.org/insects7.html). Some of the extinct species include the Atossa fritillary butterfly (*Speyeria adiaste atossa* (Edwards)) from southern California mountain ranges, the Rocky Mountain grasshopper (*Melanoplus spretus*(Walsh)) which used to migrate from the northern Rocky Mountains to Canada and the Xerces Blue butterfly (*Glaucopsyche xerces*(Boisduval)) from the sand dunes of SanFrancisco Bay area. The causes of extinction have generally been ascribed to human activities viz. habitat destruction, urban development and pollution. However, as diversity assessments have not been systematically done in India for insects, we do not even hold a draft list of extinct insect species.

The list of earth's endangered insects include some of the species, which are reported in India eg. a butterfly called the Andaman Crow *Euploea andamanensis* Atkinson, dragonfly *Burmagomphus sivalikensis* Laidlaw, pygmy hog sucking louse *Haematopinus oliveri* Mishra and Singh, Himalayan relict dragonfly *Epiophlebia laidlawi* Tillyard, 1921. In the redlist of 2018, only the pygmy hog sucking louse is included from India. The pygmy hog-sucking louse, is acritically endangered species, which is an ectoparasite on another critically endangered species, the pygmy hog (*Porcula salvania* Hodgson). It is endemic to India and now restricted to some parts of north-western Assam (Gerlach 2014). However, recent reports indicate that many more insects need to be included in the list. For eg. The Indian Moon Moth *Actias selene*Hübner, which was found in large numbers earlier is a rarity now. This gradual disappearance of the species is ascribed to habitat loss due to destruction of its natural environment (http://indiasendangered.com/rare-species-of-indian-moon-moth-spotted-in-trichy/). A recent article in Down to earth (https://www.downtoearth.org.in/news/wildlife-biodiversity/red-list-assessment-of-indian-grasshoppers-to-kickstart-74646) reports that the Grasshopper Specialist Group of the International Union for the Conservation of Nature (IUCN) has initiated the Red List Assessment of grasshoppers in India for the first time, which would also include *Tettilobus trishula* Skejo, Bhaskar et Stermšek, the new species of grasshopper discovered in the Eravikulam National Park in Kerala's Idukki district. Indian taxonomists should focus on red list assessments of other groups of Indian insects too. It would be extremely important for India to prepare a document on the extinct species of insects, the endangered (or vulnerable) ones and the critically endangered ones and get them included in the red list. Butterfly species like Queen Alexandra's Birdwing and Snow Apollo can be included in the list of wildlife species so that they are protected. The ABS regulations can then focus on conserving specifically such species and preserving the habitats, which house such species of concern.

Insects form a very large group of organisms as more than 8 lakh species of insects are reported to occur in the world. In India about 61,000 species have been recorded out of which about 20,000 species are endemic to the country. Many new species of insects are being recorded every year. Habitat destruction, over-exploitation, pesticides, pollution and introduction of exotic species are also contributing to the decline in many insect species. Thus, a fact which has to be accepted by Policy makers and regulatory bodies is that for the protection and conservation of insect species, the primary focus should be on protection of the habitat.

Some of the insect species like butterflies and beetles are traded internationally for their aesthetic values, as pets and also as food. To control illegal wildlife trade, regulations have been framed in several countries, which are unique for each country.

For eg. in India, you need Government permission to kill or collect non-domestic animals including insects, especially from biodiversity conservation sites like forests and national parks. In Japan, butterfly collection is legal, except for a few protected species. In some countries, insects and other arthropods are captive bred or sold in accordance with the law, however, in several countries, "bug smuggling" is an illegal multibillion-dollar industry. Insects, which are highly vulnerable to smuggling, are traded through illegal export and import procedures (Hall 2019). The removal of insects in large numbers from ecosystems can have a detrimental effect on pollination, food chains and nutrient recycling, while mass collections of even a single species would have a highly significant negative impact on the ecosystem. The Wildlife Protection Society of India has formally reported around 24 cases of insect poaching during the period 1995–2008, while there could be many more unreported cases. The most vulnerable states/regions in India to insect poaching are the eastern Himalayas, West Bengal, Sikkim, the Northeast and Kerala. However, there are reports which state that collections of insects purely for research have been misinterpreted as illegal trade. The reports state that entomologists from a University in eastern Europe collecting mosquitoes for their research on identifying a particular group of mosquitoes causing dengue (Hołyński 2016) and Czech entomologists collecting beetles and butterflies from a National park in West Bengal (Nature 2008) were branded as law breakers.

During earlier days, formal yet free bilateral/multilateral exchanges between Government research organisations were practised, considering any urgent research or pest management need of a country. The emphasis was then on the exporting country providing a phyto-sanitary certificate with the consignment of insects and the receiving country holding a valid import permit. The issue of ABS was not addressed till the Biological Diversity Act, 2002 and Biological Diversity Rules, 2004 (http://nbaindia.org/uploaded/act/BDACT_ENG.pdf) were brought into force. In India bilateral/multilateral exchanges of dead insect specimens for taxonomic research was the accepted practice. There have also been bilateral exchanges of live insects, for eg. between India and Kenya (*Trichogramma mwanzai* Schulten & Feijen received and *Xanthopimpla stemmator* Thunberg supplied by India) and India and Newzealand (*Cotesia kazak* Telenga received and *Helicoverpa armigera* (Hübner) supplied) purely for non-commercial research purposes.

10.3 Insect Taxonomic Studies Under the ABS Regime

International networking is a must for insect taxonomy and globally the accepted taxonomic practice involves the exchange of dead specimens between taxonomists for comparison with types housed in select museums in various parts of the world, which is imperative to arrive at the correct identity of insect species for both new and known species. The validity of descriptions of new species is in doubt unless compared with designated types. World over researchers have acknowledged that the major problem in the sound management of biodiversity is the existence of "taxonomic impediments" and one of the major impediments is the restrictions imposed on the free exchange of dead specimens between taxonomists in different countries based on Biodiversity rules and regulations. These taxonomic impediments do impact our ability to identify, conserve and utilize the biodiversity. There are millions of species of insects, mites and spiders still undescribed and there are far too few taxonomists to do the job, especially in biodiversity-rich but economically poorer tropical developing countries. Besides, a large number of museums in developed countries hold most specimens (originally from developing countries), as well as associated taxonomic information. Smetacek (2013) reported that although 11,300 moth and butterfly species are recorded in India, the museum collections are less than 50% of this number. This is the case for almost all the groups/families of insects. This has negatively impacted both research and conservation initiatives for insects. These days, a common suggestion offered to taxonomists is to depend on internet databases and digital images, which can never be substitutes for real specimens and can end up in mis-identifications. To quote Hołyński (2016) "There can be no true insect taxonomic studies without insect collections and insect specimens as there can be no archaeology without excavations, history without archives, or geography without maps."

Taxonomists in some countries (including India) have to comply with the ABS regime. If the rules under ABS regime is strictly followed, it is impossible for taxonomists to either access insect specimens from other countries or share insect specimens from their collections with other foreign taxonomists without Government approval. This rule also varies from one country to another, with some countries having strict regulations on sharing of dead insect specimens and some with no restrictions.

Indian insect museums (viz., the museums located at University of Agricultural Sciences, Bengaluru; ICAR-Indian Agricultural Research Institute, New Delhi; Tamil Nadu Agricultural University, Coimbatore; Zoological Survey of India (ZSI), Calcutta; Aligarh Muslim University, Aligarh; ICAR-National Bureau of Agricultural Insect Resources (ICAR-NBAIR), Bengaluru) hold a large number of specimens of insect and other arthropod species awaiting discovery and description. The National Insect Museum managed by the ICAR-NBAIR is the national repository for insect germplasm recognised by the Ministry of Environment, Forestry and Climate Change (MoEF&CC) and preserves around two lakh insect specimens. However, based on the national guidelines, the taxonomists at ICAR-NBAIR have to seek approval from the National Biodiversity Authority (NBA) to share specimens with foreign taxonomists. Alternately, the transfer of specimens can be done through international collaborative research projects with the approval of the concerned Departments/Ministries of the Government of India (http://nbaindia.org/uploaded/pdf/Guidelines_for_International_Collaborative_Research_Project.pdf). Point No. 3) under the guidelines for such "International collaborative research projects" states "…….**international collaborative research projects** (approved by the concerned ministry) **have been exempted from obtaining approval of the National Biodiversity Authority**…..", while point no. 4. 2) states "**the collaborators should abide**

by the provisions of existing national laws, regulatory mechanisms and international agreements or treaties…" which results in Government Departments and Ministries again seeking approval from NBA for such international collaborative research projects. Hence, currently, the method followed by majority of the Indian taxonomists is to submit Form B (http://nbaindia.org/content/26/59/1/forms.html) and obtain clearance from the NBA (which is generally granted in 45 days from the date of submission) for sharing dead insect specimens with taxonomists outside India. Formal approvals for a project period of around three years are now in place for the taxonomists from NBAIR and other Indian research organizations, enabling them to share dead insect specimens with some of the leading taxonomists in different parts of the globe. Thus, a small yet firm step has been taken towards tackling one of the taxonomic impediments. Networking between global taxonomists with formal exchange of dead insect specimens has led to strengthening of the NBAIR museum with identified specimens, updating of the databases on all groups of agriculturally important insects, mites and spiders and precise (but may not be timely) identifications of outbreak and invasive pests and potential bio control agents to target them. However, taxonomists face certain challenges while filling in few specific columns in the Form B , for eg. The Indian taxonomist has to indicate the number of specimens that will be sent to the foreign taxonomist, which is difficult to spell out clearly as the number would depend not only on earlier collections, but also on future collections from different ecosystems and also on how many specimens need identity confirmation. Besides, the Form B has to be accompanied by a written declaration by the foreign taxonomist which says "The said biological resources shall be destroyed in full after the completion of the studies/ partnership or upon completion of the studies the biological resources shall be sent back to the institution from where the biological resources were received as the case may be". Though no taxonomist would think of destroying precious insect specimens, even sending back all the specimens may not always be acceptable, especially when the Indian and foreign taxonomists jointly describe a new species and the universally accepted method is for the types to be preserved in both the museums.

ICAR-NBAIR, being the **recognized national repository for insect germplasm** in India, is responsible for the import, export and exchange of insects. Hence as per the NBA guidelines, prior to sending the material outside the country, the vouchers of the insects which are to be shared with any individual or organization outside India (after due approval from NBA), has to be deposited with NBAIR (in accordance with section 39 of the act). Further, NBAIR has to inform NBA regarding the deposition of vouchers. An open access database has been hosted on NBAIR website (https://databases.nbair.res.in//NBA/index.html), which lists out the names of researchers who have deposited voucher specimens at NBAIR museum along with additional details on (a) the organization to which the researcher belongs (b) the date of Form B approval received from NBA and the reference number (c) number of vouchers deposited and the taxonomic details of the voucher specimens (d) the registration number provided by NBAIR for the vouchers and (e) the date on which the information regarding the submission of vouchers has been conveyed to NBA. Since ICAR-NBAIR is recognized by MoEF & CC as the national repository for insect germplasm, and since the voucher specimens have to be deposited at NBAIR, the ideal system would be for this National Bureau to function as an extended arm of NBA and to function as the approving authority for the export of dead insect specimens, when it is purely for taxonomic research, with no commercial interest or monetary benefit emerging through this export.

Thus, understanding species identity of insects that we see and the interactions between diversity components and the mechanism underlying biodiversity effects would also strengthen our efforts at documenting our insect diversity and fine-tuning conservation and management initiatives. Misidentifications of either insect pests or their natural enemies have resulted in setbacks to agricultural production through adoption of wrong crop protection practices in various parts of the world, thus underscoring the fact that insect taxonomic research through international collaborations needs strong national support.

10.4 Biological Control of Insect Pests and Weeds Using Beneficial Insects Under the ABS Regime

Globally biological control strategies through augmentation, classical or conservation approaches have resulted in significant successes. India too has several bio control success stories to its credit (Singh 2004; Ballal and Verghese 2015; https://www.nbair.res.in/success-stories). Van Lenteren et al. (2011) have compiled the information on the global bio control introductions. Around 7000 introductions of bio control agents (BCAs) have been made from around 119 countries into about 146 countries involving almost 2700 BCAs. Over the last 20 years, the introduction of BCAs was based on national legislation of each country and ISPM3 (International Standards for Phytosanitary Measures No. 3) of the IPPC (International Plant Protection Convention), which specifies the responsibilities of the different players, but does not address the issue of ABS. The developed countries have taken up several classical biological control initiatives, besides being the source of bio agents. The developing countries have contributed more bio agents than they have received. When an invasive pest enters a country, in augmentative biological control, the option is to first look for effective indigenous natural enemies (as was done in India when the invasive Sugarcane Woolly aphid *Ceratovacuna lanigera* Zehntner was reported in 2005 and when the Fall Army Worm *Spodoptera frugiperda* (J. E. Smith) invaded our maize crops in 2018) and conserve them and if required augment their populations. Several very efficient indigenous bio agents were identified in India for some of these invasive pests (Joshi and Viraktamath 2004; Srikanth et al. 2012; Shylesha et al. 2018; Ballal et al. 2021). Thus, India becomes a source country for very promising bio agents, which may be required by other countries facing the onslaught of the same pests.

Besides being cost effective, biological control is an environment-friendly strategy, which also conserves biodiversity unlike chemical control strategy. The major emphasis in the augmentative and classical biological control (CBC) strategies is to reduce pesticide use, thus ensuring profitable export market for the agricultural produce with minimum chemical pesticide residue and a significant influx of foreign exchange in developing countries.

Classical biological control addresses invasive alien species which attack agriculture, forestry and the environment. When a country is faced with the onslaught of an invasive insect pest, in the CBC approach, the first step is to explore for effective natural enemies in the country of origin of the pest and this necessitates researchers to conduct surveys (which can be joint surveys by both the countries) to search for natural enemies in another country, which is followed by identifying the most effective natural enemy (which is done in the source country), export of the natural enemy (ies) (for identification and evaluation against target pest), quarantine testing for efficacy against the target pest and possible non-target effects on beneficials (by the importing country), and finally field releases against the target pest in the importing country. Here, joint exploratory surveys and bio control project implementation by bio control researchers of two countries benefit both the countries in terms of standardising survey methods, understanding biodiversity, capacity building and strengthening of research capabilities. Classical bio control programmes have been very successful when there is strong collaboration between the countries.

The current ABS regime restricts the exploration for natural enemies in another country and sharing of exotic natural enemies. The hurdles faced by bio control researchers and practitioners due to the ABS regime have been presented very clearly by the International Organization for Biological Control of Noxious Animals and Plants (Cock et al. 2009; 2010). Silvestri et al. (2020) have reviewed the experiences of six countries-Argentina, Brazil, South Africa, the USA and Canada in implementing ABS regulations relating to classical biological control of weeds. As provider countries and signatories of CBD and NP, the first three countries have faced several obstacles in collecting and exporting weed insects from their country to other countries with respect to clauses like PIC, MAT, MTA (under the national ABS regulations), while the USA (not a signatory to CBD and NP) and Canada (signatory to CBD, not to NP) faced unique difficulties in receiving and using bio agents from other countries, in spite of the fact that both the countries are free suppliers of bio agents to other countries. Besides, those countries following the ABS regulations, have categorically stated that it is highly complicated to share (in monetary terms) the benefits accrued through bio control strategies.

When a bio control agent has been successful in managing an invasive in one country, the best and most sensible option would be for the other countries facing the vagaries of the same pest to access the proven bio agent, in other words re-distribute the successful bio agent to all the affected countries. India has benefited through such an approach when the Papaya mealy bug invaded our Papaya orchards. The successful biological control approach adopted for management of Papaya mealy bug in Papua New Guinea by utilising the parasitoid *Acerophagus papayae* (Noyes and Schauff), was repeated in India. Now India has become a source country for this important parasitoid, with a responsibility to provide the same to other affected countries. Another example in the area of classical biological control of weeds is the utilisation of the salvinia weevil, *Cyrtobagous salviniae* Calder & Sands for biological control of salvinia floating fern, *Salvinia molesta* Mitchell in India, Kenya, Malaysia, Namibia, Papua New Guinea, South Africa, Sri Lanka and Zambia following the spectacular success recorded in Australia (Cock et al. 2010).

Currently, there are pending requests to India from other countries for both exotic (which have proved effective and established in the field) eg. *A. papayae, Telenomus remus* Nixon and indigenous parasitoids viz., *Tamarixia radiata*(Waterston), *Goniozus nephantidis*(Muesebeck) and *Cotesia flavipes* (Cameron). If India has to share these beneficial live bio agents with other countries, the two options available right now are either through the Form B approach or through international collaborative project mode with approval from the concerned Government Dept./Ministry. The former cannot be opted as the country receiving the live bio agents will not be able to declare that "the received biological resources would be destroyed or sent back." The latter option may not be feasible when the need is to respond to an emergency situation.

Generally, Government research organisations take up the responsibility of importing a bio agent, quarantine screening for non-target effects if any and releasing it as per the national regulations. If and when the bio agent establishes in the field and spreads (even to neighbouring countries), the organisation or the researcher has no control over the bio agent. Thus, classical biological control aims for a "free-of-charge public good" (van Lenteren et al. 2011). The bio control sector does not claim IPR rights or monetary benefit based on the benefits accrued to the farmers and consumers. The complete information on the agents imported, the results of evaluation trials and the stories on successes and failures are all made available on public platforms by several countries (as in India, eg. https://www.nbair.res.in/success-stories) so that other countries can take advantage of the strategy and the agent.

Bio control researchers do not enjoy any direct profits through the collection and utilisation of exotic natural enemies, while the global farmers are benefited. In the case of classical biological control, inoculative releases of the natural enemies by the importing organisation lead to their spread and management of the target pest, without the farmer having to intervene. The farmers and the consumers are additionally benefited through the reduced use of pesticides and lower pesticide residues in food, besides long-term economic benefits.

The long-term benefit of a biological control programme supported by ICIPE, Kenya, to target the cereal stem borers in East and Southern Africa was evaluated by Midingoyi et al. (2016). The major parasitoid which was introduced from different countries was the larval parasitoid *C. flavipes* (with India then being one of non-commercial suppliers of the parasitoid), along with another larval parasitoid *Cotesia sesamiae* (Cameron), the egg parasitoid *Telenomus isis* (Polaszek) and the pupal parasitoid *X. stemmator.* The estimated net benefit over a 20-year period was $272 million. This bio control programme was responsible for lifting 57,400 poor people per year out of poverty in Kenya, 44,120 in Mozambique and 36,170 in Zambia.

The biological control agent *Apoanagyrus lopezi*(DeSantis) was introduced into sub-Saharan Africa by IITA from South America to tackle the cassava mealybug. Zeddies et al. (2001) calculated the costs and benefits of this programme for a period of 40 years for 27 African countries. It is estimated to have saved $26/ha, with resultant $9 billion in benefits, and a benefit–cost ratio of about 200:1. In 2020, the cassava mealybug entered India and is currently creating havoc in the states of Kerala and Tamil Nadu. Based on the African success story, the solution can only be through importation and release of the exotic *A. lopezi* and the encouraging fact is that IITA, Benin is ready to share the parasitoid with India. At this juncture, we cannot ignore the nudges from other countries reminding us of our responsibility to reciprocate.

The economic returns through a classical biological control programme for the invasive papaya mealybug infesting five crops in just one Indian state, Tamil Nadu were estimated by Myrick et al. (2014). Within five months of releasing the introduced (from Puerto Rico) parasitoid *A. papayae,* the pest was controlled, chemical pesticide usage reduced and production and income of farmers increased. The annual economic benefits was estimated at $121 million to $309 million. However, the actual benefit would be much higher as the economic analysis cannot precisely quantify and take into account the health and environmental benefits, reduction in development of chemical pesticide resistance, reduction of human exposure to chemical insecticides that can lead to serious health problems and prevention of rapid spread of the pest to other parts of the country. This programme was enabled through a strong and healthy international collaboration with USDA-APHIS, thus highlighting the importance of international collaborations for managing an invasive pest through the classical biological control approach.

Augmentation of indigenous bio agents is followed to tackle serious indigenous pests and sometimes invasive pests too. Here, it is essential to establish mass production facilities to rear the insect bio agents. In the case of augmentative biological control, researchers or a company (especially in developed countries) conduct surveys to search for effective indigenous bio control agents for target pests and once the agent is identified, rearing and release techniques are formulated. The companies then sell the bio agents to growers and do generate some profit. However, besides being benefited through effective pest control and higher yield, the farmers get additional benefits of working in a pesticide free environment and higher profits for their pesticide free produce. Large commercial producers of bio agents are located in developed countries, while it is the small and medium sized entrepreneurs who take up commercial production of bio agents in developing countries. The existence of more such production units region-wise would ensure easy availability of bio control agents for farmers, besides creating job opportunities for the local people. An example is that of a study conducted by Guercia et al. (2018), which discusses on the economic feasibility of an augmentative biological control industry in Niger. This study assesses the profitability of the business taking into consideration the costs involved in production and distribution of the parasitoid *Habrobracon hebetor* Say to manage the millet head miner *Heliocheilus albipunctella* de Joannis. This small bio agent production industry assured low prices for the agent, while generating marginal profit for the producers and job opportunities for the villagers of a poor country like Niger, besides spreading a sustainable technology of plant protection in the villages. This study revealed that augmentative releases of *H. hebetor* induced a 22% yield increase per field and led to a monetary gain of \$5.94–\$121.77. The restriction imposed on commercial utilisation of insects through the ABS regulations hinders an approach for sustainable plant protection. The larger augmentative bio agent producers opine that if a natural enemy obtained from a source country becomes a commercially successful BCA, payment of 'royalties' to the country of origin can definitely be considered (Leneteren et al. 2011). However, considering the smaller profit margins (unlike chemical pesticide companies), these companies believe that benefit sharing would be more realistic if it is in terms of collaborative activities and capacity building. In India, Government research organisations and state bio control laboratories are the major producers of insect bio agents and very few commercial units are into production of parasitoids and predators. These agents would become available to farmers if more commercial units are encouraged to take up production and supply of bio agents.

10.5 Problems and Prospects

Around 104 countries are signatories to the Nagoya Protocol and have to comply with the ABS regime. The signatories understand and appreciate the fact that the Nagoya Protocol was born out of the necessity to guarantee the fair and equitable sharing of benefits arising from the utilisation of genetic resources. The major problem is that each country has put in place a different kind of a mechanism to ensure ABS and there is a total lack of clarity in accessing relevant information on the ABS Clearing House and to receive appropriate responses from the National Authorities of each country. Taxonomic research and bio control research and practice are now bound by the regulations of the Nagoya Protocol on ABS, which is being implemented independently by each country (with their own set of regulations). For eg. in countries like USA, UK, Newzealand and Australia, there is no restriction in exporting dead insect specimens for research purposes, while India has to seek NBA approval for the same. Internationally accepted guidelines on exempting taxonomic research from the ABS regime, if formulated would benefit the global taxonomists. Some countries have initiated formulating new legislations especially with respect to access to genetic resources to accommodate biological control, which may resolve some issues or can lead to more confusions. There is a dire need for an internationally acceptable legislation covering the bio control sector, which would enable fast track procedures to approve access to natural enemies, when a country is facing the onslaught of an invasive pest. The International Organisation for Biological Control has developed a biological control standard agreement under the non-commercial research umbrella, which is specifically intended as a guideline for the countries, where no regulation is in place (Martinez and Biber-Klemm 2010). Smith et al. (2018) have rightly put forth a query on what activities can be considered as utilisation, especially since biological control deals with utilisation of the biological control agents, while the benefits through biological control is non-monetary and this strategy is also the best solution to protect our genetic resources from attacks by invasive species. If biological control is recognised globally as a non-commercial strategy to preserve and use genetic resources in an equitable way, the hurdles faced by bio control researchers and practitioners would cease to exist.

To quote Cock (2010), "Biopiracy rules should not block biological control...-Global regime on benefit sharing for genetic resources should take account of non-commercial interests". Mason and Brodeur (2013) have developed a guide to help biological control scientists in Canada to provide naturally occurring biological control agents, while simultaneously providing documentation templates to ensure that the genetic resources were obtained legally and supplied to the recipient country for free use. Besides, "Best practices" for the global biological control network to enable multi-lateral exchange of biological control agents, while simultaneously protecting biodiversity have been provided by this team (Mason et al. 2017). For transfer of germplasm for scientific research and bio control releases, model concept agreements have been prepared by this team separately for countries which are under the ABS regime and which are not, which can be adopted by the countries, based on their status.

10.6 Global Taxonomy Initiative of CBD and "Special Considerations" of the Nagoya Protocol

Under CBD, ABS is defined as the Access and fair and equitable sharing of benefits arising from the "**commercial utilsation" of genetic resources** with provider contracting party, it should be noted that CBD also emphasises on **technical and scientific cooperation and coordination of global taxonomic expertise** evident through the COP decision on a Global Taxonomy Initiative, which is under active implementation now and the publication of Guide to Global taxonomy Initiative by CBD (2008).

Article 8 'Special Considerations' of the Nagoya Protocol reads as follows (https://www.cbd.int/abs/text/articles/?sec=abs-08):

"*In the development and implementation of its access and benefit-sharing legislation or regulatory requirements, each Party shall:*

(a) *Create conditions to promote and encourage research which contributes to the conservation and sustainable use of biological diversity, particularly in developing countries,'* ....
(b) *Pay due regard to cases of present or imminent emergencies that threaten or damage human, animal or plant health, as determined nationally or internationally*...........................
(c) *Consider the importance of genetic resources for food and agriculture and their special role for food security.*"

The above "Special considerations" can definitely be considered to exempt transfer of dead or live insects for non-commercial research and transfer of live insects for biological control practice from strict compliance to ABS regulations. The following set of recommendations can be considered to strengthen and fine-tune ABS regulations.

10.7 Recommendations

- Emphasis on assessment of Indian insect diversity to document the extinct, endangered and vulnerable insect species and inclusion in IUCN red list.
- Strict regulations on preventing mass collections of insects (especially from biodiversity hot spot areas) for commercial trade and export.
- Globally, multilateral exchange of insect specimens and natural enemies to be encouraged for research and bio control practice, which would ensure fair and equitable sharing of non-monetary, environmental and health benefits.
- Identification of a National Focal point in each country, which can (a) provide proper advice on ABS regulations (b) facilitate survey missions by national and international researchers (c) provide information on the endangered insect species which cannot be shared and the live bio control agents and dead insect specimens which can be shared (d) give approval for national and international linkages for taxonomic and bio control initiatives (e) provide emergency response for exchange of bio agents while facing invasive pest attacks.
- Networking between the national focal points of different countries in order to bring in uniformity in the interpretation of the ABS regulations, especially with reference to sharing of insect resources purely for non-commercial research and for biological control practice.
- ABS in relation to taxonomic research and biological control research and practice should be in terms of capacity building, shared research programmes and/or technology transfer and information exchange.
- Preparation of a document by global experts (with inputs from the developed and developing countries) on the ABS guidelines specifically with reference to sharing of dead and live insects for taxonomic research and bio control research and application, which should be applicable globally.
- International collaborative programmes to facilitate information exchange on the availability and source of invertebrate biological control agents and transfer of technologies on mass rearing of bio agents.
- Availability of free knowledge sharing platforms on insect resources and open access databases on bio control agents in each country; eg. BCANZ in NewZealand (Ferguson et al. 2007) and databases on ICAR-NBAIR website (https://www.nbair.res.in/databases).

- Formulation of agreements for non-commercial exchange of bio agents, acceptable by National biodiversity regulators of each country, which can be implemented globally, especially during emergency situations concerning invasive pest attacks.

If ABS regulations are tailored to encompass the above recommendations, it can result in conservation of our rich biodiversity along with discoveries of more new insect species and more success stories in management of notorious pests through non-chemical pesticidal approaches. It is the responsibility of global taxonomists, biodiversity specialists and the bio control researchers and practitioners to understand and appreciate the advantages of the ABS regulations in preventing illegal and commercial exploitation of our insect resources and conserving our insect diversity, especially the endangered species. However, it is equally important for this group to network amongst themselves and with their respective National biodiversity authorities and National ABS contact points to give due attention to the hurdles caused by the ABS regime and to develop a legal framework, that would facilitate the precise identification, documentation, conservation and utilization of insect diversity for the benefit of the global farmer, which is also in the best interests of humanity.

Conflict of Interest The views and opinions expressed in this article are those of the author in her personal capacity. None of the observations, opinions, or conclusions expressed herein are to be attributed to or construed as the views or positions of any organization, association, or company. No conflict of interest.

References

Ballal CR, Kandan A, Varshney R, Gupta A, Shylesha AN, Navik OP, Venkatesan T, Gracy G, Ramanujam B, Rangeshwaran R, Sivakumar G, Nagesh M, Patil J, Amala U, Shivalingaswamy TM, Subaharan K, Bakthavatsalam N, Lalitha Y, Poornesha B, Apoorva V, Ashika TR, Nayyar N (2021) Biological control for fall armyworm management in Asia—case study: India. In: Prasanna BM, Huesing JE, Peschke VM, Eddy R (eds) Fall armyworm in Asia: a guide for integrated pest management. Mexico, CDMX: CIMMYT. (In Press)

Ballal CR, Verghese A (2015) Role of parasitoids and predators in the management of insect pests. In: Chakravarthy AK (ed) New horizons in insect science: towards sustainable pest management. Springer India, Chapter 29, p 307

Cardoso P, Barton P, Birkhofer K, Chichorro F, Deacon C, Fartmann T, Fukushima CS, Gaigher R, Habel JC, Hallmann CA, Hill MJ, Hochkirch A, Kwak ML, Mammola S, Noriega JR, Orfinger AB, Pedraza F, Pryke JS, Roque FO, Settele J, Simaika JP, Stork NE, Suhling F, Vorster C, Samways MJ (2020) Scientists' warning to humanity on insect extinctions. Biological Conser 242:108426. https://doi.org/10.1016/j.biocon.2020.108426.

CBD (2008) Guide to global taxonomy initiative. CBD Technical Series No. 30, 195 p. https://www.cbd.int/doc/publications/cbd-ts-30.pdf

CBD (Convention on Biological Diversity) (1993) Convention on biological diversity (with annexes). concluded at Rio de Janeiro on 5 June 1992. United Nations-Treaty Series 1760(30619), pp 142–382

Cock MJW (2010) Biopiracy rules should not block biological control. NatureSep 23; 467(7314):369. https://doi.org/10.1038/467369a

Cock MJW, van Lenteren JC, Brodeur J, Barratt BIP, Bigler F, Bolckmans K, Cônsoli FL, Haas F, Mason PG, Parra JRP (2009) The use and exchange of biological control agents for food and agriculture. FAO Background Study No. 47, FAO, Rome: 88 p

Cock MJW, van Lenteren JC, Brodeur J, Barratt BIP, Bigler F, Bolckmans K, Cônsoli FL, Haas F, Mason PG, Parra JRP (2010) Do new access and benefit sharing procedures under the convention on biological diversity threaten the future of biological control? Biocontrol 55:199–218

Ferguson CM, Moeed A, Barratt BIP, Hill RL, Kean JM (2007) BCANZ—biological control agents introduced to New Zealand. http://www.b3nz.org/bcanz. Accessed July 2021

Gerlach J (2014) Haematopinus oliveri. The IUCN Red List of Threatened Species. 2014: e.T9621A21423551. https://doi.org/10.2305/IUCN.UK.2014-1.RLTS.T9621A21423551.en

Guercia MJ, Nortan GW, Ba MN, Baouac I, Alwang J, Amadou L, Moumouni O, Karimoune L, Muniappan R (2018) Economic feasibility of an augmentative biological control industry in Niger. Crop Prot 110:34–40

Hegde M, Bhupathy S, Sasidharan KR, Krishnakumar N (2013) Indian reptiles, amphibians and Insects in CITES appendices. Published by: institute of forest genetics and tree breeding (Indian Council of Forestry Research and Education). 15 p

Hall J (2019) Bug smuggling is big business. National Geographic. https://www.nationalgeographic.com/animals/article/bug-smuggling-big-business

Hołyński R (2016) Re: Which countries have banned insect collecting by private persons?. Retrieved from: https://www.researchgate.net/post/Which-countries-have-banned-insect-collecting-by-private-persons/57430c7aeeae391ad360f99c/citation/download

Joshi S, Viraktamath CA (2004) The sugarcane woolly aphid, *Ceratovacuna lanigera* Zehntner (Hemiptera: Aphididae): its biology, pest status and control. Curr Sci 87(3):307–316

Martinez SI, Biber-Klemm S (2010) Scientists—take action for access to biodiversity. Current Opinion Environ Sustain 2(1–2):27–33

Mason PG, Brodeur J (2013) Access and benefit-sharing and biological control. In: Mason PG, Gillespie DR (eds) Biological control programmes in Canada 2001–2012. CAB International Publishing, Wallingford, pp 6–11

Mason PG, Cock MJW, BarrattBIP KJN, van Lenteren JC, Brodeur JA, Hoelmer KA, Heimpel GE (2017) Best practices for the use and exchange of invertebrate biological control genetic resources relevant for food and agriculture. Biocontrol. https://doi.org/10.1007/s10526-017-9810-3

Midingoyi SG, Affognon HD, Macharia I, Ong'amo G, Abonyo E et al (2016) Assessing the long-term welfare effects of the biological control of cereal stem borer pests in East and Southern Africa: evidence from Kenya, Mozambique and Zambia. Agric Ecosyst Environ 230:10–23

Myrick S, George N, Selvaraj KN, Natarajan K, Muniappan R (2014) Economic impact of classical biological control of papaya mealybug in India. Crop Prot 56:82–86

Nature (2008) Entomologists jailed for illegal specimen hunt. Nature: News in Brief 454:1476–4687. https://doi.org/10.1038/454014a

Shylesha AN, Jalali SK, Gupta A, Varshney R, Venakatesan T, Shetty P, Rakshit Ojha, Ganegar PC, Omprakash Navik, Subaharan K, Bakthavatsalam N,Ballal CR (2018) Studies on invasive pest *Spodoptera frugiperda* (J.E. Smith) (Lepidoptara: Noctuidae) and its natural enemies. J Biolo Control 32(3):1–7. https://doi.org/10.18311/jbc/2018/21707

Silvestri L, Sosa A, Mc Kay F, Vitorino MD, Hill M, Zachariades D, Hight S, Weyl P, Smith D, Djeddour D, Mason PG (2020) Implementation of access and benefit-sharing measures has consequences for classical biological control of weeds. Biocontrol 65:125–141

Singh SP (2004) Some success stories in classical biological control of agricultural pests in India. APAARI Publication 2004/2, 73 pp

Smetacek P (2013) Review of Indian Lepidoptera collections and their significance in conservation. ENVIS Bulletin 14:135–139

Smith D, Hinz H, Mulema J, Weyl P, Ryan MJ (2018) Biological control and the Nagoya Protocol on access and benefit sharing—a case of effective due diligence. Biocontrol Sci Tech 28(10):914–926

Srikanth J, Singaravelu B, Kurup NK (2012) Natural control of woolly aphid by *Encarsia flavoscutellum* prevents yield and quality loss in sugarcane. J Sugarcane Res 2(1):64–68
The IUCN redlist of threatened species. http://www.iucnredlist.org/. Accessed July 2021
United Nations Environment Programme (2021). Making peace with nature: a scientific blueprint to tackle the climate, biodiversity and pollution emergencies. Nairobi. https://www.unep.org/resources/making-peace-nature
UNTC (2010) Nagoya protocol on access to genetic resources and the fair and equitable sharing of benefits Arising from their utilization to the convention on biological diversity. https://treaties.un.org/doc/Publication/MTDSG/Volume%20II/Chapter%20XXVII/XXVII-8-b.en.pdf
van Lenteren JC, Cock MJW, Brodeur J, Barratt BIP, Bigler F, Bolckmans K, Haas F, Mason PG, Parra JRP (2011) Will the convention on biological diversity put an end to biological control? Revista Brasileira De Entomologia 55(1):1–5
Zeddies J, Schaab R, Neuenschwander P, Herren H (2001) Economics of biological control of cassava mealybug in Africa. Agric Econ 24(2):209–219

Chapter 11
A Revisit to Access and Benefit Sharing (ABS) with Reference to Aquatic Systems and its Bio-Resources

K. V. Jayachandran

Abstract The Access and Benefit Sharing (ABS) related to Biodiversity is a resource-legal framework. Prior to Convention on Biological Diversity (CBD) in 1992, access to biodiversity and local knowledge relevant to utilization remained free for all mankind. The benefits went to organizations involved in the trade. Benefit sharing had started previously but remained at government level but never reached to stakeholders or custodians. There were no provisions for the fair and equitable sharing of benefits and with this background CBD was articulated in 1992 and came into effect with the ratification in 1993. The CBD has three main objectives: conservation of biological diversity; sustainable use of its components; and fair and equitable sharing of benefits arising from the use of genetic resources. Conservation of biological diversity is clear in action to protect the biodiversity. The protection of genetic resources is a multi-pronged affair and it involves protection of water bodies, the genetic component in it, creating environment for production increase and sustainable use and so on. It means that the benefit sharing should not be limited to genetic resources alone but the environment as well as the custodians. Such aspects have been omitted in the present system. Therefore, the present article is an attempt to expand the field of ABS with regard to aquatic systems and bio-resources to custodians of these resources. The areas are: riparian vegetation for bank protection; protecting water and its availability, quality, retention, renovation etc.; conservation of breeding populations of genetic resources; regulated capture of genetic resources and the people who are real custodians of these activities. All these are primary causes for the conservation of aquatic genetic resources. The industries procuring the bio-resources are producing value added products and making huge benefit by adopting precautious marketing strategies. In spite of regulations declared under CBD the fishermen and other people are not benefitted out of this. The local people who are the custodians are totally neglected from this area, though they are also eligible for such benefit sharing. This article makes a revisit to the problem concentrating on various areas, such as, bank protection, water conservation, genetic bio-resource protection, production (capture), aquaculture, ornamental fish

K. V. Jayachandran (✉)
Department of Aquatic Biology and Fisheries, University of Kerala, Thiruvananthapuram, Kerala, India
e-mail: jayachandrancmlre@gmail.com; drjayachandran@keralauniversity.ac.in

O. V. Oommen et al. (eds.), *Biodiversity Conservation Through Access and Benefit Sharing (ABS)*, https://doi.org/10.1007/978-3-031-16186-5_11

culture, marketing, industrial production of value added products, waste management, products developed from waste, ranching programmes etc.

Keywords Revisit ABS · Aquatic resources · Custodian · Riparian · Value added bio-resources

Biological organisms are integral components in nature. Abiotic and biotic factors are completely interdependent. This is essential for sustainable development. Biodiversity refers to "variability that exists among living organisms from all sources including among other things, terrestrial, marine and other aquatic ecosystems and the ecological complexes which they are part of. This includes diversity within species, between species and their ecosystems". Biological resources comprise of all genetic resources having functional units of heredity in an ecosystem with actual or potential use or value to human welfare, such as, entire organisms or any part. Current definition of genetic material is that any material of plant, animal, microbial or other biotic origin containing functional units of heredity and these form genetic resources and their uses can range from basic research leading to commercial products. The genetic resources can be from "in situ" and/or "ex-situ" conditions as well as domesticated conditions/ habitats.

The potential value of biological diversity and genetic resources is range anywhere between US$ 800 billion to US$ 1 trillion based on various estimates (Balmford et al. 2002). The potential to use these genetic resources are not direct but by employing various channels including research and biotechnological applications. Marketing strategies are also involved in the system. Even wastes generated while utilization of bio-resources are subjected to product development. The convention on Biological Diversity (CBD) is a multilateral international treaty intended to provide national governments with sovereign rights over their genetic resources and associated traditional knowledge. The CBD acknowledges the significant environmental, economic and social value of biodiversity. It also seeks to ensure that the rights of countries and communities over their biological resources and related knowledge are respected, and that the benefits derived from their use are equitably shared. In the process all are included. The CBD thus upholds commitments for maintaining the planet's ecological balance in the context of sustainable development strategies. Cardinal questions now arise here is that how the resources can be accessed? What are the criteria adopted for the access? Do we protect the interest of everybody in the link of source material and for the ecosystem protection? From the actions taken so far it can be seen that there are a large section of people who protect the environment are left out in ABS process. The ecosystems are protected (not by formulated projects and implemented alone) by native people through the adoption of many traditional knowledge and mechanisms which they have developed suitable to an area. Have we adequately addressed or envisaged this aspect in any of the ABS programmes developed so far in any country? How can we achieve wild production and protection of biodiversity without giving emphasis on the issue? The observation is that it has not been stressed anywhere. Suneetha et al. (2009) have given a comprehensive account on the guidelines of ABS adoption for India and may be suitable for other countries also. The present article is giving emphasis to aquatic ecosystems.

Man depended on biological resources initially for food and subsequently utilized for product development. Biodiversity was considered to be a common property till the end of twentieth century. It was in the Earth Submit held in 1992, at Rio that the world for first time thought of sovereign rights of states over their bio-resources and associated native knowledge systems. The local communities developed various methods for exploitation and utilization of these resources. Many of these methods became viable traditional knowledge and were often obtained from them by organizations and individuals who tried to expand it in the production sector and monopolized the benefits. This was practiced by many countries. The communities who are actual custodians of knowledge and even resources were never benefitted out of this. A few countries made legal provisions for benefit sharing from the use of genetic resources in the country of origin. Kerala's (India) kaani experience is an exception (described elsewhere). In such situations also the benefits had never reached the custodians or local community. The reason was obvious that there was no concrete legal provision for the fair and equitable sharing of benefits on a legal framework. Many precious resources under these circumstances had depleted to almost unrecoverable level. The situation was grave and on this background the notion of an international convention on bio-diversity was conceptualized with an Ad Hoc Working Group of Experts at UNEP in November 1988. In the subsequent year, the Ad Hoc Working Group of Technical and Legal Experts was formulated for framing of a legal text aiming to the conservation and sustainable use of biological diversity. It also intended for the sharing of benefits arising from their utilization with sovereign states and local communities. This was the origin of CBD. The meetings of the parties to the convention are known as Conferences of the Parties (COP), with COP 1 held in 1994 in Nassau, Bahamas; COP 2 in 1995 in Jakarta, Indonesia; COP 3 in 1996 in Buenos Aires, Argentinia; COP 4 in 1998 in Bratislava, Slovakia; COP 5 in 2000 in Nairobi, Kenya; COP 6 in 2002 in The Hague, Netherlands; COP 7 in 2004 in Kuala Lumpur, Malaysia; COP 8 in 2006 in Curitiba, Brazil; COP 9 in 2008 in Bonn, Germany; COP 10 in 2010 in Nagoya, Japan; COP 11 in 2012 in Hyderabad, India; COP 12 in 2014 in Pyeongchang, Republic of Korea; COP 13 in 2016 in Cancún, Mexico; COP 14 in 2018 in Sharm El-Sheikh, Egypt and the most recent one (COP 15) yet to be concluded in 2021/2022 in Kunming, China. In October 2010, the tenth meeting of the parties to the convention took place in Nagoya, Japan and Nagoya Protocol was ratified (archives of CBD).

International Year of Biodiversity was in 2010 with CBD as its focal point. Following the recommendation of CBD signatories during COP 10 at **Nagoya**, the UN, on 22 December 2010, declared 2011 to 2020 as the U N Decade on Biodiversity. Strategic Plan for Biodiversity for 2010–2020 and its **20 Aichi Biodiversity Targets** were adopted at COP 10 (named after Aichi, the Japanese prefecture in central Honshu Island which hosted the CBD meeting at which the targets were agreed). The First Extraordinary Meeting of the Conference of the Parties took place at Cartagena in February 1999, and this led to the adoption of **Cartagena Protocol** on

Bio-safety which became effective since 2003. COP 11 was held at Hyderabad, India and started building on the experiences on Global Warming. The meeting of COP 12 adopted the "Pyeongchang Road Map", which focused on the ways to achieve biodiversity conservation through technology cooperation, funding and strengthening the capacity of developing countries. COP 14 adopted "ecosystem-based approaches to climate change adaptation" and disaster risk reduction through voluntary guidelines "Ecological Civilization: Building a Shared Future for All Life on Earth" is the theme for COP 15 and is yet to be concluded.

CBD was finalized in Nairobi in May 1992 and came into force on 29th December, 1993 and has three main objectives: (1) biological diversity conservation, (2) sustainable use of its components, and (3) fair and equitable sharing of benefits arising from the use of genetic resources. All three are interconnected. As per the latest report, 196 parties accepted the CBD and got ratified. It took almost five years for the first countries came up with ABS legislation (the Philippines and the Andean Pact member countries) and subsequently framed legal measures by Brazil and many others. The documents of relevance are touched upon in brief here for the purpose, namely, UNCLOS, TRIPS, OECM, NAGOYA PROTOCOL, CARTAGENA PROTOCOL and AICHI BIODIVERSITY TARGETS.

The **U N Convention on the Law of the Sea** (**UNCLOS**) is otherwise called the **Law of the Sea Convention** or the **Law of the Sea Treaty**, is an international agreement to legally regulate all marine and maritime activities. The role of UNCLOS is to frame principles of international maritime law based on the experience in the ocean related activities. UNCLOS 1 introduced four treaties, namely, Convention on the Territorial Sea and Contiguous Zone (into force: 10 September 1964); Convention on the Continental Shelf (10 June 1964); Convention on the High Seas (30 September 1962); Convention on Fishing and Conservation of Living Resources of the High Seas (20 March 1966). UNCLOS II was attended by only a few parties. UNCLOS III came into force in 1994 replaces the four treatises of 1958. As of June 2016, 167 countries and the European Union are parties. Part XII broadened jurisdictional rights for enforcing international environmental regulation within their territory and on the high seas. The United Nations Sustainable Development Goal 14 additionally has a target regarding conservation of marine ecosystems. **International Seabed Authority (ISA)** also was established based on decision in Part XII and authorized exploration and mining and collect and distribute the seabed mining royalty beyond Exclusive Economic Zone (EEZ). Protection of marine ecosystems is a priority but action plans are only shaping its course of action. How far have we achieved in this respect is a matter of great concern. The **International Seabed Authority** (**ISA**) is located at Kingston, Jamaica is an intergovernmental body with 167 member states and the European Union established under UNCLOS 1982 (1994 Agreement on Implementation). The ISA's mission is: control on mining related activities and protect biodiversity in area beyond national jurisdiction (BBNJ).

Other Effective Area-based Conservation measures (OECM) is a new initiative for safeguarding the diversity of life on the planet through the protection and management of special places internationally (Jonas et al. 2014; Lopoukhine and Dias 2012; MacKinnon et al. 2015). OECM activities include: creating protected areas as the key route to delivering area-based conservation of biodiversity and ecosystem services. IUCN World Commission on Protected Areas established a Task Force in 2015 in terrestrial, freshwater and marine habitats; some generic protection work of areas likely to qualify as OECMs in the ocean were identified, to complement and supplement fisheries and other management measures to promote more sustainable use; proposed fishery managements.

The World Trade Organization Agreement on **Trade-Related Aspects of Intellectual Property Rights (TRIPS)** is the most comprehensive multilateral agreement on intellectual property (IP). It plays a central role in facilitating trade in knowledge and creativity, in resolving trade disputes over IP, and in assuring WTO members the latitude to achieve their domestic policy objectives; which deals with patentability or non-patentability of plant and animal inventions, and the protection of plant varieties. Pushpangathan (2009) proposed the need to amend the TRIPs Agreement. It is far more cost effective in the long run to establish an internationally accepted solution through TRIPs for the prevention of bio-piracy and Traditional Knowledge associated patents.

Cartagena Protocol pertains to movement and handling and use of all living modified organisms that may have adverse effects on the conservation and sustainable use of biological diversity and also risks associated with human health. This concept refers to the need to protect human health and the environment from the possible adverse effects of the products of modern biotechnology. It has 40 articles and 2 Annexure**.** The ABS cannot apply to modified living organisms.

Nagoya protocol consists of 36 articles and an annexure. It promotes the use of genetic resources and associated traditional knowledge. It intends to strengthen the opportunities for fair and equitable sharing of benefits from the use of biological resources. The Protocol is the instrument for the implementation of the convention. It will create incentives to conserve biological diversity, sustainable use its components. It further enhances the contribution of biological diversity to sustainable development and human well-being. Fair and equitable sharing of benefits is in accordance with Article 15, paragraphs 3 and 7 of the Convention, and is restated as "benefits arising from the utilization of genetic resources as well as subsequent applications and commercialization shall be shared in a fair and equitable way with the Party providing such resources that is the country of origin of such resources or a Party that has acquired the genetic resources in accordance with the Convention. Such sharing shall be upon mutually agreed terms". "Each Party shall take legislative, administrative or policy measures, as appropriate, with the aim of ensuring that benefits arising from the utilization of genetic resources that are held by indigenous and local communities, in accordance with domestic legislation regarding the established rights of these indigenous and local communities over these genetic resources, are shared in a fair and equitable way with the communities concerned, based on mutually agreed terms". "Benefits may include monetary and non-monetary benefits, including but not limited to those listed in the Annex." And "Each Party shall take legislative, administrative or policy measures, as appropriate, in order that the benefits arising from the utilization of traditional knowledge associated with genetic resources are shared in a fair and equitable way with indigenous and local communities holding such knowledge. Such sharing shall be upon mutually agreed terms".

Though Nagoya Protocol and Aichi Biodiversity targets are intended for conservation, access and fair and equitable sharing of benefits, the author believes that many vital areas are left out and a revisit may be advantageous. The Nagoya Protocol intends to create greater legal certainty, clarity, transparency and provide fair and non-arbitrary rules and procedures. It stipulates clear rules and procedures for prior informed consent (PIC) and mutually agreed terms (MAT). On this basis permits or equivalent are issued when access is granted. It opens up new vistas of R & D and emphasizes the importance of genetic resources for food and agriculture/ aquaculture/ food processing for food security and safety. The rules are beneficial for both providers and users of genetic resources by establishing more predictable conditions. Here most astonishing is that the actual custodians of environment are not identified and the safeguarding phenomen is never brought under ABS. In aquatic ecosystems water is the critical component and its health is the index of biodiversity existence. Here who is benefitted and who is denied of the benefit is a matter of serious concern.

Aichi Biodiversity targets aim to fulfill 20 targets under 5 strategic Goals by 2020. The 15th UN Biodiversity Conference (COP-15) in Kunming, China is set to endorse the "post-2020 global biodiversity framework" (GBF) for the next 10 years for strengthening CBD in the context of UN Decade of Ecosystem Restoration (2021–2030; https://www.decadeonrestoration.org/), the 26th UN Climate Change Conference (https://ukcop26.org/), the UN Food Systems Summit (https://www.un.org/en/food-systems-summit), and the UN Ocean Conference (https://www.un.org/en/conferences/ocean2020), all are concurrently convened (Rodrigo Sara, 2021/22). Time is of the essence. We are facing a global decline of ecosystems and wild species (Díaz et al. 2019). UN Secretary General Mr. António Guterres proclaimed that "humanity is waging war on nature, and we need to rebuild our relationship with it" and according to the President of the UN General Assembly, Mr. Volkan Bozkir, "Kunming must do for biodiversity what Paris did for climate change in 2015, by elevating the discourse to the mainstream and placing it firmly on the political agenda" (UN Summit on Biodiversity, New York, Sept. 30, 2020). Sir David Attenborough, at the opening ceremony of the 26th UN Climate Change Conference, provided his vision of a possible future: "A new industrial revolution powered by millions of sustainable innovations is essential and is indeed already beginning. We will all share in the benefits: affordable clean energy, healthy air, and enough food to sustain us all. Nature is a key ally. Whenever we restore the wild, it will recapture carbon and help us bring back balance to our planet" (Glasgow, UK, Nov. 1, 2021). Since we have to achieve further in biodiversity conservation, utilization and benefit sharing during the period 2020–2030, some additional and modified action plans are necessary and hence proposed below related to Aichi targets 1–20.

- Each country (different states, districts, taluks, panchayaths, wards and regions (different regions in India and equivalent terminology for other countries may be adopted) may develop a nearly complete data bank (responsibility of scientific community, academia and administration)

- Values of each biological organism in terms of ecosystem service as far as possible may be consolidated (scientific community and academia)
- Importance of the organism in the ecosystem may be scientifically proposed (scientific community and academia)
- Patterns of distribution of biota and prevailing ecological conditions of ecosystems may be arrived at (scientific community and academia)
- Consolidate the values of medium in which the aquatic biodiversity is surviving
- Collect the traditional knowledge available with local communities for conservation of water, prevailing environment, biodiversity, access of biodiversity, breeding of various species for conservation, utilization, strategies of utilization like production of various items, acceptance by society, marketing, waste generation, mitigation strategies and so on
- Make serious awareness programmes by facilitating and adopting and engaging of citizens in biodiversity facts and figures for monitoring, conservation, justifiable and sustainable use (Targets 4 to 15 ad19)
- Scientifically designed case studies on the importance of aquatic biodiversity carried out
- Technically oriented communication programmes may be developed and implemented
- Develop and adopt coherent, strategic and sustained communication systems for effective campaign depending on the type of audience and expected outcomes.
- Impress upon the target groups with nationally or internationally relevant examples with success stories
- Include subjects on biodiversity in the educational curriculum taking into consideration of the motives for sustainable development of biodiversity for human well being
- Assure that Non-governmental institutions are actively participating in the event and make it participatory
- Consolidation of periodic and consistent assessment of communication strategies and document experience generated out of the exercise and suggest relevant improvement in the effort
- Bring suitable and innovative mechanisms of access and benefit sharing to the policies for long term achievement
- The whole programme should enable some beneficial and progressive changes in socio-economic, cultural, behavioral changes in the society for meaningful implementation of access and benefit sharing of aquatic biodiversity.
- Enlisting the existing planned policies evolved across government concerning biodiversity issues.
- Identify opportunities and options for biodiversity conservation, access and utilization
- Biodiversity values and services are to be properly documented and the same may be made use of in decision making (T 19)
- Poverty is to be addressed in a meaningful and most pragmatic way
- Continuous reporting and periodic assessment and necessary strategies and action plans to be formulated for addressing the above.

- Values of biodiversity must be the key point in planning processes and based on this resource management exercises, such as, mapping of biodiversity, analysis of peculiar pattern of distribution, inter-dependence, interaction, production, ecosystem related services (targets 5,6,7) are to be framed.
- Indicator species may be identified and made use of in assessing environmental quality. These are to be documented and made available to general public so that the quality of environment can be monitored by them.
- Scientific assessment of environment must be undertaken in well planned intervals
- Natural assemblages of species in habitats must be maintained
- Evolve biodiversity strategies and action plans for the state (may be for region wise) and for the country
- Basic studies on biodiversity at regional level has to be undertaken
- Analytical mechanism has to be introduced for arriving strategies and subsequent action plans
- Always avoid negative impact factors while implementing the programmes and awareness programmes suggested/ conducted
- Incentives for conservation programmes as a sustainable mechanism. Incentives should generate voluntary conservation behavior among people (Target 2)
- Harmful alien species is to be eliminated and made aware to people
- Developing policy plans, including a prioritized list of measures, with timelines, leading to the eventual removal, phase-out, or reform of harmful incentives, including subsidies, and the introduction, or strengthening, of positive incentives for the conservation and sustainable use of biodiversity (Target 17)
- In cases where candidate incentives and subsidies for elimination, phase-out or reform are already known, taking timely policy action (Targets 6 and 7)
- Making greater use of social incentives (for example, the establishment of awards or recognition programmes promoting behaviors beneficial to biodiversity)
- Better targeting and integration of Aqua-agri-environmental schemes and other policy instruments towards desired biodiversity outcomes (Targets 4 and 7)
- Strengthen the activities related to biodiversity of all major players, such as, small to large scale industries, business entrepreneurs, planning and implementing establishments, governments and consumers.
- Interlinking all stakeholders for effective implementation of the biodiversity related issues for its sustainable development and safeguarding ecosystems and its biota
- Ecosystem based sustainable production and consumption (Target 3)
- Develop package of practices for each ecosystem development
- Discourage incentives for all activities which harm biodiversity for which guidelines are to be prepared and approved
- Action plans are to be charted out for better understanding of environmental impacts on the basis of demand driven activities
- Adulteration of Aquatic products (e.g. fish) is prevalent in many parts of the world and is to be regulated through effective mechanisms. Procurement of produce by fixing floor price by government controlled agencies and are to be cold stored.

Distribution from such system can ensure quality product distribution and ensure quality sea food products

- Ensure the job opportunities created may be continued for which employment generation plans are to be developed
- Local knowledge base on loss of natural habitat may be properly acquired and acknowledged. Such knowledge can scientifically incorporated into the policies and programmes
- Adequate benefit must be proposed for such knowledge acquisition and utilization
- Identifying factors causing habitat loss at national level and frame policies and measures to reduce such loss.
- Study the impact of alien and introduced species on the prevailing biotic organisms
- Developing a clear legal or policy framework for land and water use or spatial planning to support national biodiversity aims (Target 2)
- Prevalent use pattern of aquatic ecosystems is to be assessed and arrive at causative factors in biodiversity loss.
- Enforcement of laws relating to habitat protection and conservation has to be implemented
- Sea grasses and associated organisms provide huge wealth to resources and hence are to be protected on priority basis
- Conservation measures for depleted stocks must be enforced
- Fish exclusion devices must be introduced as a measure of conservation
- Legal measures are to be strictly enforced for conserving the aquatic stocks
- Total awareness on the fishing regulations, fishing pressures, demand driven fishing with legal binding are to be encouraged
- Size at first maturity, size at which aquatic organisms could be caught, juvenile organisms' protection, avoidance of discards etc. to be addressed on a priority basis.
- Adequate compensation by way of ABS may be provided for people protecting the biodiversity and habitat. At present people living around water bodies who are real protectors not considered for ABS.
- New candidate species may be identified, developed on a participatory mechanism for augmenting food production and income generation for custodians
- Aquaculture and ornamental fish culture be made sustainable through utilization of biodiversity.
- Organisms may be utilized for forage production and feeding relevant candidate species of aquaculture thereby aquaculture becomes an eco-friendly, organic production system. Such production systems can be replicated by local people so that they will also be benefitted.
- Value added product production should be encouraged and for this local communities may be preferred so that ABS will be extended to them also
- Research on identifying, locating new ornamental organisms for trade and also their custodians and ABS applied.
- Custodians of aquatic resources in forests are to be managed by tribes and ABS applied

- In situ brood stock development in the wild has to be promoted and custodians are to be brought under ABS
- Profit generated from Products development from Aquatic organisms are to be considered for ABS to local community from where source materials procured
- Waste materials from aquatic resource management chain has to be subjected to new product development and disposal and are to be considered for ABS to local community from where source materials procured
- Aquatic plant resources are to be brought to utilization and ABS applied
- Emphasis must be given to the culture of live feed organisms and ABS should be applied to custodian local communities.
- Maintenance of riparian vegetation is important for keeping water quality high, production systems good and improve productivity and ABS applied for custodians
- Incorporate biotic resources for water pollution control and if managed by local people for which ABS is to be applied
- Disease problems from farming systems are to be properly addressed and conservation measures to be proposed and legally promulgated
- Conservative measures like closed fishing seasons, ranching programmes, captive breeding etc. are to be encouraged and entrust the technical jobs to local bodies and communities
- New candidate species for aquaculture, forage and ornamental purposes are to be identified and biodiversity utilization encouraged for which legal framework has to be developed
- Sustainable cage culture techniques are to be promoted and ABS made applicable to local community
- Aquatic alien species are sometimes proved beneficial and in such cases R&D should identify native species to replace alien species
- Conservation measures should be stringent to avoid invasion of alien species into the natural systems
- Migratory paths of aquatic organisms along the entire water systems are to be protected and fishing regulated
- Precious resources are to be protected on priority basis.
- Area based conservation measures are to be adopted
- Promote R & D programmes for genetic improvement of highly beneficial breeds and ensure no contamination of genetic resources
- Custodian local community must be adequately benefited out of this and ABS should be modified accordingly
- Incentives must be proposed for such programmes of conservation

- Wetland ecosystems should be better put under use so that many problems can be brought under control (including diseases, health of water and well being of biota and so on)
- Increasing riparian vegetation in aquatic ecosystems for conservation of ecosystems.

- Reclamation of aquatic ecosystems should be restricted to the highest level so as to conserve aquatic biodiversity, water and micro-climatic conditions.
- Eutrophication of aquatic ecosystem has to controlled for mitigating climate changes
- Making national information available through the ABS Clearing House (refer appropriate rules)
- ABS system has to be revisited on account of local people who are responsible for maintaining biodiversity, water and mitigating climate change in aquatic ecosystems
- ABS should be extended to people inside the forest who conserve the precious biodiversity. Rich aquatic wealth is available within forest and it could be made into use on a participatory mechanism
- Fishery resource from sea, estuaries, rivers etc. are to be procured by government agencies by fixing floor prices and are to be cold stored for further sale so that the stakeholders will get their due share as ABS. This can ensure non-adulteration and well being of consumers.
- Middle men may be given alternate avenues for involvement in fisheries industries for receiving their service for conservation
- Efficient mechanisms must be introduced for the product development from waste accumulated out of biodiversity utilization and ABS applied to these efforts
- National Biodiversity Strategy and Action Plan (NBSAP) should accommodate custodians and conservators of biodiversity
- NBSAP should focus subterranean fauna as well
- Adequate funds may be made available for conservation of species and utilization in order to ensure benefit sharing in a meaningful way

ABS started in late 1980s as an innovative strategy to achieve conservation and sustainable use of biological diversity during the negotiations of the CBD. It was envisaged to get access of genetic resources and Traditional Knowledge and legally share benefits to providers. Report says that ABS clearance is facing a lot of bottlenecks thereby neither facilitate access enabling sustainable use, nor result in substantial sharing of benefits (Alves et al. 2018; Ribeiro et al. 2018; Heinrich et al. 2020; Milieu Ltd. 2020; Williams et al. 2020; Bertioli et al. 2021; Mekonnen and Spielman 2021). These constraints and their unintended consequences create barriers to academic research and commercial development. Bertioli et al. (2021) reported that bilateral ABS regimes have been shown to drastically reduce botanical collections and international seed exchange. This can negatively impact the development of new crops and technologies to ensure food security (Mekonnen and Spielmann 2021). Similar concerns have been raised by the scientific community on the Sustainable Development Goals (SDG) (Gaffney et al. 2020). There are serious arguments regarding hurdles in ABS clearance, such as lengthy processes of approval on one side and the users of genetic resources who bear the burden of regulatory compliance and high transaction costs associated with the current ABS regimes under the CBD and the Nagoya Protocol on the other side (Alves et al. 2018; Milieu Ltd. 2020).

Bonn Guidelines on ABS: These Guidelines are prepared to serve indicative suggestions when developing and drafting legislative, administrative or policy measures on ABS with particular reference to provisions under Articles 8(j), 10 (c), 15, 16 and 19 of Nagoya protocol. Some objectives are aimed at promoting conservation and sustainable utilization of biological diversity; providing transparent framework and guidance for ABS adoption to both parties and stakeholders; enabling capacity building to guarantee the effective negotiation and implementation of access and benefit-sharing arrangements; promoting awareness of appropriate provisions of CBD; strengthening the clearing-house mechanism as a mechanism for cooperation among Parties in access and benefit-sharing; contributing to poverty alleviation and be supportive to the realization of human food security, health and cultural integrity, especially in developing countries; encouraging taxonomic research, as specified in the Global Taxonomy Initiative, and providers should facilitate acquisition of material for systematic use and users should make available all information associated with the specimens thus studied.

National Biodiversity Board (NBA, India) (2021) has developed a handbook to 'Strengthening Human Resources, Legal Frameworks, and Institutional Capacities to Implement the Nagoya Protocol' (Global ABS Project)—Access and Benefit Sharing In India A Handbook For Researchers. This compendium consists of 3 parts: Part I Access and Benefit Sharing in India: A Handbook for Researchers (Chapter 1—Understanding the Concept of Access and Benefit Sharing (ABS); Chapter 2—National Framework for ABS in India; Chapter 3—Understanding the Legal Requirements for Accessing Biological Resources and Associated Knowledge for Research; Chapter 4—Regulated Activities Subsequent to Access; Chapter 5—Understanding ABS Practically—Putting Theory into Practice); Part II Ethical Code of Conduct on ABS for Researchers, Research Institutes and Organizations; and Part III Monitoring Guidelines for Research Institutions to Promote Compliance with Access and Benefit Sharing Regulations under the Biological Diversity Act, 2002.

Global freshwater systems are on the degrading process. Freshwaters are the most extensively and rapidly altered ecosystems on the planet primarily due to reclamation for agriculture use. The responses due to this are: physical structure, chemistry, ecosystem processes, biotic characteristics, and diseases. Other major changes include climate, hydrologic modification, land use, chemical inputs, invasive species, and harvest. All drivers played had substantial effects on freshwaters like: changing hydrology, channel form, land use, chemical inputs, species introductions, harvest. Climate change is projected to cause major impact on freshwaters. Temperature and precipitation patterns affected water flow, rate of nutrient cycling and toxin breakdown and so on. Variation in storm intensity and affect stream channel form, habitat destruction, transport of invasive species, inputs of non-point pollutants, alteration in habitat etc. are expected changes associated with climate change. Deforestation, urbanization, and infrastructure development often lead to increases in the host vectors of human parasites, resulting in increased disease rates. Reservoirs will also be seriously affected due to the hydrologic and biogeochemical effects. Overfishing, other anthropogenic effects on aquatic ecosystems producing decline in populations of most of the valuable fishes. Many populations may not return to their previous abundance. Subterranean bio-resources and ground water quality will also deteriorating. Aquatic system and associated fauna and flora within the forest are fast disappearing due to deforestation (Carptenter et al. 2011). Hence protection of aquatic system is of paramount importance. ABS may help in this regard.

Marine ecosystem is also facing serious threats of alterations due to climate change. Temperature gradually rising, acidity of water increasing, pollution of various sorts increasing alarmingly, water quality started deteriorating, overfishing of resources experienced, frequent decline of populations reported every now and then, migration of biotic organisms affected due to changes in water current, coral bleaching reported is alarmingly high, mangrove forest declining and all these are affecting marine ecosystems.

Only a few of the major issues in aquatic ecosystem are highlighted here. Precious aquatic fauna from forest is being fished for various industries (aquaculture, ornamental fish culture etc.). Freshwater fishes are caught for marketing, but the persons who safeguard the stream are never benefitted. Catch of fishermen are not given adequate price for their efforts. The efforts of fishermen in collecting plastics from sea are not adequately rewarded. The industries make huge profit from utilization of the bio-resource and ABS never covered for the real protectors of bio-resources. Without sharing the benefits to the real protectors of bio-resource and also the medium in which bio-resources are safely conserved, implementation of ABS will not be complete. International policy changes are needed to improve legal certainty, efficiency, and workability. In the ongoing policy discussions, alternative systems for benefit sharing have been proposed, including multilateral systems to find a solution to the present situation (https://www.cbd.int/abs/DSI-webinar/DSIPolicyOptions2021.pdf; Laird et al. 2020; Scholz et al. 2020). But care must be taken for pragmatic proposals.

Much before the CBD, Nagoya protocol and ABS came into existence, a perfect system of benefit sharing to real conservators was in force in Kerala State, India. The classical example of benefit sharing initiatives taken by the researchers for the use of TK of the Kani tribe in developing the anti-fatigue drug "Jeevani" is still considered an intelligent forerunner for ABS. A team of researchers from the Jawaharlal Nehru Tropical Botanical Garden and Research Institute (JNTBGRI), Kerala came across the revitalizing effects of a local plant, *Trichopus zeylanicus* ssp. *travancoricus* during one of their ethno botanical expeditions guided by the Kani tribes of Kerala. A standardized herbal formulation based on the active chemical compounds from the plant that could be safely used against stress and fatigue was developed. In 1995, the technology for developing Jeevani was licensed to the Arya Vaidya Pharmacy, Coimbatore, Tamil Nadu, India for commercial manufacturing. Recognizing the contribution of the TK from the Kani tribes, the JNTBGRI constituted a trust of Tribes for share of benefit received in this respect.

(Source: http://lup.lub.lu.se/luur download?func = download File & recordOId = 1,555,871&fileOId = 1,563,800) From the above discussion it is clear that some aspects not been contemplated in the ABS. Therefore, a revisit to ABS is inevitably required for which a few suggestions of importance as indications are appended below-

- The primary conservators of resource of aquatic system of forest are tribe people and their role in conservation of biodiversity by conserving ecosystem is essential and hence they must be brought under ABS. Participatory breeding programmes, packaging, marketing and accounting can help them. Non-monetary benefits can be in the form of awareness and bestowing responsibility. They must also be taught about the extent of utilizing and conserving the precious biodiversity.
- In the streams and reservoirs the local residents must also be brought into the ABS. Because they are the people who conserve ecosystems and hence they deserve to enjoy the benefit. The waste generated out of the utilization of aquatic resources may be subjected to product development. All these programmes can be on a participatory approach.
- Riparian vegetation along the banks of river system is inevitable for the keeping healthy water so that biodiversity is preciously conserved and for this local people are doing commendable service. Eco-friendly establishments, such as, mini parks can be built along the banks and the management can be entrusted to them so that their service will be adequately benefitted both monetary and non-monetary
- Aquaculture production systems should consider the engagement of local people for getting their share from benefits
- Cage culture in reservoirs can be carried out in a participatory mode
- Part of benefits from industries should reach primary conservators of source areas from where the bio-resources have been procured.
- Fish catch should be procured by governments by fixing floor price and stored under their control so that the fishermen are adequately benefitted. The middle men may be allowed to procure fish from government machinery and distributed to industries. Authorized private machinery can also be made use of. In recent times fish for sale has been preserved in unethical ways and is distributed for domestic consumption and has created serious health issues to the consumers. The mechanism proposed above satisfies fishermen, middleman and consumers alike.
- All aquatic bio-resources should be brought under a well framed policy netwrok so that the entire links in the chain will get benefited in a fair and equitable benefit sharing.

References

Alves RJV, Weksler M, Oliveira JA, Buckup PA, Pombal JP, Santana HRG, Peracchi AL, Kellner AWA, Aleixo A, Langguth A (2018) Brazilian legislation on genetic heritage harms biodiversity convention goals and threatens basic biology research and education. An Acad Bras Cienc 90:1279–1284

Balmford A, Bruner A, Cooper P, Constanza R, Farber S, Green RE, Jerkins M, Jefferiss P, Jessamy V, Madden J, Munro K, Myers N, Naeem S, Paavola J, Rayment M, Rosendo S, Roughgarden J, Trumper K, Turner K (2002) Economic reason for conserving wild nature. Science 297:950–953

Bertioli DJ, Clevenger J, Godoy IJ, Stalker HT, Wood S, Santos JF, Ballén-Taborda C, Abernathy B, Azevedo V, Campbell J (2021) Legacy genetics of Arachis cardenasii in the peanut crop shows the profound benefits of international seed exchange. Proc Natl Acad Sci USA, 118:e2104899118

Bonn, Germany: Bundesamt für Naturschutz. https://www.bfn.de/fileadmin/BfN/ina/Dokumente/Tagungsdoku/2016/ Task_Force_on_OECMs_Rp2016.pdf

Carptenter SR, Stanley EH, Zanden MJV (2011) State of world's freshwater ecosystems: physical, chemical, and biological changes. Annu Rev Environ Resour 36:75–99. https://doi.org/10.1146/annurev-environ-021810-094524

COP decisions. www.cbd.int. Archived from the original on 31 October 2020. Accessed 29 Nov 2020. Source http://www.cbd.int/doc/press/2014/pr-2014-10-06-cop-12-en.pdf Archived 13 June 2015 at the Wayback Machine

CBD Secretariat. COP 14-fourteenth meeting of the conference of the parties to the convention on biological diversity. conference of the parties (COP). Archived from the original on 17 May 2019. Accessed 8 April 2019

CBD/COP/DEC/14/5 Archived 9 September 2019 at the Wayback machine, 30 November 2018. "Voluntary Guidelines" (PDF). Archived (PDF) from the original on 4 September 2019. Accessed 4 Sept 2019

Díaz S, Settele J, Brondízio ES, Ngo HT, Agard J, Arneth A, Balvanera P, Brauman KA, Butchart SHM, Chan KMAA (2019) Pervasive human-driven decline of life on Earth points to the need for transformative change. Science 366:eaax3100

Einhorn C (2021) The most important global meeting you've probably never heard of is Now. The New York Times. Archived from the original on 19 October 2021. Accessed 19 Oct 2021

Eighth Ordinary Meeting of the Conference of the Parties to the Convention on Biological Diversity (COP 8) Archived from the original on 17 December 2013. Accessed 12 Dec 2013

Dan Laffoley D, Nigel Dudley N, Jonas H, MacKinnon D, MacKinnon K, Hockings M, Woodley S (2017) An introduction to 'other effective area-based conservation measures' under Aichi Target 11 of the convention on biological diversity: origin, interpretation and emerging ocean issues. Aquatic Conserv Mar Freshw Ecosyst 27(S1):130–137. https://doi.org/10.1002/aqc.2783

Ribeiro DS, Koopmans MP, Haringhuizen GB (2018) Threats to timely sharing of pathogen sequence data–the Nagoya protocol may impose costs and delays. Science 362:404–406

Gaffney J, Tibebu R, Bart R, Beyene G, Girma D, Kane NA, Mace ES, Mockler T, Nickson TE, Taylor N (2020) Open access to genetic sequence data maximizes value to scientists, farmers, and society. Glob Food Sec 26:100411

Dudley N (ed) (2008) Guidelines for applying protected area management categories. IUCN, Gland

Greenfield P (2021a) Major UN biodiversity summit delayed for third time due to pandemic-Cop15 negotiations to set this decade's targets on nature to be split into two, with face-to-face meetings delayed until 2022. The Guardian. UK. Archived from the original on 20 August 2021

Greenfield P (2021b) UN's Kunming biodiversity summit delayed a second time - Covid pandemic continues to hamper plans for key gathering to agree targets on protecting nature. The Guardian. UK. Archived from the original on 20 August 2021

Heinrich M, Scotti F, Andrade-Cetto A, Berger-Gonzalez M, Echeverría J, Friso F, Garcia-Cardona F, Hesketh A, Hitziger M, Maake C (2020) Access and benefit sharing under the Nagoya Protocol-quo vadis? Six Latin American case studies assessing opportunities and risk. Front Pharmacol 11:765

IISD (2022) UN Biodiversity Conference (CBD COP 15) (Part 2). International Institute for Sustainable Development. Archived from the original on 8 April 2022. Accessed 8 April 2022

Jonas H, Barbuto V, Jonas HC, Kothari A, Nelson F (2014) New steps of change: looking beyond protected areas to consider other effective area based conservation measures. Parks 20(2):111–128

Jun M (2021) Cop26 must not overshadow Kunming: we need joint climate and biodiversity goals. The Guardian. Archived from the original on 19 October 2021. Accessed 19 Oct 2021

Laird S, Wynberg R, Rourke M, Humphries F, Muller MR, Lawson C (2020) Rethink the expansion of access and benefit sharing. Science 367:1200–1202

Leadley PW, Krug CB, Alkemade R, Pereira HM, Sumaila UR, Walpole M, Mumby PJ (2014) Progress towards the Aichi Biodiversity Targets: an assessment of biodiversity trends, policy scenarios and key actions. Technical Series 78. Montreal. Secretariat of the Convention on Biological Diversity, Canada

Lewis-Lettington RJ, Mwanyiki S (eds) (2006) Case studies on access and benefit-sharing. International Plant Genetic Resources Institute, Rome, Italy. ISBN 13: 978–92–9043–721–5; https://www.cbd.int/financial/bensharing/several-case-IPGRI.pdf
Lewis-Lettington RJ, Muller MR, Young TR, Nnadozie KA, Halewood M, Medaglia JC (2006) Methodology for developing policies and legislative frameworks for access to genetic resources and benefit sharing. International Plant Genetic Resources Institute, Rome, Italy
Lewis-Lettington RJ, Dogley D (2006) Commentary on the development of the republic of seychelles access to genetic resources and benefit sharing bill (2005). International Plant Genetic Resources Institute, Rome, Italy
Lopoukhine N, Dias BF (2012) Editorial: what does Target 11 really mean? Parks 18(1):5–8
MacKinnon D, Lemieux CJ, Beazley K, Woodley S, Helie R, Perron J, Gray P (2015) Canada and aichi biodiversity target 11: Understanding 'other effective area-based conservation measures' in the context of the broader target. Biodivers Conserv 24:3559–3581
Mekonnen DW, Spielman DJ (2021) Changing patterns in genebank acquisitions of crop genetic materials: an analysis of global policy drivers and potential consequences. Food Policy 105:102161
Heinrich M, Scotti F, Andrade-Cetto A, Berger-Gonzalez M, Echeverría J, Friso F, Garcia-Cardona F, Hesketh A, Hitziger M, Maake C, Politi M, Spadafora C, Spadafora R (2020) Access and benefit sharing under the nagoya protocol—Quo Vadis? Six Latin American case studies assessing opportunities and risk. Front Pharmacol 11(765):1–19. https://doi.org/10.3389/fphar.2020.00765
Milieu Ltd (2020) Analysis of implications of compliance with the EU ABS regulation for research organisations and private sector companies. Study commissioned by the European Commission. https://ec.europa.eu/environment/nature/biodiversity/international/abs/legislation_en.htm. Accessed 2 Nov 2021. Google Scholar
NBA (2021) Strengthening human resources, legal frameworks, and institutional capacities to implement the nagoya protocol (Global ABS Project)- Access and benefit sharing in India a handbook for researchers
Pandrey K (2021) Over 100 countries sign Kunming declaration on biodiversity conservation. Down to earth. Archived from the original on 20 October 2021. Accessed 19 Oct 2021
Pushpangadhan P (2009) Access and benefit sharing: case studies and International experience. In: FICCI international conference on traditional knowledge, New Delhi
Scholz et al., 2020 Scholz AH, Hillebrand U, Freitag J, Cancio I, dos Ribeiro SC, Haringhuizen G, Oldham P, Saxena D, Seitz C, Thiele T (2020) Finding compromise on ABS & DSI in the CBD: requirements & policy ideas from a scientific perspective. https://doi.org/10.13140/RG.2.2.35180.80001GoogleScholar
Suneetha MS, Pisupati B, Kumar S (2009) Framework for benefit sharing guidelines for India. Asian Biotechnol Dev Rev 11(2):55–88
Williams C, Walsh A, Vaglica V, Sirakaya A, Silva M, Dalle G, Winterton D, Annecke W, Smith P, Kersey PJ (2020) Conservation policy: helping or hindering science to unlock properties of plants and fungi. Plants People Planet 2:535–545. Wiley Online Library Google Scholar
Woodley S, Bertzky B, Crawhall N, Dudley N, Miranda Londoño J, MacKinnon K, Sandwith T (2012) Meeting aichi target 11: what does success look like for protected area systems? Parks 18(1):23–36
Jump up to a b "About the Protocol". Convention on biological diversity. Archived from the original on 28 October 2020. Accessed 17 Sept 2020
Meeting documents: Archived from the original on 4 February 2015. Accessed 4 Feb 2015
Meeting Documents: First Extraordinary Meeting of the Conference of the Parties to the Convention on Biological Diversity, 22–23 February 1999-Cartagena, Colombia. Archived from the original on 4 February 2015. Accessed 4 Feb 2015
Meeting Documents. Archived from the original on 4 February 2015. Accessed 4 Feb 2015
Meeting documents: fifteenth meeting of the conference of the parties to the convention on biological diversity, second quarter of 2021-Kunming, China. CBD. Archived from the original on 17 September 2020. Accessed 17 Sept 2020

Welcome to COP 9. 20 January 2009. Archived from the original on 4 February 2015. Accessed 4 Feb 2015
Welcome to COP 10. Archived from the original on 4 December 2010. Accessed 29 Nov 2010
World Wide Views on Biodiversity. Archived from the original on 11 November 2017. Accessed 19 Dec 2011
http://www.cbd.int/cop2014 Archived 6 October 2014 at the Wayback Machine; Webcasting: "Archived copy". Archived from the original on 6 October 2014. Accessed 6 October 2014

Part IV
Implementation of ABS Mechanisms

Chapter 12
Strengthening ABS Implementation in India

Preetha Nilayangode, K. P. Laladhas, and Oommen V. Oommen

Abstract It is necessary to design a future ABS strategy that recognizes the results of research and promotes commercialization of the same at the same time ensuring sustainability of the resources. The present paper argues for developing innovative models of ABS based on the requirements of the community and the states implementing the same with an overarching vision for promoting the conservation and sustainable utilization of resources.

Keywords ABS · KSBB · NBA · MAT · PIC · Biodiversity conservation · Benefit sharing

12.1 Introduction

The adoption of the Convention on Biological Diversity (CBD) in 1992 brought in a new paradigm change in the manner of utilization of natural resources for research and development of products of commercial importance in a wide range of sectors including pharmaceutical, biotechnology, seed industry, crop protection, horticulture, cosmetic and personal care, fragrance and flavor, botanicals, and food and beverage. Access and Benefit Sharing (ABS) has its basic principles within the CBD and aims to ensure that Prior Informed Consent (PIC) for access is provided, Mutually Agreed Terms (MAT) are negotiated and accrued benefits are shared equitably and fairly. The Biological Diversity Act 2002 and Rules 2004 of India provides for the conservation of biological diversity, sustainable use of its components, and fair and equitable sharing of the benefits arising out of the use of biological resources and traditional knowledge. Pursuant to the adoption and ratification of the Nagoya

P. Nilayangode
Kerala State Biodiversity Board, Thiruvananthapuram, Kerala 695801, India

K. P. Laladhas · O. V. Oommen (✉)
Centre for Venom Informatics, Department of Computational Biology and Bioinformatics, University of Kerala, Kariavattom North campus, Thiruvananthapuram, Kerala 695801, India
e-mail: oommenvo@gmail.com

O. V. Oommen et al. (eds.), *Biodiversity Conservation Through Access and Benefit Sharing (ABS)*, https://doi.org/10.1007/978-3-031-16186-5_12

Protocol the Guidelines on Access to Biological Resources and Associated Knowledge and Benefit Sharing Regulations 2014 were notified under the Act of 2002. The Guidelines clarified and elaborated ABS provisions of the Act and a legal framework for the mode of benefit sharing under different scenarios were put in place in India. The basic objective of ABS is the empowerment of the communities who have conserved the resources for generations along with conservation of biodiversity. The Biodiversity (Amendment) Bill 2021 that was tabled in parliament on December 2021 has tried to bring in better clarity and address some of the issues encountered during the implementation of the Act of 2002 in India. The present paper argues for developing innovative models of ABS based on the requirements of the community and the states implementing the same with an overarching vision for promoting the conservation and sustainable utilization of resources.

12.2 Challenges, Responses, and Effectiveness of ABS Regime of India

Being a signatory of CBD and Nagoya Protocol on Access and Benefit Sharing, India has put in a robust legal and institutional framework for the operationalization of ABS with statutory, autonomous bodies at the National, State, and Local level mainly National Biodiversity Authority (NBA), the State Biodiversity Boards (SBBs) and Biodiversity Management Committees (BMCs) respectively. In India, multiple regulatory authorities at the National, State, and Local levels are mandated with regulating access to bio resources by different user groups and they are also involved during different stages of access with permit based access, and benefit-sharing agreement as a condition of access. The ABS system implemented in the country considering its varied diversity in natural resources, culture and associated knowledge envisages a sectorial approach with different approaches for Research and Development, Commercialization, and non-commercial research for Indians and foreign entities. The benefit-sharing was visualized in both monetary and non-monetary manner including participating in the research and analysis, providing technical skills and experience, sharing of technology, and building the capacity of the local institution.

Humphries, Benzie and Lawson (2021) in a review of literature relating to Access and Benefit sharing had concluded that India has the largest body of literature relating to the ABS of all the countries examined (47%), and most of the work focused on issues relating to the effective implementation of ABS measures. It was pointed out that none of the articles stressed the role of ABS directly to conservation of biological diversity. Twenty years down the line India has come a long way to ensure compliance with CBD by establishing national focal points and other institutional structures at the national, state, and local level but whether the benefits has percolated down to the real benefit claimers need to be evaluated. The State Biodiversity Boards (SBBs) have spent time and energy in discussing and negotiating with the key players, but with how much success is a matter which requires introspection. Immense human and financial resources have been expanded to negotiate and sign ABS agreements. The benchmark for this should be how much benefits have percolated down to the local community and whether the users of biodiversity have adopted the concept of "putting people and natural capital along with profits" and how much it has benefited conserving the rich diversity of this mega diverse country. Strengthening implementation of ABS in India in the post-2020 Global Biodiversity Framework requires an analysis of the Strengths, Challenges, Options, Responses, and Effectiveness of the entire process from Access of bio-resources to the development of a product and the method of sharing of benefits with the real custodians of the resources and the impact on conservation of biodiversity. The discussion in the Post 2020 Global Biodiversity Framework makes this an opportunity to review the implementation of ABS in India to identify the challenges/issues faced by the stakeholders at different levels including the NBA, the SBBs, BMC, Traders, bio resource based industries and research community.

12.3 Biodiversity Management Committees–The Last Mile Connectivity

The NBA and the SBB are body corporates and there is no overlap in the functions of NBA and SBBs. Under the Nagoya Protocol, NBA being the national focal point, if a permit or its equivalent is granted for accessing genetic resources the permit—an internationally recognized certificate of compliance is made available to the ABS–CH. The BMCs although established across the country at all three levels of local bodies consequent to the National Green Tribunals order in OA No 347 of 2016 dated 18.03.2020, is neither financially nor technically competent to address these issues. The institutional capacity of BMC to levy collection charges the primary step in the access mechanism as envisaged in the Act is negligible and there are very few successful cases which can be quoted in this instance. A time-bound action plan to empower the BMC technically, administratively, and financially to function is needed. The Biodiversity Bill 2021 has rightly addressed this issue and the amendments proposed provides flexibility to the states to decide the structure of the BMC and also to increase the number of members to ensure better participation from a range of user groups.

As mandated by Section 41 of the Act the approval of the concerned BMC is needed to provide access to bio-resources for commercial utilization both for entities covered under Section 3 and Section 7. But in majority of the cases the geographical location of collection of resources will not be specified in the application form or if at all specified it will be accessed through traders thus weakening the negotiation power of the SBBs as companies claim this as an excuse for not complying with the legislations. At the national level, an e-filing portal for ABS approvals is functioning but the same is not the case with SBBs. The stakeholders well aware of the intricacies of dealing with multiple agencies not only within the state but across different states of the vast country prefer a centralized, single authority that can ensure a standardized ABS process and also bring down administrative difficulties and transaction costs. In many instances although the state is providing the resource much of the processing and value addition is undertaken elsewhere so that the benefits do not directly percolate to the community. This has made the access regime more of a top-down approach with NBA or SBBs taking the lead for negotiating ABS agreements for non-Indian or Indian entities as the case may be.

Many of the states have attempted to document the tradable and commercially potential bio-resources for ABS successfully. The ABS measures for states like Kerala where the industry is dominated by MSMEs which is a major income generating and employment providing sector with relatively lower investment and lower turnover needs to be explored. One of the Best Practises of ABS documented (Shreyas Bhartiya 2019) is the conservation breeding project of Golden Mahaseer one of the twenty mega fish species of the world and a game fish. The activities of BMC Hirabambai, Maharashtra demonstrates that apart from the management of biodiversity, a BMC can also play a key role in providing livelihoods and generate its own finances through efficient conservation and sustainable use of biodiversity.

12.4 Scope of Access and Benefit Sharing Measures

The doctrine of Access and Benefit Sharing, since its inception has become one of the most debated among the three major objectives as visualized in the CBD, but in spite of this there is still no clarity about what is included within the scope of ABS especially amongst the SBBs. This is mainly because the activities falling under the mandate of the ABS has been left to be defined by the national legislations and in a country with diverse economy ranging from traditional village farming to modern seed industry, home remedies toadvanced pharmaceutical industries there is no one stop solution. Even though there is a set of international principles laid down in the CBD and Nagoya Protocol, the specificities of the activities which fall within its mandate are left largely to be defined by the national legislation and regulation. One of the key impediments in implementing ABS is the lack of awareness and misconceptions regarding terminology and what is coming under the purview of Bio-resources, Commercial utilization, Value addition, By-products, Normally traded commodities, etc.

The point of debate among users and regulatory authorities at the regional level is what can be considered under the purview of the Act and whether it is to be made applicable in the case of cultivated crops, bio-resources from forest, public or private land. It has been suggested that the national ABS framework should include the protected area ABS policies as well (Nelliyat and Meenakumari 2018). It has also been argued that national forest and ABS legislation shall be designed in a harmonized and mutually supportive manner to ensure that ABS and forest considerations are taken into account. Certification in the forestry sector has received attention in recent years as a mechanism that can contribute to sustainable forest management.

Integration of the Fair Wild certification approach with the ABS regiment Bhima Shankar Wildlife Sanctuary (BWLS) in Pune district and Sangameshwar in Ratnagiri district of Maharashtra is another best practise cited (Bhartiya 2019). The certification has led to premium price for the collected fruits from the certified trees resulting in appreciation of the value of biodiversity and conservation and sustainable use of the resources.

Apart from protected areas the ABS measures in cultivated landscape also needs attention. In a bio diverse but rapidly urbanizing state like Kerala, Trees outside the Forest are more closely connected with the community and play a significant ecological, environmental and economic role to sustain households and household income. The extent of Trees outside forest in Kerala has been estimated to be 14,394 sq km which is 37% of the total geographical area of the state (India State of Forest Report 2021). The chunk of the timber needs of the state is met from trees outside the forest and the unsustainable extraction of these will ultimately affect the ecological conditions of Kerala. Khan et al. has suggested a species selection tool for bio resource selection for ABS based on six criteria- Availability, Price, Market availabilities, Traditional knowledge and diversified uses, Industrial demand, labour cost, Time-bound, Accessibility, Collateral damage to regeneration and Storage issues (Salman Khan et al 2019). A similar work is in developing a BCG matrix to identify bio-resources with ABS potential has been reported (Mehta et al. 2019). The states shall work on developing such regionally adaptable methodology for identifying bio resources with ABS potential with the twin objective of sustainability and benefit sharing.

12.5 ABS and Bio-Economy

Globally nearly 40% of the world's economy is derived from the direct use of biodiversity. India's Bio-economy during 2019 is valued at $62.5 B with BioPharma accounting for $ 36 B (58% of total Bio Economy), Bio agriculture accounted for $12.0 B (19%), bio industrial $ 5.0 B (8%), and Research $ 9.8 B. The contributions of the enzymes and biofuels sectors are estimated at 46 and 54% share of the total Bio Industrial segment respectively (The India Bio Economy Report March 2020). Advancements in biotechnology is contributing to economy in a large way. The ABS mechanism shall ensure that in the larger interest of the humanity it is necessary to improve biological data access mechanism and how benefits are shared. The India Bio Economy Report March 2020 has stated a case of exemption to all cultivated biodiversity from the provisions of the National Biodiversity Act as such activities automatically ensures their conservation. At the same time access to bio-resources from the wild is dwindling as the dynamics of forest management have evolved from production-oriented to sustainable forest management, encompassing environmental, economic, and social dimensions. The Indian forestry sector has also adapted to the growing sustainable forestry industry through certification schemes consequent to restrictions on import of non-certified timber, non-timber forest products, etc. by several countries.

India ranks sixth in essential oil production and export of products derived from medicinal plants (Natarajan 2020). The total production of essential oils in the world is estimated at about 0.10 to 0.11 million tons and in terms of value India stands in second position with a world share of 21–22 per cent. India is the largest producer of mint (*Menthaarvensis*) and basil (*Ocimumbasilicum*).

The highly traded species also include the threatened medicinal plant species like. *Aconitum heterphyllum, Cosciniumfenestratum, Decalepishamiltonii, Nardostachysgrandiflora, Oroxylumindicum, Picrorhizakurroa, Saracaasoca, Swertiachirayita, Vateriaindica e*tc. (Gopi et al. 2018).Plant based raw material has always been the backbone of Ayurveda and traditional medicines and have guided the development of potential drugs. The sector wise analysis of ABS applications received by the NBA until October, 2017 reveals that Pharmaceuticals occupy 54.5%, Research 18.9%, Nutraceutical 15.1%, Cosmetics 2.6 and other 8.9% (Shanbhag Gayathri and Rana (2018).The plant part as raw material in trade includes leaf, flower, seed, stem, bark, root, rhizome, tuber, heartwood, grain, gall, exudate, extracts and even as a whole for certain preparations. The Ministry of AYUSH, has estimated there were 8954 manufacturing units in India during 2019–20, out of which 7718 units were into Ayurveda products, 625 into Unani and 397 Homeopathy. (https://ayush.gov.in/). The CII in a study concluded that only around 50 companies have revenue above INR 100 crores which account for over 85 per cent of the revenue generated by this sector. Top 50 companies (across both food and Ayurveda non-food products) reported revenue of around INR 22,500 crores for the financial year April 2017-March 2018 from sale of Ayurveda products only. The size of the natural/Ayurveda veterinary and feed supplement market is estimated to be INR 700 crores with nearly INR 150 crores of exports (Confederation of Indian Industry (CII) 2018). CII estimated, the total market size of Indian Ayurveda industry as USD3 billion, which includes both Ayurveda products (ethical, classical, over the counter (OTC), personal care and beauty products) and services (medical, well-being, and medical tourism services). During financial year 2015, the Indian wellness industry was estimated at INR850 billion (USD13.9 billion). It is reported that the gross market size was of $ 4.4 billion, or roughly Rs 30,000 crores during 2018. The report also pointed out that 77% of Indian households are using Ayurvedic products as against 69 percent in 2015. The industry whose total worth was $3.4 billion in 2015 is expected to reach $9.7 billion by 2022.

Ayurvedic medicines falls under two main categories—classical products and proprietary products. The market share of proprietary products is ever increasing with the decline in classical medicinal products. Most of the classical drugs are manufactured traditionally and are highly labour-intensive with fewer profit margins. The turnover of proprietary products is Rs. 5500 crores compared to Rs. 500 crores for classical medicines. The major share of exports are raw materials at 56 TMT (Thousand Metric Tonnes) when compared to finished products export which was 16 TMT in 2010. The corresponding figures for 2000 was 18 TMT for raw materials and 14 TMT for finished products. The export earning in total has increased from

US $ 39 million to US $ 150 million during this period even though finished product share has increased only 2% (39 to 43%) (Chaturvediet al 2014)This preference for raw materials over finished product is leading to huge fall in revenue for India.

The Kerala State Scheduled Caste Scheduled Tribe Development Cooperative Federation Ltd. (Federation) are entrusted with the collection and marketing of NWFPs, which include medicinal plants. There are weekly markets for the sale of raw drugs in four districts in Kerala, at Vadakara in Kozhikkode, Velanthavalam in Palakkad, Perumbavoor in Ernakulam and Chala market in Thiruvananthapuram. The medicinal plants collected from outside the forests, private gardens, homesteads, wastelands and common properties are sold directly or through agents to the dealers and medicine manufacturing firms. As per the latest official data 794 herbal and ayurvedic industries are registered in Kerala and 144 cosmetic industries are listed in Kerala. (Official communication form Drugs Controller Authority, Kerala) It is reported that 12% of the nation's formal ayurvedic industrial units are established in Kerala. These companies have an annual turnaround of about Rs. 650 crores. Of these, the majority are privately owned units and one of the noted government sector companies is Oushadhi. Oushadhi supplies the Kerala Government for their 90% needs, which is 70% of Oushadhi's total production, 20% goes to other states and remaining 10% is marketed through retail outlets of the company.

Medicinal plants trade raises issues of equity, the way benefits are distributed along the supply chain. Often the supply chain often extends to 3–4 tiers without much value addition but with increase in sale price at each level. There is, on an average, 70 to 100% increase in sale price of crude drugs from primary collector's level to pharmacy level. Many times the same crude drug is available in various grades with major traders having considerable difference in price. The final prices of the most of the raw drugs are significantly higher than the collection charges leading to overexploitation of the natural resources much beyond the natural regeneration capacity of the species. The ABS guidelines of 2014 stipulates that the benefit sharing obligations on the trader shall be in the range of 1.0 to 3.0% of the purchase price of the biological resources and the benefit sharing obligations on the manufacturer shall be in the range of 3.0 to 5.0%of the purchase price of the biological resources. But as the trading of raw drugs is largely unorganized and there is no traceability of the source of collection, ABS has not been implemented at this level of access in most of the Indian states. The feasibility of benefit sharing measures to be implemented in the case of raw drug traders need to be explored.

Kerala is among the top producers of agro-based products such as spices, coconut, tea, as well as marine products in the country. Marine fish, including sardines, shrimps, lobsters, cuttlefish, squid and tuna, account for a share of 76 percent in the total fish production of the state. The main sources and products of primary interest for marine nutraceuticals and ingredients include omega-3 fish/algal oil, phospholipids (bound omega 3-fatty acids), micro/macro algal nutrition supplements, fish proteins and peptides, hydrolysates, shellfish chitin, fish collagen. The total fish production in Kerala in 2018–19 was 8.01 lakh tonnes, of which marine fish production was 6.09 lakh tonnes and inland fish production was 1.92 lakh tonnes. In 2018–19, Kerala accounted for 5.4 percent of the country's total exports of processed food. Kerala currently accounts for nearly 75 percent of India's EU-Certified Seafood Units, and has 5 state-of-the art food processing parks including 2 mega food parks. (EXIM Bank of India 2020). For the ten maritime states of India marine resources are the major sources of livelihood for the artisanal fishers but ABS measures for marine products is largely an unexplored area.

12.6 Bio Resource Monitoring Tool for ABS Compliance

The ABS system currently being practiced at the state level is a top-down approach where SBB had played the active role in initiating ABS agreements which is contrary to the bottom-up approach as visualized in Biological Diversity Act. Even in this top-down approach benefits received from the user can be distributed to the BMC, wherever identified only if the geographical origin of the biological resources accessed is known. For sharing of benefits, it is necessary to identify the players in the entire supply chain from providers of raw materials to finished products. This is often not possible since the resources are usually transferred many times along the value chain from the provider to the end-user, and it may not be able to trace back the geographical location of origin of a resource used in the development of a product. Understanding where the value is added along the value chain and understanding market dynamics and transactions is critical to establish benefit-sharing. Bio-resources trade is extremely complicated involving intermediates at various stages whether it be domestic trade or export of bio-resources. In 2019, a total of 1.10 million tonnes of spices and spice products valued US$ 2.80 billion were exported from the country and India accounts for half of the global trading in spices. At the same time, raw spices are imported into India and are used as raw materials in the food processing units which are then either exported or consumed within the market. Many of the industries need resources in certain stages of maturity for the extraction of specific products and prefer to import raw materials at competitive prices also. Many SBBs had conducted a study on the supply chain of tradable bio resources and it was observed that industries invariably access genetic resources through intermediaries. The industries find difficulty in declaring the geographical area from the bio resources is sourced.

Multiple legislations existing in India promote traceability of resources to the origin of the collection. The Indian System of Medicine has adopted the GMP for herbal products manufacturing units, intending to ensure that the raw materials should be of good quality for GMP certification. As per Drugs & Cosmetics Rules, 1945 manufacturing units of Ayurveda Siddha and Unani drugs are required to maintain a record of the raw material used in the manufacture of their products to the State Drug licensing authorities and to the National Medicinal Plants Board on an annual basis vide notification No. K. 11,020/2/2006-DCC (AYUSH), dated 09th July 2008. Although the legal system is in place, the multiplicities of enforcement agencies and reluctance in data sharing defeat the very purpose of the Act i.e. the up-liftment of the local community.

The NBA can act as an umbrella agency and develop a Bio-Resources Monitoring Tool- a single centralized window that covers production, trade, and price indices related statistics including domestic, export, and import of bio resources ranging from crops, livestock, forestry produces, marine bio resources, and microorganisms of India etc. Identification of checkpoints of regulation is one way of ensuring compliance. Any new ventures need the required clearance from the Local Self Government and hence this can be taken as a primary checkpoint whereby only if the necessary access permission has been provided by the concerned BMC the clearance is provided. The starting link on this chain can be a certificate of origin for the accessed bio-resource by the concerned BMCs. This database can also act as a checkpoint in ensuring that the terms of access as per the benefit-sharing agreement are being met by the users. Up-to-date statistics will help in highlighting trends on use of Indian bio-resources, which can also be linked with ePBR database, taxonomic databases and information on bio-resource based commercial products. The Nation mission on Biodiversity and Human Health of Government of India is one such initiative and aims to transform biodiversity science by linking it to the peoples' economic prosperity.

12.7 Integrating ABS and Welfare Schemes

The major benchmark of SBBs for the successful implementation of ABS is the monetary benefits accrued as a result of ABS agreements. SBBs can think out of the box and develop innovative models of non-monetary benefits. India is one of the first countries in the world to make corporate social responsibility (CSR) mandatory, following an amendment to the Companies Act, 2013 in April 2014. Companies that have an annual turnover of Rs. 1,000 crores or more **or** a net worth of Rs. 500 crores or more or a net profit of Rs.5 crores or more as part of CSR implement programs to ensure environmental sustainability, ecological balance, and conservation of natural resources. It has been reported that 42% image of a company is stemmed from its CSR activities. CSR was perceived as a measure of being socially responsible but increasingly it has gained importance as a competitive advantage for the companies. It helps to build the equity of a brand by creating a positive image in the mind of consumers. But one interesting observation is that the corporates are reluctant to enter into such association with SBB as they want to go beyond the legal obligations of CSR and see this as an investment for creating Brand image among customers thus showing that the company cares more than just profits (Mahmood 2020). Although CSR and ABS are under two regulatory systems SBB can think of Certifications and other measures by integrating schemes that enable corporates to be identified as socially responsible and use appropriate benefit-sharing to develop a high-brand reputation.

12.8 A Sectoral Approach to ABS

The value additions on biological resources in industrial sectors such as pharmaceutical, food, and nutraceutical or cosmetic vary substantially and a sectoral approach has been advocated. A review of the applications for access to bio resources received at KSBB over the years from Section 3 entities and Section 12.7 entities brings out the fact that more than 90% of the applications are for plant based resources. India is a leading producer of natural base ingredients, catering to 60% of the global spice oleoresin demand and 80% of the globalmint extracts demand. 75% of India's base ingredient production is being exported. Over 75% of this market is constituted by natural ingredients. The global base ingredients market for both flavours and fragrances and nutraceutical ingredients is close to USD 13.5bn, The Indian base ingredients market is currently sized at about USD 2.8bn, growing at a CAGR of 11%. Within this segment, spice extracts and oleoresins is the fastest growing sub-segment and is expected to grow at a CAGR of 16%. Globally the F&F market is USD 27.1bn in size, expected to grow at 6.4% over the next 5 years. The organized Indian F&F production market is USD 1.1bn in size, close to 30% of which is exported (Singh and Bhattacharyya 2016) The global market for nutraceuticals is estimated at USD 198.7bn in 2016. The organized Indian market stood at ~ USD 550 mn in 2016, and is expected to grow at 20%.

The endophytic bacteria, fungi population and the marine algae and archaebacteria are reservoirs of bio-actives/metabolites of pharmaceutical and industrial importance, which remain untapped. The secondary metabolites obtained from seaweeds demonstrate a broad spectrum of bioactivity. Apart from these, marine polysaccharides, such as agar, alginate, carrageenan, fucoidan, ulvan, laminarin, and furcellaran from seaweeds are important in marine phycocolloid industry as it constitutes the most abundant source of polysaccharides. The maritime states of India can promote development of value added products from marine resources and work out a sectoral approach to ABS.

The SBB instead of merely looking out for a graded percentage of the turnover of the company as benefit sharing obligation shall work on a case to case basis sectoral approach defining the goals. Criteria for selection of benefit sharing shall falls on the SBB based on the needs of the area as this may vary from an urbanized area to a remote tribal area or from a maritime state to a hilly area.

12.9 Creation of Access and Benefit Sharing Facilitation Cell

ABS is based on prior informed consent being granted by a provider of resource or associated knowledge to a user of such resources, and on negotiations between both parties to develop mutually agreed terms, to ensure the fair and equitable sharing of benefits arising from the use of the resources. In many a case a research project based on bio-resources and associated knowledge may lead to the discovery of a patentable invention. Most of the universities and research organizations of India have a Patent Facilitation cell and the responsibilities are restricted to facilitating the process of filing patents. Although research per se by Indian entities does not come under the purview of the Biological Diversity Act, the premier research organizations of the country shall take a proactive role in establishing ABS cell for facilitating PIC at the initial stages of research itself so that traditional knowledge holders are benefitted if the results are patented and the cell can also provide an enabling environment for technology transfer for commercialization of the results of research.

12.9.1 Creating an Enabling Environment for ABS

An enabling environment for implementing ABS can be developed only if the investment in R & D and innovation and utilization of the knowledge generated by public research is utilized for the welfare of the community. The Indian Patents Act, 1970 was amended in harmony with the Biodiversity Act and requires the applicant to disclose the source and geographical origin of the biological material used in an invention in the specification to comply with Section 10(4)(D) of the Patents Act. To limit unauthorized use of biological resources and associated traditional knowledge the applicant for patent needs to obtain NBA approval before the grant of a patent (Section 12.6). IPRs can be strong tools for successful and regulated commercialization of products based on genetic resources and associated traditional knowledge. The institutional framework for the protection of traditional knowledge and sharing of benefits accrued through commercial utilization of bio resources and products developed is put in place. The Traditional knowledge digital libraryaims to protect Indian traditional medicinal knowledge and prevent its misappropriation at foreign patent offices.

There have been suggestions to exempt non-commercial research from the international ABS regime altogether and that a distinction shall be made between basic research with academic intent and applied research and development with commercial intent (Jungcurt 2011). It has been suggested to reduce the legal barriers to accessing genetic resources for non-commercial research purposes (Rourke 2018). The Biodiversity Amendment Bill 2021 has attempted to address this issue and facilitate research by making the process of obtaining approval more user friendly.

Much research is being carried out by the research institutes and universities in India but only a handful of these research findings find commercial applications. A quick screening of a total of 203 theses produced in Kerala during the last few years revealed that 179 are related to floral studies and 13 only are related to faunal studies. A large number of theses focus on Health ailments and treatments (109), while some focus on Food and Agriculture (87) and others on Ecology and Environment (16). Apart from this large numbers of papers are being published relating to traditional knowledge. Of the total 173 papers screened, 96 represent T.K related researches in Kerala on wild edibles and agricultural crops having medicinal values and the rest 77 are the clinical studies done on traditional and edible varieties of plants. Of the 96 T.K studies, 36% of the total papers are researches on wild edible fruits and vegetables and 30% of the studies are on tubers and rhizome. The bacterial resources found in Kerala have been a study material in 80% of research for developing therapeutic base than nutraceuticals out which only 2% went for in vivo trials. Most academic research results are at the stage of laboratory prototype or preclinical trials and require substantial further research and development before they can be turned into a commercially viable product. In India, not more than 5% of patents reach the market, and two-third of the research originates from Government labs. Experiences elsewhere have shown that the model has worked well when the research community are engaged with startups to commercialize the product.

Biotechnology Industry Research Assistance Council (BIRAC) IP & Technology Management Cell is making considerable efforts to identify translatable technologies for further commercialization. The Make In India Facilitation Cell at Biotechnology Industry Research Assistance Council (BIRAC) and the Department of Biotechnology (DBT) of the Government of India have been targeting biotech startups in the country and the number of startups in the country has reached 2,669 at the end of 2018. Coordination between such initiatives of Government shall work proactively for commercialization of results of research so that ultimately the benefits flow back to the community. Similar initiatives in the field of life sciences will pave the way for creating value from research and develop new products and services from bio resources leading to socio-economic development of the area and community.

ABS can serve to mobilize resources to conduct scientific research on TK for developing new scientifically validated herbal drugs/nutraceuticals etc. and enhance the livelihood of knowledge providers. It is necessary to design a future ABS strategy that recognizes the results of research and promotes commercialization of the same at the same time ensuring sustainability of the resources. The "ABS potential" of the country needs to be studied and understood in a better way.

References

Bhatia N (2018) Ayurveda industry market size, strength and way forward, 2018, Confederation of Indian Industry (CII)). https://www.eximbankindia.in/

Bhartiya S (2019) Good practices of access and benefit sharing Indo-German Biodiversity programme access and benefit sharing partnership project ISBN 978-81-935619-1-1. Published by Deutsche Gesellschaftfür (GIZ)

Chaturvedi S, Ladikas M , Guo L ,Srinivas KR (2014) The living tree: traditional medicine and public health in China and India. Academic Foundation, New Delhi

Export-Import Bank of India (2020) Working paper no. 96 promoting exports from Kerala: Insights and Policy Perspective Export-Import Bank of India June 2020. https://www.eximbankindia.in/

Gopi DK, Mattummal R, Narayana SKK, Parameswaran S (2018) IUCNRed listed medicinal plants of siddha. IUCN Red List 1(1):15–22

Fran H, Benzie JAH, Lawson C (2021)A review of access and benefit-sharing measures and literature in key aquaculture-producing countries 2021. Rev Aquacult 13(3):1531–1548. Published: 26 Jan 2021, https://doi.org/10.1111/raq.12532,June

https://ayush.gov.in/
India Bioeconomy Report March (2020) Biotechnology Industry Research Assistance Council. https://birac.nic.in/
India State of Forest Report (2021) Forest survey of India, Ministry of environment forest and climate change
Jungcurt S (2011) Access and benefit sharing under the CBD and access to materials for research. In: Uhlir PF (ed) Designing the microbial research commons: proceedings of an international symposium. National Academies Press. https://www.ncbi.nlm.nih.gov/books/NBK92715/
Khan S, Seal S, Sharma S, Joshi U (2019) Species selection index (SSI) a novel tool designed for bio-resource selection under access and benefit sharing mechanism archives of agriculture and environmental science. https://doi.org/10.26832/2456632.040206
Mahmood JB (2020) How does corporate social responsibility transform brand reputation into brand equity? Econ Noneconom Perspect CSR Int J Eng Busin Man 12:1–13. https://doi.org/10.1177/1847979020927547journals.sagepub.com/home/enb
Michelle RF (2018) Access and benefit-sharing in practice: non-commercial research scientists face legal obstacles to accessing genetic resources. J Sci Pol Gover 13(1). Policy analysis: abs in practice. www.sciencepolicyjournal.org JSPG
Natarajan S, Venkateswaran K, Rao P, Reddy M, Rajasekharan PE (2020) Threatened medicinal plants of eastern ghats and their Conservation, pp 2–7. https://doi.org/10.1007/978-3-030-39793
Prakash N, Meenakumari B (2018) CEBPOL, NBA. Protected areas and access and benefit sharing (ABS): a review, 2018. published by the Centre for Biodiversity Policy and Law, National Biodiversity Authority, India, p 91
Pradeep M, Mukherjee A, Joshi A (2019) Conserving bioresources for better livelihoods, identifying tradeable and potential bioresources for ABS in Kumaun, Uttarakhand Deutsche Gesellschaftfür (GIZ)
Rourke MF (2018) Who are "indigenous and local communities" and what is 'traditonal knowledge' for virus access and benefit sharing? A traditional analysis of the convention on biological diversity and its Nagoya protocal. J Law Med 1–37. Griffth University, Australia
Shanbhag Gayathri N, Rana RS (2018) Regulation of access to biological resources and benefit sharing in India: an analytical study, National Biodiversity Authority, Chennai. Internationale Zusammenarbeit (GIZ) GmbH
Singh P, Bhattacharyya K (2016) Flavours, fragrances and nutraceutical ingredients in India Avendus Capital Private Limited

Chapter 13
The Biological Diversity Act and Enforcement: Challenges in India

Pushpa Kumar Lakshmanan

Abstract With the rapid growth of biotechnology and increasing demand for nature-based health remedies, access to biological resources and traditional knowledge has become a paramount necessity in the present context. The Convention on Biological Diversity and the Nagoya Protocol on Access and Benefit Sharing have rolled out legally binding obligations to the state parties. India being one of the pioneering parties in promoting access and benefit sharing (ABS) mechanism, has enacted the Biological Diversity Act (BD Act) to regulate the ABS process to conserve biodiversity and to protect the interests of traditional knowledge holders. The National Biodiversity Authority and the State Biodiversity Boards have taken several measures to implement the BD Act. The Biodiversity Management Committees have been set up at the local level in most of the states. Nevertheless, there are many compliance issues with the industry stakeholders and research institutions. Enforcement of the provisions of the BD Act need a fresh look to cope with the modern-day requirements of the law. The article focuses on the implementation and enforcement issues relating to the Biological Diversity Act in India.

Keywords ABS · BMC · Enforcement · Regulation · BD Act · Health remedies

13.1 Introduction

The Convention on Biological Diversity (CBD), 1992[1] brought out a paradigm shift in the global biodiversity governance. It recognised the sovereign rights of nation-states to protect and regulate the biological diversity and biological resources found within their national jurisdiction. This was necessitated to provide economic advantage of the bio-wealth by the biodiversity rich countries, to prevent biopiracy through

[1] The Convention on Biological Diversity (CBD) adopted at Rio de Janeiro on June 5, 1992, entered into force on Dec. 29, 1993, reprinted in 31 ILM 818 (1992). Almost all countries in the world (196 State Parties) are Party to the CBD except only Holy See (Vatican City) and the United States.

P. K. Lakshmanan (✉)
Law Centre-I, Faculty of Law, University of Delhi, Delhi, India
e-mail: lpkumar@lc1.du.ac.in

O. V. Oommen et al. (eds.), *Biodiversity Conservation Through Access and Benefit Sharing (ABS)*, https://doi.org/10.1007/978-3-031-16186-5_13

misappropriation of genetic resources and traditional knowledge, and to strengthen the sense of ownership to the nation-states to protect biodiversity and traditional knowledge available within their countries and to enable their sustainable use. This move gave the state parties new opportunities to devise legal and policy strategies to protect biological diversity from adverse activities. The CBD's broad objectives of conservation of biological diversity, sustainable use of its components and equitable sharing of benefits arising out of utilisation of genetic resources and traditional knowledge rolled out new strategies in favour of conservation as well as facilitated access to genetic resources and associated knowledge for sustainable use of biological resources for developmental purposes.[2] The Nagoya Protocol on Access and Benefit Sharing (ABS)[3] crystallized the rules for access and benefit sharing and obligated the state parties to come out with appropriate legal, administrative and policy measures to facilitate access and benefit sharing. The coming into force of Nagoya Protocol has shifted the global debates on ABS towards its implementation at the national levels.[4]

13.2 The Biological Diversity Act in India

India had been a major player in the negotiations of the CBD as well as the Nagoya Protocol on ABS. India made an early legislative response as early as in 2002 by enacting the Biological Diversity Act (BD Act) to implement the objectives of the CBD. Now this law is implementing the provisions of Nagoya Protocol also. The BD Act is implemented with the help of the Biodiversity Rules, 2004, the Guidelines on Access to Biological Resources and Associated Knowledge and Benefits Sharing Regulations, 2014 and a spate of notifications issued under the Act.[5] The BD Act established a three-layered institutional arrangement for regulating biological resources and associated knowledge in the country.

---

[2] See Article 1 of the CBD.

[3] The Nagoya Protocol on Access to Genetic Resources and the Fair and Equitable Sharing of Benefits Arising from their Utilization to the Convention on Biological Diversity, 2010. www.cbd.int/abs/text/. Accessed 09 April 2022.There are 135 parties to the Nagoya Protocol as on 25 May 2022. 63 countries are the non-parties, these include Australia, Bangladesh, Canada, Columbia, Costa Rica, Italy, Jamaica, New Zealand, Nigeria, Russian Federation, Singapore, Sri Lanka, and the USA. Morocco and Saint Lucia have already ratified, and the Protocol has not yet come into force for them.

[4] Balakrishna Pisupati, Access and Benefit Sharing: Issues and Experiences from India, Jindal Global Law Review, (2015) 6(1):31.

[5] See the Compendium of all the notifications issued under the BD Act, http://nbaindia.org/uploaded/pdf/Compendium_Book.pdf.

The National Biodiversity Authority

The National Biodiversity Authority (NBA) is the apex body in the hierarchy that functions at the central level in Chennai, Tamil Nadu.[6] Besides being the national level regulatory authority of access and benefit sharing mechanism, the NBA is an advisor to the Central Government on matters relating to conservation of biological diversity, sustainable use of its components and equitable sharing of benefits arising out of the utilization of biological resources. The NBA grants approval to the biological resources and knowledge as per sections 3, 4 and 6 of the BD Act for persons who are non-Indians and intuitions that have non-Indian participation. It grants prior approvals to applications pertaining to intellectual property rights involving biological resources or traditional knowledge. It advises the state government in the selection of areas of biodiversity importance to be notified as heritage sites and devising strategies for their management. It coordinates with the State Biodiversity Boards (SBBs) and Biodiversity Management Committees (BMCs) in the process of approving access and determining benefit sharing terms. NBA is also empowered to take necessary measures to oppose the grant of intellectual property rights in any country outside India on any biological resource obtained from India or knowledge associated with biological resource derived from India.

State Biodiversity Boards

The State Biodiversity Boards (SBB) are the state level regulatory bodies established by state government as per section 22 of the BD Act. The SBBs advise the state government on matters relating to conservation of biodiversity, its sustainable use and benefit sharing aspects. It grants approvals to the request for commercial utilization or bio survey or bio utilization of any biological resource for commercial utilization by Indian applicants covered under section 7 of the BD Act.

Biodiversity Management Committees

In every state and union territory, the local bodies constitute the Biodiversity Management Committees (BMC) as per section 41 of BD Act for the purpose of promoting conservation, sustainable use and documentation of biological diversity. The BMCs also engage in preservation of natural habitat, conservation of land races, folk varieties and cultivators, domesticated stock and breeds of animals, microorganisms and documenting of knowledge relating to biodiversity. The BMCs prepare, maintain, and validate Peoples Biodiversity Registers (PBR) in consultation with the local people. The BMC maintains a Register giving information about the details of access to biological resources and traditional knowledge and the details of the collection fee imposed by the BMC and the details of benefits derived from each case and the mode of sharing. The BMCs also advise on any matter referred to it by the SBB or NBA. It is expected to maintain the data of local *vaids* and practitioners using biological resources.

[6] Section 8 of the BD Act.

13.3 Enforcement Challenges of the Biological Diversity Act

As there are other articles touching upon the functioning of the NBA, SBBs and BMCs, and the number of approvals granted so far under the BD Act, this article does not analyze those aspects. However, in the enforcement angle, one may appreciate how India has effectively enforced the provisions of the BD Act to implement the Nagoya Protocol by uploading maximum number of Internationally Recognized Certificates of Compliance (IRCC). As on date, 3815 IRCCs have been uploaded by the state parties to the ABS Clearing House of the Nagoya Protocol. Even though there are 135 parties implementing the Nagoya Protocol, only 25 countries have uploaded IRCCs on the ABS Clearing House of the Protocol. Out of these, India alone has issued 2755 IRCCs distinguishing itself as a leader of ABS facilitation among all the countries. The other countries that come after India in terms of issuance of IRCCs are France (553), Spain (135), Kenya (78) and other countries issuing less than fifty IRCCs. From this one may infer that those 110 countries have not yet uploaded the IRCCs or not able to implement the provisions of the Protocol within their countries.

Although the best efforts have been made to implement the BD Act in India since its inception, there have been many critiques to this law. There are many reasons for criticisms. These include lack of legislative acumen in drafting the law in the first place that led to vagueness as well as lack of clarity in some provisions of the BD Act, requirement of formal prior approval for all sorts of research activities, andbenefit sharing obligations for cultivated bioresources.

1. Lack of Clarity in Law: One of the major enforcement challenges of the BD Act is the vagueness and lack of clarity in the law. For example, section 7 of the BD Act simply mentions intimation to SBBs but when we read it with sections 23 and 24 of the Act, it will signify that prior approval is required for access to biological resources by Indian applicants or non-section 3 entities. This provision gives the scope to the unwilling parties to argue that intimation does not mean prior approval. Vagueness in the legal provision has given much headache to the SBBs in many states. This issue has been challenged in the Divya pharmacy case,[7] wherein the Uttarakhand High Court stated that prior approval and benefit sharing are mandatory if we read sections 7 along with 23 and 24. If the law would have been drafted properly, such confusions would not occur.

2. Problem for researchers: Another major issue is the difficulty faced by the researchers (Indians as well as non-Indians) when they want to carry out academic research without any commercial motive. Many University research scholars as well as scientists from different research organizations felt it a real hurdle to get prior approvals for conducting research on genetic or biological resources or bio-survey or bio-assay. As most of the research programmes involve inter-country collaborations

[7] Writ Petition (M/S) No. 3437 of 2016, High Court of Uttarakhand, India; 2018 SCC Online Utt. 1035.

between public and private organisations, sharing of biological resources and transfer of research results after research from one researcher to the other were considered difficult as one has to get another approval for this purpose. This is considered as a major hurdle for research activities stifling research.

3. One time Exemption: Considering the large number of violations of the BD Act in different forms by the Indian industries and researchers in complying with the BD Act, the MoEFCC has issued a notification to allow a one-time exemption to the violators. Even though this notification has no backing of the BD Act, the Central Government has opted this route to bring the Indian entities in compliance with the law. The NBA has effectively handled this and several cases have been resolved.

4. Exempting Cultivated biological resources: A major claim by the users of biological resources is that when they cultivate the resources in their own gardens or farms, they are not obtaining the resources from anyone or not accessing them from the forest or wild. Hence, the argument is there is no harm to conservation of biodiversity and there is no need for obtaining access from any person and there is no requirement of benefit sharing. This has been a major demand of the industries that use biological resources for commercial utilisation. As of now there is no provision in the BD Act to exempt the cultivated resources from the purview of ABS process. When a review committee to amend the Act, Rules and Regulations deliberated this aspect, the committee recommended exemption of cultivated resources from the ambit of the Act. However, there could be a challenge for enforcement agencies in this context. If any person or company cultivates biological resources within its premises and uses it partially and procures a substantial quantity of the same biological resources from non-cultivated sources through traders or collectors from wild clandestinely, if they are not caught or proved, one can very well claim that all the biological resources utilized by the person are from cultivated biological resources. It is easy to circumvent the law in this way. How will the National Biodiversity Authority or the State Biodiversity Boards monitor such activities? The enforcement agencies should keep this aspect in mind, and they have to device proper mechanisms to investigate and monitor such activities.

5. Benefit sharing for access and utilisation of associated knowledge: The objective of Nagoya Protocol as well as the BD Act is to provide regulated access to genetic/biological resources and associated/traditional knowledge. Under the Indian BD Act, a non-Indian entity under section 3 needs to get prior approval from NBA for accessing biological resources as well as associated knowledge. Nevertheless, in the case of applicants falling under section 7, prior approval is required only for accessing biological resources for commercial utilization. Associated knowledge is not covered under section 7. However till 2018, even for the applicants under section 3, benefit sharing was determined only for the access and utilisation of biological resources even if the traditional knowledge was accessed. This scenario was changed when the ABS Committee recommended 1% of benefit sharing for accessing associated knowledge.

6. Regulating Access to genetic resources under the Nagoya Protocol and genetic resources relating to food and agriculture under ITPGFRA

The BD Act is a comprehensive legislation dealing with all kinds of biological resources, including agro-bio resources. When the ITPGFRA rolled out the multilateral system of exchange for international sharing of genetic resources relating to food and agriculture for research, the BD Act had to respond. Based on the recommendation of the NBA, the MoEFCC has issued an order in December 2014 with the effect that the crops covered under Annex-I of the ITPGRFA will not be governed by the BD Act and the Ministry of Agriculture that implements ITPGRFA will deal with those crops based on the multilateral system of exchange.[8]

13.4 A fresh look needed to enforce the BD Act

1. The seventeen years of functioning of the BD Act has gathered many valuable lessons that calls for a fresh look at the enforcement of the Act with new initiatives. E-filing of applications for approvals is a robust step taken by the NBA. In addition, the approval process should also be linked with the online services. One of the major delays for approval process is signing of ABS agreement after formal decision to give approval. Sending the physical agreement copy by post/mail and delay in signing or not responding for long time by the applicants have led to lot of delays in conclusion of ABS agreements. To avoid such scenarios, the online approval based on specific conditions can be devised that will effectively ease the approval process and benefit better implementation of the law.
2. Another concern is that the benefit sharing amount deposited in the gene fund does not reach the benefit claimers or the states for conservation purpose in time. An amendment in law or an administrative direction may be given to the effect that once the benefit sharing is determined by the NBA or SBB, the amount may be deposited directly to the BMC/identified benefit claimers or to the State where the biological resource or associated knowledge is obtained. This will provide immediate transfer of the benefit sharing amount and will avoid delay.
3. The non-commercial research activities by Indian or non-Indian parties must be permitted through online approvals without delay with specific conditions of prior approval in case of IPR protection or commercial application networking with the implementing agencies of the countries wherein the research is being conducted or the party resides by using user country measures.
4. Even after the years of coming into force of the BD Act and after the one time exemption given to the non-compliers of the law, there is a resistance among the

[8] Notification F. No. 28–5/2008-CS (NBA), dt. 17 Dec 2014, Ministry of Environment, Forests and Climate Change, Government of India.

industrial players in abiding by the law. Massive awareness needs to be undertaken for different stakeholders to encourage compliance of the law. Moreover, the good industries that comply with the provisions of the BD Act need to be recognised with some suitable awards or monetary incentives for their good conduct and compliance. This will encourage enforcement of the law effectively. The NBA may take up this issue in the Authority meeting and take a suitable decision on this.

5. At present, NBA, MoEFCC and UNDP are jointly recognising the good efforts of individuals and institutions for their contribution by awarding Biodiversity Awards. Best BMCs and best People's Biodiversity Registers are also recognised with the awards. In the same way, suitable recognition and awards shall begiven to different experts and stakeholders such as government officials, advocates, judges, academicians, NGOs, etc., who strengthen the implementation of the law directly or indirectly in through their innovative work and writing. This will hugely encourage them to contribute more in terms of implementation of the BD Act.

13.5 The Biological Diversity (Amendment) Bill, 2021

The Biological Diversity (Amendment) Bill, 2021 was introduced in the Lok Sabha on 16 December 2021. The Bill was referred to the Joint Parliamentary Committee on 20 December 2021. The Bill seeks to introduce fundamental changes in the law. It relaxes many requirements in favour of Indian companies and research activities. Section 3 of the Act borrows the definition of the Companies Act, 2013 to define non-Indian entities as the one incorporated or registered in India which is a foreign controlled company.[9] This provides a great relief to the industries that have small or namesake participation of foreign shareholders or board members that add value to their board.

The Bill also introduces a term 'codified traditional knowledge' that is neither defined in the Bill nor codified comprehensively except that are mentioned in the classical texts of Ayurveda, Siddha, Unani and the like.[10] The Bill reiterates that the 'codified traditional knowledge' is only for Indians.[11] However, with the lack of a comprehensive codification or an authentic database, it will be quite a challenging task for the enforcement agencies to differentiate between codified and uncodified traditional knowledge at the time of providing prior approval and determination of

[9] Section 5 of the Biological Diversity (Amendment) Bill, 2021.

[10] The term 'codified traditional knowledge' is being intruded in the definition of benefit claimers, section 4 relating to transfer of research results, and in section 7 of the Act.

[11] See section 3 of the Biological Diversity (Amendment) Bill, 2021 that substitutes a modified definition of 'benefit claimers' and section 6 of the same Bill that modifies section 4 of the BD Act.

benefit sharing. Again, Peoples Biodiversity Registers developed by the BMCs codify the locally available traditional knowledge. If this can be qualified as a codified knowledge, then this new provision will preclude any kind of approval or benefit sharing in favour of the BMCs or local communities as per section 7 of the Act. This might also hamper codification of traditional knowledge at the BMC level because once codified, it may be available only for Indian users and section 3 entities will not be able to use them for research and the results of research cannot be transferred as per amended section 4 of the Act. Under section 4, the Bill substitutes a provision to the effect that one may share or transfer any results of research on any biological resources or associated traditional knowledge without prior written approval of the NBA but that cannot include codified traditional knowledge which is only for Indians.[12]

The persons covered under section 7 of the BD Act need not get prior approval of the NBA before getting any intellectual property rights which is a mandatory requirement in the present BD Act. They need to only register with the NBA before the grant of the intellectual property rights. Only when they commercialize the intellectual property right, they need to get prior approval of the NBA.[13]

Indian companies and persons covered under section 7 of the BD Act are required to give prior intimation and obtain prior approval as per sections 23 and 24 for accessing biological resource and its associated knowledge for commercial utilisation. They are not required to give prior intimation nor to get prior approval from SBBs while using codified traditional knowledge or cultivated medicinal plants and its products for commercial utilisation. In addition to the already exempted persons, the registered AYUSH practitioners including those practicing Indian systems of medicine for sustenance and livelihood are out of the purview of section 7 of the Act.[14]

Section 58 of the Act that made any offence under the Act cognizable and non-bailable is going to be omitted.[15] The effect would be that the offences under the Act will be no longer cognizable and easily bailable which is a major relief for the industry and researchers. The provisions prescribing imprisonment for violations of the provisions of the Act have been removed totally and instead penalty or fine has been suggested.[16]

A new provision is added to appoint an officer by the Central Government not below the rank of Joint Secretary to the Government of India or a Secretary of the State Government as an adjudicating officer to impose penalty. The adjudicating officer is given the quasi-judicial powers to conduct the inquiry, take evidence, pass an order imposing penalty.[17]

---

[12] Section 6 of the Biological Diversity (Amendment) Bill, 2021.

[13] Section 8 of the Biological Diversity (Amendment) Bill, 2021.

[14] Section 9 of the Biological Diversity (Amendment) Bill, 2021.

[15] Section 39 of the Biological Diversity (Amendment) Bill, 2021.

[16] Section 38 of the Biological Diversity (Amendment) Bill, 2021 modifying section 55 of the BD Act.

[17] See Section 38 of the Biological Diversity (Amendment) Bill, 2021 introducing section 55A.

The Bill also provides power to the authority or officer empowered by the Central Government to enter, inspect and survey any activity to implement the provision of the BD Act.[18]

As the Biological Diversity (Amendment) Bill, 2021 introduces many new provisions as discussed above; this move may trigger many enforcement issues. One has to wait and see how many of these provisions see the light of the day when it comes into effect as an amendment.

Acknowledgements The author gratefully acknowledges the support provided by the Institution of Eminence, University of Delhi under the FRP that helped in developing this research article.

---

[18] Section 38 of the Biological Diversity (Amendment) Bill, 2021 introducing section 55B.

Part V
ABS and Its Economics

Chapter 14
Access and Benefit Sharing: An Innovative Financial Mechanism for Biodiversity Conservation

Prakash Nelliyat

Abstract Biodiversity is a critical element for providing ecosystem services and the basis for good health, food and livelihood security, economic growth, and moderation of climatic conditions. Besides, biodiversity has significant commercial importance, as it provides the source materials for different industries. However, biodiversity degradation is one of the major global challenges. The funding gaps in meeting biodiversity strategies and targets are huge and their achievements through public financing alone is extremely difficult. We need a shift towards a new investment and policy paradigm incorporating the economic value and financial benefits of biodiversity. In this regard, companies and the financial sectors should come forward for biodiversity management. After the advent of the Convention on Biological Diversity (CBD), we have seen a quick transition of biodiversity from a global public good/heritage to a resource of those countries where the biodiversity occurs (sovereign right). In this context, innovative financial mechanisms like Access and Benefit Sharing (ABS) have a good scope. As a party to CBD, India initiated legal and institutional measures for the implementation of the ABS and mobilized an amount of Rs. 130 crores for biodiversity conservation. However, the ABS process in the country faces huge challenges, as different stakeholders are not well aware about the ABS principles and their operation. At present, only very few cases come under the ambit of ABS and the amount collected also not been effectively utilized for biodiversity conservation. ABS is a techno-legal issue. Hence, its proper understanding and enforcement requires special attention with a more scientific approach. No doubt in a mega bio-diverse country like India, which possesses substantial bio-wealth in its rich ecosystems, the scope of the ABS potential is huge and its effective administration will be a reliable source for mobilizing finance for the conservation of biodiversity.

Keywords Financial mechanisms · Access and benefit Sharing · Biodiversity conservation · Bio-entrepreneurs · Tradable bio-resources

---

P. Nelliyat (✉)
Former Fellow (ABS), Centre for Biodiversity Policy and Law, National Biodiversity Authority, 5th Floor, TICEL Biopark, Taramani, Chennai 600113, India
e-mail: prakashnelliyat@gmail.com

O. V. Oommen et al. (eds.), *Biodiversity Conservation Through Access and Benefit Sharing (ABS)*, https://doi.org/10.1007/978-3-031-16186-5_14

14.1 Introduction

Biological diversity is a global asset, with tremendous value to the present and future generations. However, the species and ecosystem are under greater threat now than ever before. Considering this fact, arresting the decline of biodiversity is a major objective of global environmental policies with initiatives at national and local levels. With this goal in mind, the Convention on Biological Diversity (CBD) was initiated on 5th June 1992 at the United Nations Conference on Environment and Development (the Rio "Earth Summit") and came into force on December 1993 as an international instrument for comprehensively addressing biological diversity. 196 nations, including India, are parties in the CBD. The objectives of the Convention include: the conservation of biological diversity, the sustainable use of its components, and the fair and equitable sharing of benefits arising from the utilization of genetic resources.

The overwhelming evidence of the recent Intergovernmental Science-Policy Platform on Biodiversity and Ecosystem Services (IPBES) Global Assessment, presents an ominous picture. It stated that "the health of ecosystems on which we and all other species depend is deteriorating more rapidly than ever. We are eroding the very foundations of our economies, livelihoods, food security, health and quality of life worldwide". However, the report also tells us that it is not too late to make a difference, but only if we start now at every level from local to global. Through 'transformative change', nature can still be conserved, restored and used sustainably. This is also important to meetmost other global goals including Sustainable Development Goals (SDGs). By transformative change, the study means a fundamental, system-wide reorganization across technological, economic and social factors, including paradigms, goals and values (United Nation 2019).

For biodiversity management considerable volume of finance is required and the studies proved that there is a huge funding gap, which is an emerging challenge. Generally, the Governments' budget allocation for the environment sector is a small portion compared to other sectors. More funds are allocated for the developmental sectors. However, considering the gravity of biodiversity degradation, the Government as well as the business sector needs to come forward with innovative financial mechanisms for biodiversity conservation. Access and Benefit Sharing (ABS), one of the objectives of CBD, is an emerging financial (incentive to community) option for biodiversity conservation and its sustainable use.

For the successful operation of the ABS, countries need to implement exclusive legal and institutional measures. Further, the significance of ABS with respect to the conservation of biodiversity and sustainable use of its components has to be demonstrated to various stakeholders, including the industrial communities. Biological resources are inputs or raw-materials for manufacturing different consumer products and profit options for bio-entrepreneurs. India is rich in biodiversity or biological resources and also the home of a vast repository of traditional knowledge (TK) associated with biological resources. For the effective implementation of ABS, India enacted the Biological Diversity Act (2002)—Here after BD Act–and Rules (2004) and made decentralized institutional arrangements.

According to the CBD and the BD Act, the providers of particular bio-resources are entitled to avail a share of the benefit arising from the commercial utilization of that resource from the agencies or entrepreneurs through ABS. The fund derived through ABS shall be used, by the community, for the conservation of biodiversity. India made some progress in ABS, but its potential is huge. If we succeeded in recap our full ABS potential and mobilized the money, it becomes a substantial amount for biodiversity conservation.

This chapter "ABS an Innovative Financial Mechanism for Biodiversity Conservation: India's Experiences, Challenges and Way Forward" examines the Paradigm Shift in Biodiversity Management, Finance for Biodiversity Conservation, ABS Principle and its Significance as a Financial Solution for Biodiversity Conservation, India's ABS Experiences and Challenges, ABS Money distribution and Pre-conditions for Strengthening the ABS Process in India.

14.2 Paradigm Shift in Biodiversity Management

Biodiversity represents the variety of life on earth, which include species diversity (the numbers and kinds of living organisms), genetic diversity (genetic variations within species) and ecosystem diversity (the variety of habitats, biological communities and ecological processes). The services of ecological systems are critical to the functioning of the earth's life-support system. They contribute to human welfare, both directly and indirectly, and therefore represent a significant part of the total economic value of the planet (Costanza et al. 1997). Bio-diverse ecosystems provide vital services such as; the regulation of water flows and levels, protection against extreme weather conditions, the purification of air and water, the prevention of soil erosion, and opportunities for recreation and spiritual reflection. Besides, biodiversity offers essential resources and goods, such as food, fibre, and medicines (CBD 2011).

However, biodiversity faces multiple challenges from various factors that include: land-use change habitat fragmentation, degradation and loss, over-exploitation of resources, shrinking genetic diversity, spread of invasive alien species, declining forest resource base, climate change and desertification, and impacts of various development projects including pollution. The loss of biodiversity constitutes a concern for human welfare, especially for the well-being of the poorest, since it acts as a major livelihood option for them. Biodiversity loss presents significant economic challenges to the society and nation. Hence, arresting the further destruction of biodiversity and its conservation is an agenda in the environmental management policies of the government.

Generally, biodiversity and biological resources are unequally distributed in the world and their supply is restricted in certain regions. On the other hand, their demand is escalating universally particularly in the globalized era. Broadly, biological resources collection, transfer, and exchange are progressing at an alarming rate in biodiversity rich areas of the world. This business trends on biodiversity has led to the transformation of biodiversity more from a global public good to a regional or local public good or as state property and viewed as national sovereignty. In this context, the CBD insisted their parties to follow ABS through legal and institutional arrangements for the conservation and sustainable use of their biodiversity.

In brief, historically biodiversity has been considered as a 'global public good' and its conservation and management are recognized as a global responsibility. However, the scenario is changed drastically in recent period, particularly with the advent of CBD and the Nagoya Protocol. According to Pisupati (2012), the insufficient understanding of public and private goods leads to challenges, particularly for managing public goods like biodiversity.

14.3 Finance for Biodiversity Conservation

The Biodiversity Finance Initiative Workbook (2018) on "Finance for Nature" stated that awareness on 'biodiversity underpins sustainable development' has increased among finance decision makers and practitioners. Biodiversity finance is growing in value and sophistication. However, all of this has not yet resulted in a major shift or boom in finance flows towards biodiversity. Effective governance and partnerships between finance and environmental actors are essential to guarantee the up-scaling and sustainability of biodiversity financing.

Broadly, financing biodiversity is a shared responsibility of governments and the private sectors. Quantification of Biodiversity Finance Needs, past expenditures and the value of natural capital are the prerequisites to frame sound Biodiversity Finance Solutions. Biodiversity finance is not only about mobilizing new resources, but it is concerned with delivering better on what is available, reallocating resources from where they harm to where they help, and acting to reduce the need for future investments (UNDP 2018).

Biodiversity is in a severe decline due to a combination of conflicting private and public interests, incoherent policy and governance, and insufficient financing. Although at least US$ 52 billion is spent on biodiversity per year globally, this is against an estimated annual financing need of between US$150 billion and US$440 billion. This funding gap is a major challenge, hampering the achievement of both the CBD's Strategic Plan and the SDGs. However, it can be overcome, as it is between just 0.2 percent and 0.6 percent of global GDP (UNDP 2018). In this context, we need a shift towards a new investment and policy paradigm that incorporates the economic value and financial benefits of biodiversity. At present, even if companies and the financial sector are benefiting from biodiversity, they are not investing adequately in biodiversity conservation. Globally, ABS might be a legally bounded principle to overtake the above hurdles.

The BIOFIN Project (India) assessed the expenditure being incurred on biodiversity in the country and the requirement of funds for implementing the National Biodiversity Action Plan (NBAP). Based on this exercise, the annual average public finance for the period 2017–18 to 2021–22 is assessed to be around Rs. 70,121 crores (USD 10 billion) and the financial needs for implementing NBAPs for the same period to be around Rs.1,15,970 crores (USD 16.5 billion) (NBA 2019).

India's Biodiversity Finance Plan (BFP) suggests twelve potential finance solutions to bridge the finance gap, taking into account all existing financial as well as innovative instruments being tried out in the country. The BFP considered ABS as one of the financial solutions for mobilizing the fund for biodiversity conservation.

14.4 ABS Principle and Its Significance as a Financial Solution for Biodiversity Conservation

Biological or genetic resources are the fundamental source for bio-prospecting, which has been described as 'the exploration of biodiversity for commercially valuable genetic and biochemical resources'. Generally, bio-prospecting is primarily taking place in the developed nations or elsewhere with their technical support. Since, developed nations are advanced in biotechnology knowhow they are reaping considerable benefits of biological/genetic resources' (which are mainly confined to developing nations) commercial utilization. However, the commercial value and benefits obtained from genetic/biological resources need to be distributed to the primary stakeholders, and communities in such a manner that they act as a positive force or incentive for the conservation and sustainable use of biodiversity (Bavikatte and Morten 2015). This mechanism is generally known as ABS. Ultimately, through this mechanism a substantial fund required for biodiversity conservation can be mobilized by nations.

ABS, one of the key objectives of the CBD, refers to the way in which genetic resources may be accessed, and how the benefits that result from their use are shared between the people or countries using the resources (users) and the people or countries that provide them (providers). ABS is an incentive mechanism for local communities in biodiversity conservation and its sustainable use. ABS provides for the use of natural resources in an efficient and sustainable manner for various purposes, including commercial ones. It is a tool which simultaneously enables the national mandates of biodiversity conservation and economic development in a sustainable way (Nelliyat and Balakrishna 2014).

The ABS framework provides a formal guidance for the way in which biological or genetic resources are accessed, and the way benefits are shared between people or countries using the resources (users) and the people or countries that provide them (providers). The ABS philosophy proposes that providers of bio-resources are entitled to receive fair benefits from the users. In this context, ABS balances the rights of the users of bio-resources with the rights of the providers of such resources. It is vital that both users and providers of bio-resources understand and respect the legal, administrative and policy frameworks at the national and local levels, as well as those outlined in the CBD and the Nagoya Protocol. The ABS is based on prior informed consent (PIC) being granted by a provider to a user, and negotiations between both parties that result in mutually agreed terms (MAT).

Now, globally, ABS is emerging as an innovative approach and an effective financial (incentive) mechanism for biodiversity (even protected areas) conservation and its sustainable utilization (Nelliyat and Meenakumari 2018a). The rational for the ABS is much broader (Fig. 14.1). Hence, it is assumed that, if biodiversity conservation and sustainable use materialize through ABS the biodiversity stock of the country is intact. Hence, it can provide precious services including carbon sequester, moderate weather, hydrological cycle, nutrients cycle etc. along with the supply of bio-genetic resources, as raw-materials to the commercial sectors. This could be continued in a cyclical manner for the harmony of nature and human wellbeing.

Biological resources based industrial units are using wild as well as domesticated biological resources, which include: plants, animals, microorganisms and genetic materials, available in divergent ecosystems such as: forests, agriculture, wetland and marine spots. Industries use biological resources in different quantities based on the nature of products manufactured by them. Further, the application of modern biotechnology in the production process also varies substantially among the industries.

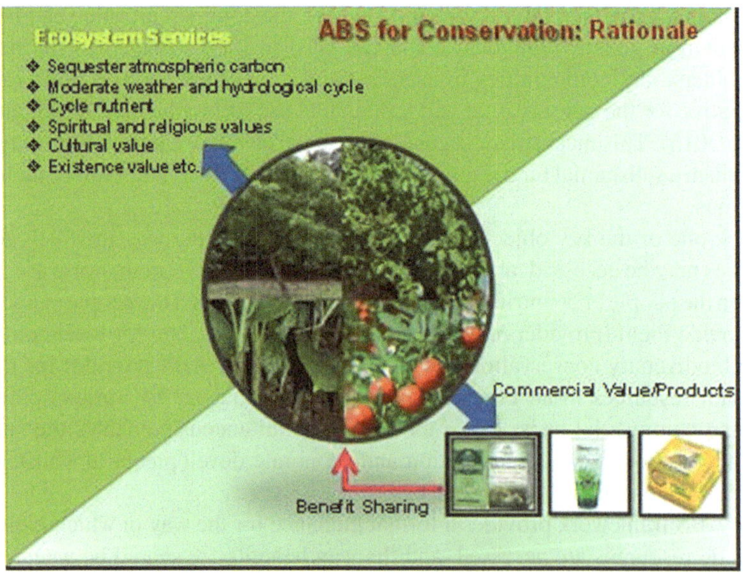

Fig. 14.1 Rational for ABS

For example: pharmaceuticals (modern drugs) and agricultural seeds industries collect genetic/biological resources in limited quantities for research, particularly for identifying the biochemical compound or genetic modifications with the application of modern biotechnological devices. They have provisions for their culture/multiplication. On the other hand, botanical medicines (AYUSH) and food processing industries need biological resources as raw-materials in bulk quantities. Compared to the pharmaceutical and seed industries, the research and development (R&D) and application of modern biotechnology devices in botanicals and food processing industries are limited. In certain cases, similar biological resources may generate different values in different industrial sectors, which make the ABS process more cumbersome. In this scenario, sector specific approach on ABS would facilitate in framing a convincing argument for different industries and other stakeholders (Nelliyat and Meenakumari 2018b).

14.5 India's ABS Experience

India is one of the 17 mega bio-diverse countries of the world. India has only 2.4% of the world's land area. But the country possesses 16.7% of the world's human population and 18% of livestock. It contributes about 8% of the known global biodiversity. The country has 7–8% of all recorded species, including over 45,000 species of plants and 91,000 species of animals. Of the 34 global biodiversity hotspots, four (the Himalaya, the Western Ghats, the North-east, and the Nicobar Islands) are present in India. Considering the outstanding universal values and exceptionally high levels of endemism in the Western Ghats, 39 sites in the states of Kerala, Karnataka, Tamil Nadu and Maharashtra have been considered on the United Nations Education Scientific and Cultural Organization (UNESCO) World Heritage List in 2012 (MoEFCC 2014).

The rich biological/genetic resources and the heritage of traditional knowledge in the country obtained global attention long ago. Large sections of the rural population are heavily dependent on natural resources for their livelihood. The genetic/biological resources of India are raw-materials for manufacturing different consumer products or the benefit/profits options for large number of bio-entrepreneurs. India's Gross Domestic Product (GDP) is expected to reach US$ 6 trillion by 2027. In this context, nature-based solutions such as Access and benefit-sharing (ABS) are certainly necessary for a more inclusive, sustainable, resource efficient and greener economy.

As a party in the CBD, India's initiatives in fulfilling its objectives (the conservation of biological diversity, the sustainable use of its components and, the fair and equitable sharing of benefits arising from the utilization of genetic resources) were appreciable. India ratified the Nagoya Protocol on ABS on 12th October 2014, and implementing the protocol's provisions through well assigned legal and decentralized institutional measures. India enacted the Biological Diversity Act in 2002 and notified the Rules (Biological Diversity Rules) in 2004 to give effect to the provisions of the Convention including those relating to ABS. The Nagoya Protocol on ABS is also being implemented through the BD Act at the national level.

The BD Act is implemented through a three-tier institutional mechanism: The National Biodiversity Authority (NBA) at the national level; State Biodiversity Boards (SBBs) at the provincial (State Government) level; and the Biodiversity Management Committees (BMCs) to be set up by the elected bodies at the local level. While all the states (28) have set up SBBs, setting up of BMCs is an ongoing process. So far around 2.77 lakh BMCs have been constituted by the local bodies in the country.

The objectives of the BD Act are similar to the CBD objectives. The salient features of the BD Act are, it.

- Provide sovereign rights of the country over its biological resources
- Stops misappropriation of biological resources and associated TK (bio-piracy).
- Regulates access and use of biological resources and / associated knowledge
- Ensure sustainable utilization of biological resources and / associated knowledge and equitable benefit sharing.
- Provide legal recognition and support to the biological resources and associated TK.
- The Key provisions of the BD Act and Rules are given in Table 14.1

Exemption in ABS provision as per the Act include: (a) human genetic material, (b) value added products (products which may contain portions or extracts of plants and animals in unrecognizable and physically inseparable form), (c) 421 biological resources notified as ‘normally traded as commodities’, and (d) local people and communities, including growers and cultivators of biodiversity, and *vaids* and *hakims,* who have been practicing indigenous medicine. Contravention or abetment of contravention of the provisions of Sections 3, 4, 6, 7, 20 or 24 of the Act amounts to a cognizable non-bailable offence.

For the implementation of various provisions of the BD Act, several notifications have been issued so far. The Guidelines on the Access to Biological Resources and Associated Knowledge and Benefit Sharing Regulations (2014) (hereafter Guidelines–2014) prescribe the scheme of processing the applications, along with templates and terms for benefit sharing.

Accordingly, ABS can be availed of from the biological resources based industries either: based on the biological resources’ purchased price by the industries, based on the raw-material cost, prescribed as 3–5% or based on the ex-factory sale value of the product minus government taxes, where biological resources are involved in production fully or partially (0.1–0.5%). When transferring the results of research, the benefit sharing obligation is 3.0–5.0% of the monetary consideration. In the case of Intellectual Property Rights, the benefit sharing obligations are: (a) If the applicant himself commercializes the process/product/innovation, the monetary sharing shall be 0.2–1.0% of the annual ex-factory gross sale (minus govt. taxes) and (b) If the applicant assigns/licenses the process/product/innovation to a third party for commercialization, the monetary sharing shall be 3.0–5.0% of the fee received in any form and 2.0–5.0% of the royalty. When accessed bio-resources and TK are

Table 14.1 Provisions of the biological diversity act, 2002 and rules, 2004

Section	Persons	Activity	Purpose
Section 3 (NBA)	Foreign citizens, Non-Resident Indians (NRIs), body corporates, associations or organisations not incorporated or registered in India or incorporated or registered in India which has any non-Indian participation in share capital or management	Obtainment of any biological resource occurring in India or knowledge associated thereto	Research, Commercial Utilization, Bio-survey and Bio-utilization
Section 4 (NBA)	Indian citizens, foreign citizens, NRIs, body corporates, associations or organisations incorporated or registered in India with or without any non-Indian participation in share capital or management and body corporates, associations or organisations not incorporated or registered in India	Transfer of results of any research relating to any biological resource occurring in, or obtained from India, to any person covered under Section 3	Transfer of research results for monetary consideration or otherwise
Section 6 (NBA)	Indian citizens, foreign citizens, NRIs, body corporates, associations or organisations incorporated or registered in India with or without any non-Indian participation in share capital or management and body corporates, associations or organisations not incorporated or registered in India	Application of any IPR in or outside India for any invention based on any research or information on a biological resource obtained from India	Obtaining IPR, by whatever name called, in or outside India
Section 20 (NBA)	Any person who has been granted approval under Section 19	Third party transfer of any biological resources or associated knowledge there to which is the subject matter of an approval granted by the NBA under Section 19	Transfer of biological resources or associated knowledge

(continued)

Table 14.1 (continued)

Section	Persons	Activity	Purpose
Section 7 (SBB)	Indian citizens, body corporates, associations or organisations which are registered or incorporated in India and not covered under Section 3	Obtaining any biological resource	Commercial utilization, bio-survey and bio-utilization for commercial utilization

Source: NBA (2010)

Table 14.2 ABS fund mobilized in NBA

ABS	Approximate amount (INR in Crore)
ABS fees on red sanders	100
Benefit sharing, upfront fees and application fee	5
Total	105

Source: NBA (2020)

transferred, the benefit sharing obligation is 2.0–5.0% (sectoral approach) of any amount and / or royalty received from the transferee.

The NBA and the SBBs mobilized Rs.130.6 crore through ABS. The following tables (Tables 14.2, 14.3 and 14.4) provide a comprehensive picture about the ABS money collected till 2020 in India.

BIOFIN assessments revealed the high potential of ABS as a profitable financial solution. A rough estimate of the ABS revenue of SBBs in India (carried out based on the experience/data of Uttarakhand Biodiversity Board) for the Biodiversity Finance Plan through the BIOFIN project, would at least be about Rs.1500 crore in terms of resources available for biodiversity conservation and utilisation in a sustainable way, which would contribute to a greener economy and sustainable livelihood. (This might be a small portion of the real ABS potential of the States in India). There is also high potential for job creation and other socio-economic benefits. The BD Act provides that the ABS money collected could, in addition to conservation and promotion of biological resources, be spent on socio-economic development of the area from which the resources have been harvested. Assuming that 50% of the annual potential ABS revenues estimated would be spent on wages, the potential for employment generation is about 50 million working days for agricultural related workers per year (according to the national daily rate for agricultural labourers declared by the Central Government in 2017) (NBA 2019).

Table 14.3 ABS amount collected by different SBBs

S. No	States/SBBs	ABS Collected (INR)
1	Andhra Pradesh	12,79,98,985
2	Uttrakhand	5,93,43,986
3	Karnataka	2,91,24,110
4	Madhya Pradesh	75,49,700
5	West Bengal	2,97,892
6	Telangana	3,13,714
7	Himachal Pradesh	76,18,056
8	Kerala	96,657
9	Uttar Pradesh	91,84,221
10	Maharashtra	44,24,350
11	Sikkim	8,500
12	Gujarat	94,10,134
13	Tripura	7,01,577
14	Goa	2,000
	Total	25,60,73,882

Source: Concerned SBBs (2020)

Table 14.4 Total ABS Fund Mobilized in India

Institution	Amount (INR in crore)
NBA	105
SBBs	25.6
Total	130.6

Source: NBA and SBBs (2020)

14.6 ABS Money for Biodiversity Conservation: Strategies Initiated for Fund Utilization

In ABS process, along with the mobilizing the money, its utilization for biodiversity conservation is also important. In this regard various sstrategies and approaches followed in the country includes: (a) Benefit Sharing Proposals of the Expert Committee on Red Sanders; (b) NBA's Instructions to SBBs for Handling the ABS Applications; and (c) Ordinance of the Kerala Government on Kerala Biodiversity Fund.

14.6.1 Benefit Sharing Proposals of the Expert Committee on Red Sanders

As huge money is acquired through the red sanders related ABS into the NBA as well as the Andhra Pradesh State Biodiversity Fund, its distribution for biodiversity conservation was a priority. Since the red sanders case is unique and specific, NBA decided to administer the distribution of ABS money through the suggestion of an Expert Committee (Box 14.1).

Box 14.1
Expert Committee on Red Sanders

It is very clear that the red sanders (*Pterocarpus santalinus*) play a significant role in India's ABS, particularly to the NBA (National Biodiversity Fund) and in the Andhra Pradesh State Biodiversity Board. The Expert Committee on Red Sanders came up with a "Comprehensive Policy for Conservation, Sustainable use and Fair and Equitable Sharing of Benefits Arising from the Utilisation of Red Sanders under the Biological Diversity Act, 2002".

The red sander is a species of tree which has high-economic value and is endemic to the South Eastern Ghats of India. It is used for furniture making, manufacturing musical instruments, pharmaceuticals, extraction of dye, cosmetics industry etc. It has high commercial value both in the domestic and global markets due to its rich red colour and hardness. Red sanders are a unique natural heritage of India and endemic biological resource. The species is listed as an endangered species in the IUCN Red List and included in Appendix II of the CITES. Its export is either prohibited or restricted under India's foreign trade policy. The Ministry of Commerce and Industry recently relaxed the restrictions in the export of seized/confiscated red sander wood to enable the Government of Andhra Pradesh and DRI to export 9784.1363 MT in the log form lying in their custody.

High demand in international markets coupled with lack of lawful supply of wood from India, explains its rampant illegal harvest and smuggling it out of India. 5% of the sale amount (excluding taxes and other Government dues) has to be deposited with the NBA / SBB as a benefit sharing component. Successful bidders can proceed with the formalities in accordance with Regulation 3 (3) of the ABS Guidelines, 2014.

Accordingly, the NBA received benefit sharing amount from 15 successful bidders, who have sought for approval to access red sanders wood auctioned by the Andhra Pradesh Forest Department. As the benefit sharing amount was obtained from the sale of confiscated red sanders wood, no individual benefit claimers could be identified. Hence, the NBA, as per Section 27 (2) of BD

Act, 2002 decided to defray the benefit sharing amount for the purpose of the conservation of red sanders.

The Committee recommended that benefit sharing amount realised from the sale of red sanders by the NBA/SBB may be shared among various stakeholders at prescribed percentages for various activities subject to certain conditions. The SBBs concerned should initiate effective conservation measures, including the prevention of illegal felling and smuggling, framing of simplified, streamlined and standardized procedures for raising and harvesting of large scale plantations by farmers and further creating a conducive marketing ecosystem in order to achieve the long term goal of conservation and sustainable use of red sanders. In this context the committee has come up with detailed modalities to utilise the benefit sharing component realised from the sale of red sanders.

14.6.2 NBA's Instructions to SBBs for Handling the ABS Applications

After the introduction of the Guidelines (2014), the SBBs in the country initiated the ABS as per Section 7 of the BD Act. As there is some confusion and flaws in processing the applications, the NBA decided to come up with the Guidelines for handling different steps in ABS, including the identification of the geographical locations of the bio-resources (Box 14.2). This guideline provides a comprehensive picture about each of the steps to be followed by the SBBs in processing the ABS applications and the points insisted upon in identifying the biological resources' origin with respect to the benefit sharing. However, based on the experience of different SBBs, they realized that, despite the concerned officers' repeated instructions or reminders, the applicants are not in a position to reveal the origin of the bio-resources. According to SBB officials in the present circumstances, it is practically impossible for the applicants (bio-resources based industries) to reveal the bio-resources origin as they are obtaining it from the markets or traders.

Box 14.2

Operational Guidelines to the State Biodiversity Boards for Processing of Applications for Access to Biological Resources received under Section 7 of the BD Act, 2002.

- The officer or person concerned shall verify whether the biological resources to be accessed are clearly specified in the application along with their common and scientific names, their parts and quantity.

- The officer or person concerned shall also verify if the biological resource to be accessed is from the wild (forest area) or cultivated or sourced from traders or communities or Joint Forest Management Committees.
- In case of any discrepancy or lack of information, the same shall be clarified from the applicant.
- The SBB is expected to entertain an application for access to biological resources only if the biological resources specified occur within the territorial jurisdiction of the State concerned.
- The officer or person concerned needs to verify whether the exact geographical location like the name of the Village, Panchayat, Block, Taluk and district is within the State from where the particular biological resource is proposed to be accessed.
- If the biological resource is proposed to be bought from a trader or from the local market, institution or communities, etc. the officer or person concerned needs to verify that the complete details (Name, complete address, contact details, etc.) of such sources are also provided by the applicant. This is important while sharing benefits to the concerned.
- In case a person other than the applicant is authorized for collecting the biological resources, the name and contact details of the person so authorized have to be verified. The authorization letter in favor of the person so authorized needs to be obtained.
- The officer or person concerned shall verify the exact quantity of biological resources and their parts proposed to be collected from each geographical location.
- This exercise may also involve a judicious assessment of the availability of some of the biological resources depending on the particular parts to be gathered in the light of sustainable harvesting.
- When the biological resources are sourced from the jurisdiction of two or more BMCs, the total amount of the accrued benefits shall be shared among them in proportion as decided by the governing body of the SBB with due diligence.

14.6.3 Ordinance of the Kerala Government: Kerala Biodiversity Fund

In the Kerala State Biodiversity Boards' (KSBB) case, out of the 190 Form I applications they received, none of them mentioned the origin of the bio-resources or the geographical locations. Most of the applicants had written 'un-known', as they have obtained them from the market or traders. The Board also decided that keeping the ABS money idly in the State Biodiversity Fund is inappropriate. Hence they set up, a Government Ordinance (Box 14.3) for the 'Establishment, administration, utilization and auditing of Kerala Biodiversity Fund' with multiple biodiversity conservation options. The Board thinks that this Ordinance might be an effective intermediate option till the applications are received with the details on the origin of bio-resources, which may not be possible now.

Box 14.3
Ordinance of the Kerala Government

The Government of Kerala has come up with an Order for the establishment, administration, utilization and auditing of funds relating to the Kerala Biodiversity Fund on 18th January, 2020. The fund is sourced from 12 heads, which include: (a) any benefit passed onto the KSBBas per the benefit sharing agreement entered into by the NBA under Section 21 of The BD Act 2002 and Rule 20 of The Biological Diversity Rules 2004, (b) cess or fee or royalties on benefits acquired through benefit sharing agreements with persons referred to in Section 7 of the Biological Diversity Act 2002.

The Kerala Biodiversity Fund shall be utilized for the following activities:

- Management and conservation of Biodiversity heritage sites
- Compensation or rehabilitation of any section of the people economically affected by notification under sub Section 37 of the Act;
- Conservation and promotion of sustainable utilization of biological resources;
- Socio-economic development of areas from where such biological resources or knowledge associated therewith has been accessed, in consultation with the local bodies concerned.
- Support an existing or a new program or project in biodiversity conservation, including operational expenditure.
- Capital works on sites of local biodiversity conservation or enhancement may be considered provided the biodiversity value of the site is clearly demonstrated in the application and established in site verification.
- To site management works that are part of a larger conservation management plan, or part of a larger scheme established for the benefit of the biological diversity of the state.
- Documentation, monitoring and for in situ and ex situ conservation of marine, fresh water and terrestrial species, indigenous varieties and breeds.
- Documentation and monitoring of critical habitats and ecosystems and biodiversity rich areas/community conserved areas.
- Inventorization, and cataloguing of biodiversity and monitoring the rate of depletion of biodiversity
- Conservation programs in public and also in private lands;
- Eco-restoration of degraded ecosystems

- Projects or programmes of the KSBBor the BMCs, except for the preparation of PBRs.
- Implementation of the state and local biodiversity strategies and action plans
- Funding of Governmental and NGOs, and other research organizations for location specific research related to conservation, sustainable use and fair and equitable benefit sharing of biological diversity and traditional knowledge associated with biological resources.
- Support to biological prospecting, value addition and bio-survey of bio resources
- Capacity building of all stakeholders of biological resources on the conservation, sustainable use and fair and equitable benefit sharing of biological diversity.
- Scouting and documentation (including digitization) and conservation of TK
- Recognizing and rewarding major contributions including innovations in conservation and sustainable utilization of bio-resources.
- Any other innovative schemes/programme/projects aimed at biodiversity conservation and its sustainable use.
- For meeting the expenses incurred for the purposes authorized by this Act.
- All expenses of administration of the fund
- Monitoring and Evaluation

It is very clear that the above approaches are situation specific and circumstantial. According to Section 32 (2) of the BD Act, the State Biodiversity Fund shall be applied for biodiversity conservation strategies in general. Here, the purpose of ABS may be mitigated partially, not fully. If this strategy follows, it might be a discouragement to the ABS stakeholders, particularly the enforcement agencies (NBA and SBBs) to track the efforts to explore the geographical location of the bio-resources which have entered into the ABS, and its conservation. We cannot come up with the guidelines for each bio-resource involved in the ABS process. What is needed is a generic approach which is applicable for all the bio-resources used by different industrial sectors. Not knowing the origin of the bio-resources cannot be an everlasting excuse for a country like India, which is a pioneer in the ABS.

14.7 Challenges in the ABS Mechanism in India

Even if India enacted the BD Act and the Rules 20 years before, the ABS glory of the country is not much appreciable. A large number of bio-resources based commercial enterprises and research organizations that come under the ambit of the BD Act are still not onboard. Hence, we have failed miserably in enforcing the Act and mobilizing the funds for the conservation of the biodiversity, which is the need of the hour. Based on the experiences so far with respect to the enforcement of the BD Act, the major risk factor is, that in most cases, industries go to court and get a stay, which leads to delay and the entire process becomes more complicated.

At present, most of the SBBs in the country don't have any clarity on the biological resources based industrial units or traders in their respective states, which come under the ambit of the BD Act and the ABS. Further they are facing huge challenges in handling the industries as their understanding of ABS is very vague. Most of the industries failed in evolving the philosophy of the ABS and its overall business scope. Industries and their Associations generally argued that the BD Act is not for Indians who use biological resources, but only for foreigners. Lack of transparency and cooperation from bio-resources based industries; misunderstandings on the various concepts of the BD Act, inadequate clarity and methodology for distributing the ABS money to the providers of bio-resources (community) are the major challenges.

A majority of the industries felt that the BD Act is a burden. For them, profit is the only objective and they are not bothered about the resources' (raw materials) stock or their sustainability. Generally, industries are submitting their application in Form 1 with details indicating their turnover. However, these industries are not coming forward for signing the ABS agreements. According to them, submission of the Form 1application is the intimation of the bio-resources' access, and that is sufficient for them and there is no need for signing the ABS agreement. Bio-resources traders (that also come under the ambit of ABS) and their businesses are highly unorganized.

Bio-resources come under the normally traded commodities (NTC) list and the associated ABS issues are a concern. At present, broadly the SBBs are focusing only on the wild bio-resources for ABS. However, bio-resources other than wild items' trading and manufacturing need to be captured. All states need a policy on handling the cultivated bio-resources and their ABS. A majority of the cultivated items are NTCs and its ABS scope is a challenge.

Traditional practices, especially practicing indigenous medicines, are excepting from the BD Act. In this context, many village and household entrepreneurs are claiming their manufacturing belongs to indigenous practice. Similarly, value added products (products which may contain portions or extracts of plants and animals in unrecognizable and physically inseparable form) are also excepting from the AB Act. However, many manufactures are using secondary bio-resources including extracts as raw materials and manufacturing final products. According to them extracts are the value added products, hence required exemption from the Act, but it is controversial.

Huge volumes of biological resources are transferred by individuals/entities that have accessed the resources from the providers, including the local communities. Unfortunately, we don't have any mechanism to trace these resources and bring them under the purview of the ABS. The resources that go to the overseas markets can be followed through the customs (ports/airports) departments. It is also important to understand the complete value chain of the bio-resources.

Many bio-tech companies (seed, modern drug, cosmetic etc.) are signing agreements (MoU) with universities and research institutions. After the research, the results which are capable of being developed into commercially viable (consumer) products are used by the companies for manufacturing different products (commercialization). However, when the research communities or organizations transfer the research results to companies for commercial purpose or manufacturing products, they should voluntarily approach the NBA and sign the ABS agreements, which are not taking place. For NBA tracing these cases with respect to its commercial applicability is a huge challenge.

As NBA's prior permission is required before applying for the biological resource–based patent, all the applicants are approaching the NBA and complying with the BD Act provisions. However, all the IPR cleared cases are not going for commercialization. Experts revealed that, only very limited cases (may be around 3% of the patent cleared cases) go in for commercialization. Unless the patent holders reveal the status of commercialization, one cannot understand the current development.

Even where the law (BD Act) is being complied with and money is being generated under ABS the funds are not being ploughed back to the providers of bio-resources. Some States such as Andhra Pradesh, Uttrakhand, Karnataka, Uttar Pradesh, Gujarat, and Madhya Pradeshhave made good progress in terms of collecting fees from users of bio-resources, but the funds collected are still deposited with the concerned State Biodiversity Fund. NBA also mobilized some money through ABS, which too not successfully transferred to the community. Thus, the ultimate purpose of ABS for equitable sharing of benefits and conservation of biodiversity has not been realised.

14.8 Pre-conditions for Strengthening the ABS Process in India

There is a need to unlock the full potential of ABS and, in the process, bring users that currently do not comply within the ambit of ABS. For strengthening the ABS mechanism of the country and mobilizing more funds for biodiversity conservation, following strategies proposed:

Generally, the total stock of the bio-resources available in the Nation/State may not come under the purview of trade or ABS (first box in the Fig. 14.2). Some of the bio-resources don't have *'use value'* and are not being used by humans for their own-consumption or other purposes. Hence, these resources are untouched from nature or not traded or not commercialized at all.

Normally, the tradable bio-resources (which are having demand from domestic and industrial sectors) are a small part of the total stock of the bio-resources (second box in the Fig. 5.2). Even if the genetic/bio resources have 'use *value* 'their entire utilization will not come under the domain of ABS. Generally, only the bio-resources used with commercial intention come under the purview of the ABS and it will be only a portion of the total tradable bio-resources (third box in the Fig. 5.2) (Nelliyet 2020).Further, BD Act provides exemption for: human genetic material, 421 biological resources notified as 'normally traded as commodities', and resources

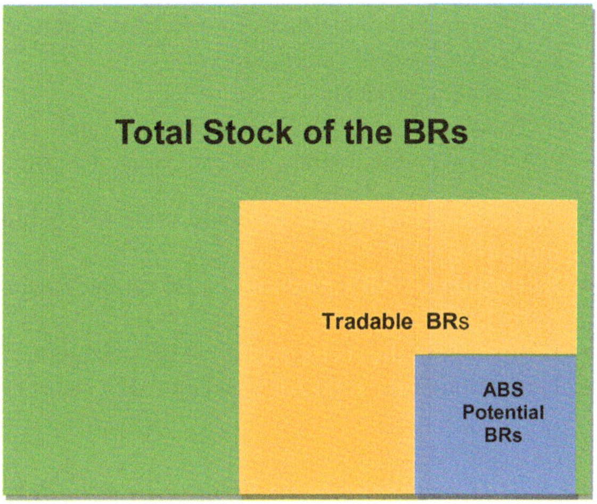

Fig. 14.2 Bio-resources at glance (Categorization)

used by local communities (for self-consumption) and *vaids* and *hakims* (for practicing indigenous medicine). Appropriate database on the above discussed categories of bio-resources is a prerequisite, which is possible only through a comprehensive documentation of the bio-resources in the nation/state.

14.8.1 Inventory of Bio-Resources Based Industries and Traders and Estimation of ABS Potential

This inventory contains key economic indicators of each industry includes: investment, gross fixed assets, raw-materials used and their costs, annual turnover, products manufactured, marketing, value of the products, export etc. Information may have obtained from; Directorate of Industries and Commerce, District Industries Centre, Department of Statistics, State Planning Commission, Industrial Association and concerned industry.

Enforcement agencies should come up with (a) the list of industries and traders who come under the ambit of the ABS within their jurisdiction, and (b) what will be the ABS potential. This will reveal the gravity of the need for ABS money mobilization. In this regard following steps to be followed:

- Identification of major bio-resources based industries who come under the ambit of BD Act and the ABS,
- Categorization of bio-resources based industry under the following heads:
  - Pharmaceuticals (modern drugs)
  - Botanical Medicines (AYUSH)
  - Agricultural seeds
  - Ornamental horticultural products

  - Crop protection products (bio-fertilizers and pesticides)
  - Health/personal care products
  - Cosmetic products
  - Food and beverages (food processing) and
  - Others (cotton textiles, leather, paper and pulp, jute etc.)

- Collect and analyse the purchase price of the bio-resources and turnover from industries.
- Estimation of ABS potential based on the norms prescribed in the Guidelines (2014).
- Estimation of the sector specific ABS potential.
- Estimation of the total ABS potential of a states or nation.

14.8.2 Comprehensive Strategies for Identification of Bio-Resources Origin

As per the current situation of the bio-resources, the information asymmetry, understanding its movement from the collection/cultivation points to the end user is a difficult task. Most of the time, the end users (industries) are getting the bio-resources from the wholesale markets or traders. The Traders may have some ideas about the bio-resources' origin, but generally are not willing to reveal it. In this regard, reliable methodology for tracking the bio-resources origin or identifying the provider is significant. Some of the draft approaches suggested is.

14.8.2.1 Supply Chain Analysis

The supply chain represents the steps it takes to get the product or service from its original state to the customer. Broadly, the functions in a supply chain include product development, marketing, operations, distribution, finance, and customer service. However, in the ABS context, our concern is the movement of bio-resources from the providers (local communities) to the user (industry) for commercial utilization or product development (Fig. 14.3). Generally, the ABS amount will be fixed based on the ex-factory sale value by the enforcement agencies. Hence, the product movement from the industry to the customer is insignificant in the ABS based supply chain.

Fig. 14.3 Supply chain of the bio-resources

For understanding the supply chain of particular bio-resources, the bottom-up approach (provider to the user) need to be followed. In this regard, the movement of bio-resources from the collector/cultivators (community) to the local/village trader or SHGs (if any) to big traders to wholesale marketers, then to bio-resources based manufacturing industry (which uses bio-resources as a basic raw-material) is required to be explored in a systematic manner with the support of the different stakeholders (Fig. 14.4). Similarly, the backward chain (from the user to the primary provider) of the bio-resources is also to be explored.

Further, the assessment of the bio-resources based manufacturing companies' production steps and R&D with complete cost details would be helpful in understanding the value addition on the bio-resources. Estimating the value addition in the production stage is a complicated task. A research carried out on a microbial resource based crop protection produces value addition estimation. The cost escalation is mainly under 3 heads: research and development, manufacturing, and marketing. Under each head different cost distribution criteria were indicated (Fig. 14.5).

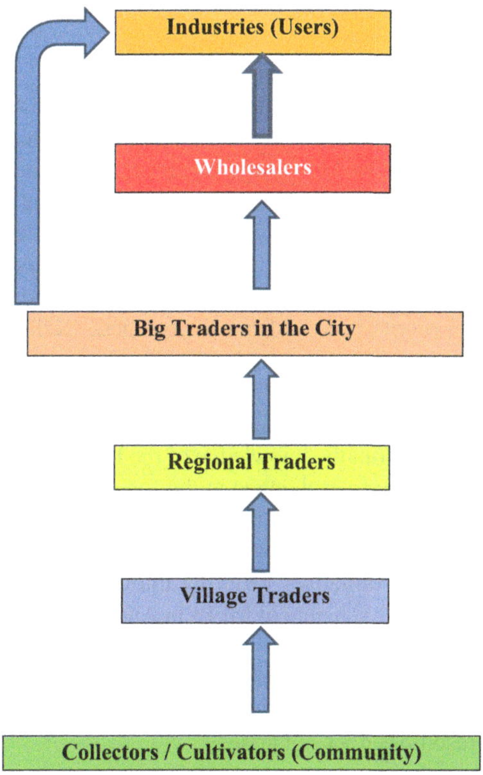

Fig. 14.4 Supply chain of biological resources

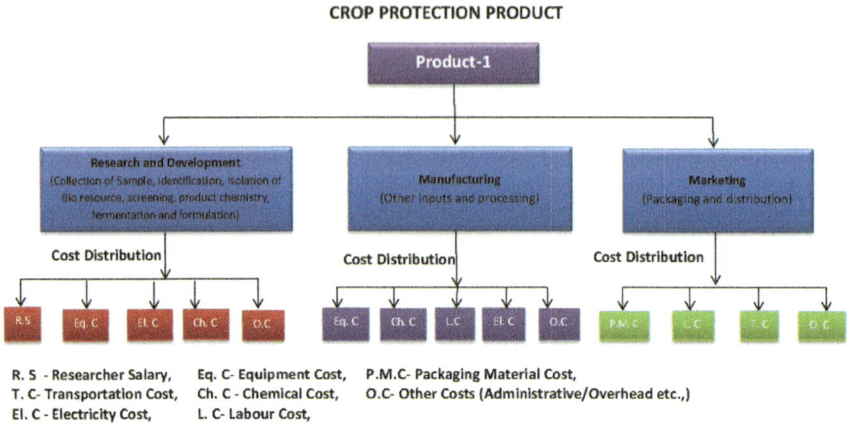

Fig. 14.5 Value addition in manufacturing

14.8.2.2 Formalization of the Bio-Resources Market and Direct Supply to Users

Most of our ecosystems or biodiversity spots such as forests, rivers, estuaries, oceans, etc. are common properties. Hence, goods like bio-resources coming out from these areas experience market failure/distortion. Due to market imperfections, bio-resources are not properly exchanged and the current market price at their collection point does not represent their real/true value, but only an exchange rate which is arbitrarily assessed or fixed; hence, the resources are always 'underpriced'. The providers (local communities) are being exploited by the traders/companies, who may make substantial profits/benefits from the business (Nelliyat and Balakrishna 2013).

For example: from the forests, a buffer zone of biodiversity, the tribal communities are collecting huge volumes of bio-resources, which is transferred to commercial hubs by traders with or without the knowledge of the Forest Department. Generally, non-timber forest resources are exchanged at the forest gate in traditional practices like auction. However, formalization of the market through an appropriate agency is required, which also facilitate in implementing the ABS mechanism in an effective manner. In this regard, the initiatives taken by the Girijan Co-operative Corporation(GCC) in procuring and exchange of MFPs in Andhra Pradesh are an interesting learning case (Box 14.4).

If the bio-resources are directly transferred to the users (industry) through authorized institutions (corporations or BMCs) the issue of identifying the geographical location of the bio-resources does not arise.

Box 14.4
Girijan Co-operative Corporation (GCC)

GCC is one of the largest tribal development co-operative agencies in India for procuring Minor Forest Produce (MFPs) in the forest areas of Andhra Pradesh. This state undertaking Corporation (functions through a wider network in the state even in remote and inaccessible tribal habitats with a large transport fleet) serves nearly a 4 million tribal population spread over 32,000 km2 of forests. For the past 60 years, the GCC safeguards the tribes from exploitative middlemen, petty traders and establishes a mutually beneficial relationship with the rest of the world. GCC's functions include: purchasing MFPs from tribes at reasonable and fair prices from the very doorstep of the tribes, imparting training to the tribes in the collection of the MPPs- so that forest yields can be increased without endangering the trees and environment, with a proactive search for adding new MFPs to the existing list. In brief, GCC aims for 'protecting the environment through the regeneration of species of MFPs and ensuring pure and natural products to the consumer at reasonable prices. In brief, GCC's efforts may considered as a first step towards operationalzing the ABS.

Source: (GCC 2018).

14.8.2.3 Empowerment of the BMCs and Fulfilment of their Responsibility

In accordance with the provisions of Section 41 of the BD Act, every local body shall constitute a BMC within its area for the purpose of promoting conservation, sustainable use and documentation of biological diversity including preservation of habitats, conservation of land races, folk varieties and cultivars, domesticated stocks and breeds of animals and microorganisms and chronicling of knowledge relating to biological diversity. The BMCs are statutory bodies at the local level with specific roles and responsibilities.

At present the BMCs of the country are not having the capacity to handle most of their responsibilities or facilitating ABS implementation. However, the BMCs' capacity building and empowerment, PBR preparation and its updating and the maintenance of the bio-resources register and collection of levy fee will help in tracking the ABS potential bio-resources. No doubt that, if we handle the bio-resources movement from the provider to the users through BMCs, the channelization of the ABS money for that particular resource's conservation who contributed in ABS could be possible and become a success model. BMCs can come up with a data base on the status of the bio-resources in their jurisdiction and innovative proposals for their conservation.

14.8.3 Bio-Resource Inventory Based Conservation Strategy

The bio-resources inventory/database induced conservation strategy should be a realisable method for biodiversity conservation. ABS attributed bio-resource base conservation strategy is meaningless in certain cases. For example, if the ABS is obtained from a bio-resource (neem), which is plenty in a region, making more efforts for its conservation is meaningless. Instead of that, the conservation of some other rare and endangered species in the region is important. Use of the ABS money for the conservation of resources having high market demand (otherwise the possibility of it's over extraction and resource depletion is high) and bio-resources sustainable harvesting practice should also be encouraged. In this regard species-wise harvesting techniques should be developed and training should be given to the harvesters.

14.8.4 Strict Enforcement

NAB and SBBs should identify the bio-resources based industries in their jurisdiction and issue notices and take stringent legal action against those who violate the rules. NBA needs to monitor all the universities and research organizations and identify those projects, which deal with collaborative research based on Indian bio-resources. Without the co-operation from other countries, it is difficult to understand the complete scenario about the bio-resources' movement and its utilization. Apart from bio-resources based industries, the biological resources based research activities in India also should closely observe and understand its objectives, the overseas support and collaboration (finance and technical), possibilities of transfer of the research results etc. The basic information can be gathered through UGC, ICAR, CSIR, ICMR etc. and the in-depth information from the concerned research institutions.

NBA is supposed to monitor the bio-resources based patent holders very closely. What is needed is to chase each case which has been approved. IT based monitory tools were developed by experts. With the application of this methodology one can get a clear picture about the Indian biological resources based patents abroad. Further, all CBD partners / member countries' cooperation in this regard is required. The user country's measures should be strictly followed by all the member countries of the CBD, particularly those ratified by the Nagoya Protocol.

14.9 Conclusion

Biodiversity underpins human well-being and various economic activities. Biodiversity provides the essentials like clean air, food security and fresh water necessary for a healthy and productive human life. Hence, investments in biodiversity are considered as investments in achieving the SDGs, contributing directly to poverty reduction, resilience and long-term economic growth and sustainability. Nearly half the human beings, particularly the most vulnerable people depend directly on biodiversity for their livelihood. The recent phenomena like climate change- induced frequent floods and droughts and pandemics like Covid-19 have strong connections with biodiversity degradation. Rich biodiversity can absorb the Green House Gases and regulate the climate change impacts and stabilize the ecosystem integrity or control the growth and domination of single species like the Corona virus. By preserving biodiversity and ecosystems, we are retaining the ability of the planet to sustain our prosperity (UNDP 2018).

Here, the management of biodiversity is crucial while adequate finance is a prerequisite. At present the share of the Government budget allocation for biodiversity is inadequate in meeting the funding gaps for biodiversity conservation. Hence, the required fund should be mobilized in the private and corporate sectors, where ABS plays a crucial role. However, at present ABS is in its infant stage. ABS is a package. Hence, different components/articles of the BD Act should operate simultaneously rather than in isolation for achieving its overall objectives. All three enforcement agencies (NBA, SBBs, and BMCs) staff strength and capacity building have to be enhanced. In the ABS process, what is needed is a workable solution, rather than a complicated one.

Policy makers and enforcement agencies need to view ABS from a holistic perspective (as indicated in Fig. 14.1) and convey the essence of the philosophy to the commercial users (industry and research organization) in an amicable and mutually agreed manner. For bio-resources based industries, bio-resources are basic raw-materials with no substitute. Through the ABS mechanism biodiversity conservation and sustainable use can be achieved, which guarantees raw-material security to them. If this fact is admitted by all users of bio-resources and they come forward to obey the BD Act and contribute the prescribed share of their benefit to the local communities, there is no doubt that, ABS will become a strong innovative financial mechanism for biodiversity conservation.

Acknowledgements This paper/chapter has been prepared based on the scientific inferences and knowledge, on different ABS issues, that I have accrued from the research I carried out in NBA, Centre for Biodiversity Policy and Law (CEBPOL) and the UNDP.I wish to thank Dr. Balakrishna Pisupati Former Chairman NBA, Shri. T Rabikumar IFS, Former Secretary NBA, Dr. V. Rajagopalan IAS, Former Secretary MoEFCC and Sr. Technical Consultant UNDP-BIOFIN India, Dr. K.P. Raghuram, Technical Officer (NBS) NBA, and Dr. T. Narendran, Technical Officer (IPR) NBA for the insightful discussions on various ABS matters I have carried out on different occasions. The views expressed in this paper are personal and do not reflect those of the organizations I worked for.

References

Bavikatte, KS, Morten WT (2015) Beyond the thumbrule approach: regulatory innovations for bioprospecting in India. 11/1 Law Environ Dev J. http://www.lead-journal.org

CBD (2011) Nagoya protocol on access to genetic resources and the fair and equitable sharing of benefits arising from their utilization to the convention on biological diversity. Convention on Biological Diversity, United Nations

Costanza R, D'Arge R, de Groot R, Farber S, Grasso M, Hannon B, Limburg K, Naeem S, O'Neill RV, Paruelo J, Raskin RG, Sutton P, van den Belt M (1997) The value of the world's ecosystem services and natural capital. Nature 387(6630):253–260. https://doi.org/10.1038/387253a0

Girijan Co-operative Corporation (2018) Girijan co-operative corporation–GCC: profile and mission. http://www.apgirijan.com/

Ministry of Environment Forests and Climate Change (2014) India's fifth national report to the Convention on Biological Diversity (CBD). Ministry of Environment and Forests, Government of India

National Biodiversity Authority (2010) Biological diversity act, 2002, NBA, Chennai

National Biodiversity Authority (NBA) (2019) Biodiversity Finance Plan (Working Document), GoI-UNDP project on Biodiversity Finance Initiative (BIOFIN)

Nelliyat P, Balakrishna P (2013) Biodiversity economics from access and benefit sharing perspective. Dissemination Paper-1 in 'Valuation of Biodiversity', National Biodiversity Authority, Chennai

Nelliyat P, Balakrishna P (2014) Economic Valuation of Bio resources for access and benefit sharing. Report, National Biodiversity Authority, Chennai

Nelliyat P, Meenakumari B (2018a) Protected Areas and Access and Benefit Sharing (ABS): a review. Centre for Biodiversity Policy and Law, National Biodiversity Authority

Nelliyat P, Meenakumari B (2018b) Compliance of Access and Benefit Sharing (ABS): a sector specific review. Centre for Biodiversity Policy and Law, National Biodiversity Authority

Nelliyat P (2020) Tradable Bio-resources' documentation (Database) and identification of its ABS potential with supply chain: a manual. UNDP-India Sponsored High Range Mountain (Munnar) Landscape Project, Kerala State Biodiversity Board

Pisupati (2012) Lessons for international environment governance. National Biodiversity Authority, Chennai

United Nation (2019) UN Report: nature's dangerous decline 'Unprecedented'; species extinction rates 'Accelerating'. UN Report: Nature's Dangerous Decline 'Unprecedented'; Species Extinction Rates 'Accelerating'-United Nations Sustainable Development

UNDP (2018) Biodiversity Finance Initiative (BIOFIN) workbook. United Nations Development Programme, New York

Chapter 15
Economic Valuation for ABS—A Paradigm to Capture Multiple Priorities and Values

M. S. Suneetha, Balakrishna Pisupati, Prakash Nelliyat, Ishwar Poojar, Geoffrey Burton, Paul Oldham, and G. Haripriya

Abstract The Convention on Biological Diversity (CBD) urges its member countries and other stakeholders to engage in plural values approach to implementing its core mandate of conserving biodiversity (CBD, 1992; Aichi Target 1). This is a key principle for the CBD's third objective of ensuring equitable share of benefits arising from the commercial use of genetic resources with all those involved in the value chain (Access and Benefit Sharing–ABS). Valuation for ABS therefore entails ensuring equity amongst the different stakeholders, which then is dependent on the factors such as knowledge associated with use of a species, the contribution of this knowledge towards increasing the probability of developing a successful product and/or in reducing costs of development, potential income that may be generated over a reasonable period of time which again is dependent on the type of sector that utilizes the species. This then links valuation for ABS to more plural approaches as it looks at economic values and intangible values (e.g., associated traditional knowledge). Issues of species population status and their endemicity also need to be factored in. In this paper, we highlight how a more systematic plural approach to undertaking valuation for ABS can set a reasonable baseline for negotiations between providers

M. S. Suneetha (✉)
UNU Institute for the Advanced Study of Sustainability, Tokyo, Japan
e-mail: suneethams@gmail.com; subramanians@unu.edu

B. Pisupati
Law Division, United Nations Environment Programme (UNEP), Nairobi, Kenya

P. Nelliyat
Former Fellow (ABS), Centre for Biodiversity Policy and Law, National Biodiversity Authority, 5th Floor, TICEL Biopark, Taramani, Chennai 600113, India

I. Poojar
UNEP-GEF-MoEF ABS Project. National Biodiversity Authority, Chennai, India

G. Burton
Jean Shannon and Associates Pty Ltd, Canberra, Australia

P. Oldham
One World Analytics, Lancaster, UK

G. Haripriya
Department of Humanities and Social Sciences, Indian Institute of Technology, Bombay, India

O. V. Oommen et al. (eds.), *Biodiversity Conservation Through Access and Benefit Sharing (ABS)*, https://doi.org/10.1007/978-3-031-16186-5_15

and users of genetic resources. For this we have used actual estimates of costs, returns and business models from the traditional medicine (Ayurveda, Siddha, Unani) sector in India. Our analysis also indicates that uncertainties relating to resource availability should be factored in entry restrictions to users.

Keywords Access and benefit sharing · Economic valuation for ABS · Plural valuation · Bio-prospecting · Equity

15.1 Introduction

The Convention on Biological Diversity (CBD) is unique in obligating its member countries to undertake conservation action in a manner that is equitable to different stakeholders, especially those who relate in close proximity to biodiversity (UN 1992). Such stakeholders could be indigenous peoples and local communities (IPLCs) or countries that are endowed with rich biodiversity. Either way, they are custodians of resources and associated knowledge related to resource use and management (Laird and Wynberg 2008).

By design the CBD urges its member countries and other stakeholders to engage in a plural values approach to implementing its core mandate of conserving biodiversity (CBD 1992; Aichi Target 1). While values related to nature are primary to its mandate, it recognizes that in order to achieve conservation, focusing on the relational values that people hold with nature and between social groups is important (Duraiapp et al. 2014; MA 2005). Relational values imply how people relate to nature that may be instrumental or non-instrumental and tangible or intangible, meaning that biodiversity may be considered to provide some form of gain or it may just be an important part of one's identity, socio-cultural contexts and ways of life (IPBES, 2016; Pascual et al 2017; Klain et al 2017).This pluralistic approach considering different social and ecological values related to nature and attributable to different actors are often not considered in valuation exercises undertaken to capture biodiversity values (IPBES, 2016). This is especially the case in economic values ascribed while determining a fair share of benefits accruable to actors in the downstream of a supply chain of a biological resource. Commonly referred to as Access and Benefit Sharing (ABS), it is the third objective of the CBD to ensure that access to resources is facilitated and on utilization, a fair and equitable sharing of benefits generated from use is ensured amidst the different actors, especially the providers of resources and associated traditional knowledge (TK) (Laird and Wynberd, *ibid*; Suneetha and Pisupati 2009). This explicitly recognizes the stewardship roles and innovative capacity of such providers. While various attempts to ensure equity in transactions have been attempted at, so far,a plural approach towards valuing the development of a product from resource to end- use has not been sufficiently defined (Zafra et al. 2020).

Here, we attempt to highlight how this gap in valuation may be addressed taking the case of valuation for Access and Benefit Sharing (ABS), using an example from India. The research was undertaken as part of efforts to strengthen implementation of

ABS in India under a UNEP-GEF project implemented by the National Biodiversity Authority of India.

15.2 Dimensions of Access and Benefit Sharing

The concept of ABS involves 2 parts: *Access to genetic resources/biological resources*and associated traditional knowledge and Fair and equitable *Benefit sharing.*

Access to genetic resources/ biological resources is determined by: ***ownership claims over the resources*** (whether individual ownership, joint ownership of resources, public resource, resources available in transboundary areas or in areas beyond national jurisdiction), ***stakeholder priorities*** (whether for profit, for conservation, for livelihoods, for cultural reasons) and ***management regulations*** (responsible institutions to determine management of resources) (Suneetha and Pisupati, *ibid*).

Benefit sharing is determined by:

Who is involved (sector and actors): This is an important aspect to consider as benefits that can be generated from resource utilization varies considerably between sectors owing to different costs, including time involved in the development of product, variances in the types of markets and consequent differences in the benefits generated (see Table 15.1).

What is involved (type of resource): Whether a resource accessed is endemic to an area, and itspopulation status would determine access conditions and a premium placed on access fees. Species that are endemic or showing high degree of threatened status have a higher rarity value requiring regulated access including quantitative restrictions and higher access fees.

Inputs used in product development: Depending on whether only a biological resource is accessed or if associated traditional knowledge (TK) is also accessed, the relative costs of product development and terms of benefits to be shared would vary. If TK is used, the extent to which it has contributed to the development of the product–whether directly informing the use from public sources (e.g databases, pharmacopeia or commonly dispersed information) or private sources of information (e.g., held amongst communities or individuals) or indirect contributions such as identification of resource and location also need to be considered as they influence the probability of developing a successful product, at least in the lead development phase. A higher probability of success should translate to higher benefits accruable to the providers of resources and associated TK.

Clarity of terms of partnership: This is the crucial part of an agreement where the users and providers of resources know well what to expect from either side, and

Table 15.1 Average Market value and costs of development (USD Billion) of products across sectors highly dependent on biodiversity

Sector	Market Value USD Billion	Cost of production USD	Prob of success	Business model
Biopharmaceuticals (incl. biotechnology)	270 (Businesswire, 2019)	0.8–2.5 Billion (Farid et al. 2020)	0.5–0.001 (increases with ATK)	Shuttle development, small companies develop parts of product
Botanicals/Nutraceuticals	285.2 (IMARC 2019)	<10 million	~0.5	Many suppliers, many developers
Agricultural biotechnology	278 (Medgadget 2020)	100–200 Million	0.5–0.001	Shuttle development, small companies develop parts of product
	Conventional seeds: 50 (Grandview Research 2019a)			
Personal products and natural cosmetics	34 (Grandview Research 2019b)	<10 Million	~0.5	Many suppliers, many developers

if what has been agreed to can be considered fair and equitable by third parties (Schroeder and Pisupati 2010).

15.3 Economic Valuation for ABS

An economic valuation exercise for ABS should therefore inherently attempt to capture multiple values–those relating to species, cultural relationships and to economic value from trade. It is not an exercise to capture value of a resource, but to capture a fair and equitable share of benefits accrued from the development and use of it.

An attempt was made through the UNEP-GEF funded project in India to develop a methodological basis to undertake valuation for ABS. Four major sectors using

Table 15.2 Important sectors using biological resources and methodological approaches identified

Category of bio-resources	Possible methodological approach	Payment detail
Bio Pharmaceuticals (modern drugs) (*Population status, Rare Endangered and Threatening (RET), Abundant, Endemic*)	Scarcity Rent (SR), Information Rent (IR)–share a proportion attributable to the product Endemic Rent (ER)	Initial payment + payment at the time of product development + payment at marketing stage Monetary + Non- Monetary (for endemic and RET)
Bio-technology (Seed/Agriculture Related), Land races, Microbes	Information Rent (IR)–share a proportion attributable to the product	Initial payment + payment at the time of product development + payment at marketing stage Monetary + Non- Monetary (for endemic and RET)
Crop protection products	Information Rent (IR)–share a proportion attributable to the product	One time
Botanicals (AYUSH)	Based on the proportion of Net Present Value (NPV) of the profit x the contribution of input to the out put	One time
Nutraceuticals/Personal Products cosmetics	Based on the proportion of NPV of the profit X (times) the contribution of input to the out put	One time
Academia/R&D (non-commercial scientific research)	Onetime fee + renegotiation change in intent	One time

SR–Scarcity Rent refers to the value ascribed to a resource depending on its population status: It refers to the cost imposed due to extraction of a finite resource and its impact on availability to future generations

IR–Information Rent refers to costs ascribed to various dimensions involved in the production of a product

ER–Endemic Rent refers to the exclusivity of an ecosystem or resource that arises from the prevalence of species that cannot be found anywhere else

NPV–Net Present Value refers to the difference in present value of cash inflows and expenses over a period of time

biological resources were identified in India, and Table 15.2 highlights the methodological approach that was finalized through an expert consultation process to determine a fair monetary share to providers.[1]

In order to illustrate how this may be operationalized, a detailed analysis of benefits accruable and R&D costs (average) was worked out with industry experts from the

[1] National Consultation meeting on Economic Valuation of Bio-resources for Access and Benefit Sharing, 9–10 Dec 2013.

Table 15.3 Benefits and costs of product development (average) in AYUSH sector

Sector	Likely returns per product per annum (INR)	Life of product (average) in years	R&D cost (total) INR	Development period in years	probability of success	Source of traditional knowledge
A,S,U	55 million	15	30 million	7	0.04	From documented sources primarily
Herbal cosmetics	55 million	4	3 0 million	4	0.01	From documented sources primarily
Herbal Veterinary	55 million	10	2,500,000 to 20 million	4	0.2	From documented and direct sources

AYUSH sector (traditional medicine industry–Ayurveda, Siddha and Unani sectors also referred to as A,S,U), following a workshop on Economic Valuation for ABS in April 2017 at Chennai, India. Experts from A,S,U sector (Dr N Brindavanam former Chief of R&D, Dabur and Dr Amit Agarwal from Natural Remedies Ltd) were available to discuss the various details involved in R&D in the sector (in 2018). It was decided that it would be good to focus on the A,S,U sector to illustrate the methodological approach being undertaken in this project, as the sector is a major user group and further due to better availability of data with the experts (see Table 15.3).

The Net Present Value was calculated at the rate of 8% per annum (optimum expected rate of return on an investment) over a period of 15 years, which was considered an average Market life of a product. The value of the product was then weighed by the probabilities of success in developing a product, which ranges from 0.04 in the case where no traditional knowledge (TK) on use is involved, to 0.01 when TK is accessed from public sources and 0.5 when specific TK is accessed. This captures the changes to the value of a product when relevant information enhances the degree of success of developing a product (Table 15.4; Sources of info–Personal communication, A,S, U experts, 2018).

Difference in probability weighted NPV and unweighted NPV gives us a minimum level of benefits that account for the contributions from traditional knowledge (from documented to undocumented TK).

Table 15.4 NPV of a product in AYUSH sector at various probabilities of success with TK

Sector	NPV (over 15 years at 8% discount rate) INR (without accounting for 'p' of success)	p1(without TK)	p2 (with TK in open sources)	p3 (specific TK)
A,S,U		0.04	0.2	0.5
NPV*p (INR)	78,81,043	3,15,242	15,76,209	39,40,521

15.4 Access Conditions

An important consideration that needs to be factored involves weighing in access conditions to species. This needs to be carefully nuanced and addressed under the following categories.

1. Population status of species (Whether Abundant, Rare, endangered or threatened and if being sourced from the wild or cultivated sources and which part of the resource is being utilized would determine the kind of access arrangements that can be entered into. It would also have implications of the kind of levy fees that may be determined to access a resource.
2. Endemicity of species: Endemic species need to be protected from over exploitation and hence regulated with clear access conditions (eg quantitative restrictions, access fees, parts collected and similar aspects oriented towards conservation and sustainable use practices).
3. Cultural values associated with species–this implies respecting local knowledge of use and identity and location and beliefs (such as sacred species, species of high cultural value) related to species and ensuring that access to their exploitation is accorded based on the free, prior and informed consent of stakeholders who hold such cultural values.
4. Form of utilization of species- This implies the form in which the user accesses a species- whether in raw, modified, or synthetic form–whether in bulk supplies, semi-processed form or in synthetic forms also has implications for terms of ABS partnerships. It would have implications for the kind of benefits (in terms of livelihood security, conservation priorities, supply linkages, R&D capacity building, co-ownership of product) that can be determined from such partnerships.

   All the above are intangible values and have stewardship effects and preserve options of biodiversity for future. Therefore, accessing resources of this nature requires respecting ecological and cultural values (SCBD 2011). They are not amenable to be factored into a valuation matrix, but are important dimensions to be considered while deciding on granting access to resources and terms of engagement. These categories need to be considered especially while determining type of partnerships to be arranged between the providers and users and further, look at types of milestone benefits that may be accrued. These benefits may be monetary (for e.g. as levy fees)and/or in-kind (support for conservation activities, livelihood activities, other development support, etc).

15.5 Conclusions

The paper illustrates how alternate methodological approaches that take into account multiple values could enable devising more equitable outcomes. This is because such approaches attempt to discover feasible solutions that address concerns of multiple actors in a value chain. Such approaches build on existing valuation approaches but at the same time bridge across disciplinary rigidities.

In the case of valuation for ABS, the purpose is to ensure that equitable benefits are transferred to the custodians of biological resources and associated knowledge that are used in the development of new products. Essentially, the valuation therefore needs to be oriented not just to the market value of the product, but also to availability of a species, and associated traditional knowledge/ cultural values related to use of the resource in addition to undertaking a relative assessment of costs and benefits of products. Such an approach, as seen through the example of the formal traditional medicine (Ayurveda, Siddha and Unani) sector in India, provides a more robust and equitable basis for negotiations between the users and providers of biological resources. Unlike a typical resource valuation approach, this factors in conservation concerns (through access regulations based on indicators of species abundance, endemicity), cultural concerns (through contributions from traditional knowledge to development of the product in terms of probability of success) and industry concerns (through annualized potential net present value of a product based on sectoral peculiarities). Such an approach, as the example shows, is not difficult to pursue and requires bringing together expertise and information from different disciplines and sectors. By the same token, this could also nudge better policy coherence across various sectors.

Acknowledgements The research was supported by the United Nations Environment-Global Environment Facility through the Ministry of Environments, Forests and Climate Change, India and implemented by the National Biodiversity Authority, India in collaboration with the United Nations University-Institute for the Advanced Study of Sustainability, Japan

Conflict of Interest The views and opinions expressed in this article are those of the author in his personal capacity. None of the observations, opinions, or conclusions expressed herein are to be attributed to or construed as the views or positions of any organization, association, or company. No conflict of interest.

References

Businesswire (2019) The global biopharmaceuticals market is projected to grow by 8.7% through to 2025 and reach $446 Billion-ResearchAndMarkets.com, June 24, 2019. https://www.businesswire.com/news/home/20190624005782/en/Global-Biopharmaceuticals-Market-Projected-Grow-8.7-2025

Coffman (2020) Benchmarking biopharmaceutical process development and manufacturing cost contributions to R&D, mAbs, 12:1, 1754999. https://doi.org/10.1080/19420862.2020.1754999

Duraiappa AK, Asah ST, Brondizio ES, Kosoy N, O'Farrell PJ, Prieur-Richard A-H, Subramanian SM, Takeuchi K (2014) Managing the mismatches to provide ecosystem services for human well-being: a conceptual framework for understanding the New Commons. Curr Opin Environ Sustain 7:94–100

Farid SS, Baron M, Stamatis C, Nie W, Coffman J (1992) United Nations, convention on biological diversity. https://www.cbd.int/doc/legal/cbd-en.pdf

Farid SS, Baron M, Stamatis C, Nie W, Coffman J (2020) Benchmarking biopharmaceutical process development and manufacturing cost contributions to R&D. MAbs. Jan–Dec 12(1):1754999. https://doi.org/10.1080/19420862.2020.1754999. PMID: 32449439; PMCID: PMC7531566

Grandview Research (2019a) Commercial seeds market size, share & trends analysis report by product (Maize, Soybean, Vegetable, Cereals, Cotton, Rice, Canola), By Biotechnology/GM Seeds, By Region, and Segment Forecasts, 2019–2025

Grandview Research (2019b) Natural cosmetics market size analysis report by product (Skin Care, Hair Care, Fragrance, Color Cosmetics), By Distribution Channel (Supermarket/Hypermarket, Online), and Segment Forecasts, 2019–2025

IMARC (2019) Nutraceuticals market report. https://www.imarcgroup.com/global-nutraceuticals-market

IPBES (2016) Preliminary guide regarding the diverse conceptualizations of the multiple values of nature and its benefits, including biodiversity and ecosystem functions and services, deliverable 3(d), IPBES/3/INF/13

Klain SC, Olmsted P, Chan KMA, Satterfield T (2017) Relational values resonate broadly and differently than intrinsic or instrumental values, or the New Ecological Paradigm. PLoS ONE 12(8): e0183962. 10.1371

Laird S, Wynberg R (2008) Access and benefit-sharing in practice: trends in partnership across sectors. Secretariat of the Convention on Biological Diversity, Montreal

Medgadget (2020) Agricultural biotechnology market report. https://www.medgadget.com/2020/02/agricultural-biotechnology-market-report-2020-global-size-research-study-industry-share-regional-growth-competitive-landscape-top-company-profile-technology-trends.html

Millennium Ecosystem Assessment (2005) UNEP, Nairobi

Pascual U, Balvanera B, Dı´az S, Pataki G, Roth E, Stenseke M, Watson RT, Başak E D, Islar M, Kelemen E, Maris V, Quaas M, Subramanian SM, Wittmer H, Adlan A, Ahn S, Al-Hafedh YS, Amankwah E, Asah ST, Berry P, Bilgin A, Breslow SJ, Bullock C, Ca´ ceres D, Daly-Hassen H30, Figueroa E, Golden C, Go´ mez-Baggethun E, Gonza´ lez-Jime´ nez D, Houdet J, Keune H, Kumar R, Ma K, May PH, Mead A, O'Farrell P, Pandit R, Pengue W, Pichis-Madruga R, Popa F, Preston S, Pacheco-Balanza D, Saarikoski H, Strassburg BB, van den Belt M, Verma M, Wickson Fand Yagi N (2017) Valuing nature's contributions to people: the IPBES approach current opinion on environmental sustainability, vol 26, pp 7–16

SCBD (2011) Nagoya protocol on access to genetic resources and the fair and equitable sharing of benefits arising from their utilization to the convention on biological diversity: text and annex. Secretariat of the Convention on Biological Diversity

Schroeder D, Pisupati B (2010) Ethics justice and the convention on biological diversity. UNEP, Nairobi

Suneetha MS, Pisupati B (2009), Benefit sharing in ABS: options and elaborations. UNU-IAS Policy Report

United Nations (1992) Convention on Biological Diversity, https://www.cbd.int/doc/legal/cbd-en.pdf

Zafra-Calvo N, Balvanera P, Pascual U, Merçon J, Martín-López B, fvan Noordwijk M, Heita Mwampamba T, Lelei S, Ifejika Speranzaj C, Arias-Arévalo P, Cabro DM, Cáceres D, lPO'Farrell P, Subramanian SM, Devyi S, Krishnan S, ilqCarmenta R, Guibrunet L, Kraus-Elsin Y, Moersberger H, Cariño J, Díaz S (2020) Plural valuation of nature for equity and sustainability: insights from the global South, Global Environ Change 63,102115. https://doi.org/10.1016/j.gloenvcha.2020

Chapter 16
Institutions and Incentives to Facilitate Indigenous and Local Peoples' Participation in the Access and Benefit Sharing Regime in India

U. Sankar

Abstract The important role of indigenous and local peoples (ILPs) in conservation and sustainable use of biodiversity is getting greater recognition in recent years. This policy shift from exclusion of the stakeholders to inclusion and integration in biodiversity conservation and development would correct not only the 'historical injustice' to the ILPs but also enhance their livelihood opportunities and increase biodiversity. The success of this policy shift would depend on legal measures such as security of right to livelihood and rights over biological resources, lowering of transaction costs of exchange of biological resources and creation of enabling institutions and incentives for efficient participation in the access and benefit sharing regime of the Convention on Biological Diversity. Community forest resource management institutions require capacity building as well as partnership with biodiversity research institutions and firms to derive benefits from the access and benefit sharing regime.

Keywords Indigenous and local people · Stakeholders · Institutional structures · Transaction costs · Traditional knowledge

16.1 Introduction

Compared with developed countries, developing countries are richer in biological resources and traditional knowledge. But as these resources were largely under common-pool open access system of management for centuries, they were over-exploited and bio piracy was widely prevalent. Also, many colonial governments were interested primarily in revenue maximization from forests and viewed the custodians of biological resources– indigenous and local peoples (ILPs)-as "encroachers"/threats to wildlife conservation and marginalized them by denying ownership/management rights over these resources. These factors reduced the livelihood opportunities of the ILPs and also contributed to biodiversity loss.

U. Sankar (✉)
Madras School of Economics, Chennai, India
e-mail: usankar@mse.ac.in

O. V. Oommen et al. (eds.), *Biodiversity Conservation Through Access and Benefit Sharing (ABS)*, https://doi.org/10.1007/978-3-031-16186-5_16

The UN Convention on Biological Diversity (CBD) (1992), says it is 'conscious of the intrinsic value of biological diversity and of the ecological, genetic, social, economic, scientific, educational, cultural, recreational and aesthetic values of biological diversity and its components' (Preamble). It reaffirms that States have sovereign rights over their biological resources. Article 6 says that each contracting party shall 'develop national strategies, plans or programmes for the conservation and sustainable use of biological diversity' (p. 5). The Convention's three objectives are: the conservation of biological diversity, the sustainable use of its components and the fair and equitable sharing of the benefits. The three objectives are inter-linked. It may be noted that if the right institutional structures and incentive mechanisms are created for the fair and equitable sharing of these benefits from the transactions on biological resources and the associated traditional knowledge, the first two objectives can also be realized.

The Biological Diversity Act was enacted in India in 2002. The National Biodiversity Authority was set up in 2003. State Biodiversity Boards have been constituted all states and Biodiversity Management Committees have been set up at local levels. Preparation of Peoples' Biodiversity Registers (PBRs) is nearing completion. The Forest Rights Act was passed in 2006 to correct the "historical injustice" done to the ILPs. Its implementation has been slow. For the benefits to reach the providers, some impediments in the transaction process have to be removed. The impediments arise from the nature of good/service transacted lack of security of tenure, ill-defined/absence of property rights, unorganized markets and high transaction costs in devising, monitoring and enforcing contracts between the providers and the users. The impediments can be overcome with right institutional structures, governance reforms and incentive schemes.

The plan of the paper is as follows: Sect. 16.2 deals with the roles of ILPs in conservation and sustainable management of biodiversity. It defines the roles of ILPs as given in CBD, Nagoya Protocol, UN Declaration on Rights of Indigenous People (UNDRIP) and the Forest Rights Act 2006, and reviews the progress made so far in restoring the rights to ILPs in India. Section 16.3 reviews the implementation of the Biodiversity Act in India, focusing on the access and benefit-sharing (ABS) regime. It considers the forms and guidelines used for different purposes. Section 16.4 is devoted to the management of community forest resources. It also indicates what needs to be done to enable the providers to get the potential benefits. Section 16.5 contains concluding remarks.

16.2 Recognition of and Empowerment of ILPs

Article 8 J of the CBD says: 'Subject to its national legislation, respect, preserve and maintain knowledge, innovations and practices of indigenous and local communities embodying traditional lifestyles relevant for the conservation and sustainable use of biological diversity and promote their wider application with the approval and involvement of the holders of such knowledge, innovations and practices and encourage the equitable sharing of the benefits arising from the utilization of such knowledge, innovations and practices'(p. 6).

The Nagoya Protocol on Access to Genetic Resources and the Fair and Equitable Sharing of Benefits (NP) arising from their utilization to the CBD notes 'the interrelationship between genetic resources and traditional knowledge, their inseparable nature for indigenous and local communities, the importance of the traditional knowledge for the Convention on Biological Diversity and the sustainable use of its components and for the sustainable livelihoods of these communities' (p. 3) out of the utilization of genetic resources. UNDRIP (2007) says that it is 'concerned that indigenous peoples have suffered from historic injustices as a result of, inter alia, their colonization and dispossession of their lands, territories and resources, thus preventing them from exercising, in particular, their right to development in accordance with their own needs and interests' (p. 3). Article 26 of UNDRIP says, 'indigenous peoples have the right to own, use, develop and control the lands, territories and resources, that they possess by reason of traditional ownership or other traditional occupations or use as well as those which they have otherwise acquired. States shall give legal recognition and protection to these lands, territories and resources. Such recognition shall be conducted with due respect to the customary tradition and land tenure system of the indigenous peoples concerned.'

The Annual Report of the Ministry of Tribal Affairs, 2019–20, says that 'the rights of forest dwelling Scheduled Tribes and other traditional forest dwellers on their lands and habitats had not been recognized despite their being integral part to the very survival and sustainability of the forest ecosystem,' (p. 53). It notes that 'the conservation processes for creating wilderness and forest areas for production of forestry somehow ignored the bona-fide interests of the tribal community from legislative framework in the regions where tribal communities primarily inhabited. The simplicity of tribes and their general ignorance of modern regulatory frameworks precluded them from asserting their genuine claims to resources in areas where they belonged and depended upon' (p. 53). It attributes the lack of recognition to past conservation policy based on "exclusive" approach rather than integration of these people.

The Forest Rights Act, 2006 was passed to correct the historic injustice and empower these peoples, Gram Sabhas and local level institutions are given the right to protect, regenerate, conserve and manage any community forest.

16.2.1 Empowerment of ILPS

The empowerment of ILPs requires facilitating roles for governments at various levels to remove the impediments ILPs face in the participation in ABS regime. ILPs need help in getting easier access to biological resources, in getting secure rights over the resources, in lowering the transaction costs of the exchanges and in designing and enforcing the contracts with the users of biological resources and traditional knowledge.

16.2.2 Nature of Biological Resources and Traditional Knowledge

If a resource is a private good having the properties of exclusion and rivalry, markets can exist or can be created and the market prices would reflect their social values. A pure public good has properties of non-rivalry and non-exclusion. We need non-market institutions for supply of public goods. Some biological resources yield non-consumptive benefits e.g. scenic beauty, eco-tourism. A few biological resources may come under the category of 'merit goods' and hence their access be made available to everyone, especially the poor. A few biological resources may come under inter-generational public goods and would be useful for development of new seed varieties and new pharmaceutical products. Some biological resources have quasi-option values and they have to be preserved for future use as and when the demand arises.

Ecosystems services may also be classified under provisioning services, regulating services, supporting services and cultural services. See Millinnieum Ecosystems Assessment (2005). Perfect markets do exist only for some provisioning services. Hence the valuation problem remains in entering contracts between the providers and the users in ABS. An attempt is being made to value ecosystem services and capture the values by methods such as payment for ecosystem services.

Traditional knowledge about medicines is available in some Indian languages. Concerned with bio-piracy in patent cases abroad, Council of Scientific and Industrial Research (CSIR) (2021) in cooperation with the Ministry of AYUSH translated traditional knowledge in Indian medicines to prevent bioprivacy. TKDL is a pioneering initiative of India to prevent misappropriation of country's traditional medicinal knowledge at International Patent Offices. NBA tries to document this information in Peoples Biodiversity Registers.

Traditional knowledge is also unwritten and takes the forms of oral traditions, folklores, songs, conventions passed on from generation to generation. Gadgil et al. (1993) observed that 'indigenous peoples with a historic continuity of resource-use practices often possess a broad knowledge base of the behavior of complex ecosystems in their own localities. This knowledge has accumulated through a long series of observations transmitted from generation to generation. Such "diachronic" observations can be of great value and complement the "synchronic" observations on which western science is based…It is vital, however, that the value of the knowledge practice-belief complex of indigenous people relating to conservation of biodiversity is fully recognized if ecosystem and knowledge would be most appropriately accomplished through promoting the community-based resource management system of indigenous people' (p. 1).

Traditional knowledge is valuable for co-existence with wild animals, ensuring food, water and other requirements of wild animals, evolving coping strategies for weather—related changes, understanding cultivation of location—specific traditional crops under different climatic conditions, identifying genetic resources for developing new medicines/new seed varieties, and guiding tourists in a safe manner.

16.2.3 Right to Development and Security of Tenure

In the past, government policies for conservation resulting in eviction of some ILPs from forests caused not only loss of livelihoods and hardships in resettlement of these people but also biodiversity loss. Article 10 of UNDRIP states that 'Indigenous peoples shall not be forcibly removed from their lands or territories. No relocation shall take place without the free, prior and informed consent of the indigenous peoples concerned and after agreement on just and fair compensation and, where possible, with the option of return'.

The Forest Rights Act 2006 contains the following provisions:

- a resettlement or alternatives package has been prepared and communicated that provides a secure livelihood for the affected individuals and communities and fulfills the requirements of such affected individuals and communities given in the relevant laws and the policy of the Central Government;
- the free informed consent of the Gram Sabhas in the areas concerned to the proposed resettlement and to the package has been obtained in writing;
- resettlement shall take place until facilities and land allocation at the resettlement location are complete as per the promised package.

Despite these provisions, forced evictions of ILPs occur due to reasons such as conservation, preservation of wild life, development etc. Such cases should be avoided in future.

16.2.4 Rights Over Biological Resources

Section 3 of the Forest Rights Act 2006 lists the following rights to ILPs:

(a) Right to hold and live in the forest land under the individual or common occupation for habitation or for self-cultivation for livelihood by a member or members
(b) Community rights such as nistar, by whatever name called, including those used in erstwhile Princely States, Zamindari or such intermediary regimes
(c) Right of ownership, access to collect, use, and dispose of minor forest produce which has been traditionally collected within or outside village boundaries
(d) Other community rights of uses or entitlements such as fish and other products ofwater bodies, grazing (both settled or transhumant) and traditional seasonal resource access of nomadic or pastoralist communities;
(e) Rights including community tenures of habitat and habitation for primitive tribal groups and pre-agricultural communities
(f) Rights in or over disputed lands under any nomenclature in any State where claims are disputed
(g) Rights for conversion of pattas or leases or grants issued by any local authority or any State Government on forest lands to titles

(h) Rights of settlement and conversion of all forest villages, old habitation, un surveyed villages and other villages in forests, whether recorded, notified or not into revenue villages
(i) Rights to protect, regenerate or conserve or manage any community forest resource which they have been traditionally protecting and conserving for sustainable use
(j) Rights which are recognized under any State law or laws of any Autonomous District Council or Autonomous Regional Council or which are accepted as rights of tribals under any traditional or customary law of the concerned tribes of any State
(k) Right of access to biodiversity and community right to intellectual property and traditional knowledge related to biodiversity and cultural diversity
(l) Any other traditional right customarily enjoyed by these people as the case may be which are not mentioned in clause (a)–(k) but excluding the traditional right of hunting or trapping or extracting a part of the body of any species of wild animal
(m) Right to in situ rehabilitation including alternative land in cases where these people has been illegally evicted or displaced form forest land of any description without receiving their legal entitlement to rehabilitation prior to the 13th day of December, 2005.

The rights include individual rights, community rights or both and user rights. Individual property rights-right to own, right to use, right to exchange and right to exclude others from using this property provide the owner an incentive to invest and develop his/her land as he/she can appropriate the higher returns from land and other resources.[1] Community rights for common-pool closed access system can yield benefits to the members and the community as a whole if an incentive-based institutional mechanism can be created (See Sect. 16.4). Forest dwellers sometime get user rights for non-timber forest products (NTFPs). Section 20 of the FRA Act defines ‘minor forest produce as all non-timber forest produce and includes, bamboos, brushwood, stumps, canes, tussar, cocoon, honey, waxes, lac, tendu/kendu leaves, medicinal plants and herbs, tubes and the like’.

The NTFPs sector is unorganized and its sustainability is being questioned. The report of the Sub-group on NTFP and their sustainable management in the 12th Five Year Plan notes,‘ one of India’s largest unorganized sectors having a dependent population of 275 million and with a business turnover of more than Rs. 6,000 crore per annum, the NTFP sector has however and unfortunately been neglected since the pre-independence period’ (p. 2). It notes depletion of the resource base because of the diversion of forest land for non-forest purpose and unsustainable harvesting

[1] ElinorOstrom in her Nobel Prize lecture 2009 notes five property rights that individuals using a common-pool resource might cumulatively have: (1) Access—the right to enter a specified property, (2) Withdrawal—the right to harvest specific products from a resource, (3) Management—the right to transform the resource and regulate, (4) Exclusion—the right to decide who will have access, withdrawal, or management rights, and (5) Alienation—the right to lease or sell any of the other four rights (pp. 419–420).

practice. It says that poor R&D, inadequate post-harvesting practices, insufficient funds and infrastructure, and unorganized nature of the trade have made the sector financially vulnerable. The Sub-group recommends resource management through conservation of all genotypes, development of sustainable harvesting practices and zone wise inventory of NTFPs. It also recommends minimum support price mechanism, certification, value chain development. India's experience with MSP is that it has been effective only in cases of major crops like paddy and wheat. In case of NTFP, problems arises first in fixation of minimum support prices because of quality variations in each of NTFP, and second in creating infrastructure for procurement, storage and development of value chain. The community owned Van DhanVikas Kendra constitutes 15 tribal self-help groups, each comprising upto 20 tribal NTFP gatherers or artisans. The Tribal Cooperative Marketing Development Federation of India (TRIFED), established in 1987, collaborates with government departments to get geographic indication (GI) tag for tribal products. GI is a constituent of TRIPS.

There are delays in granting individual and community rights under the Forest Rights Act. As on 30 September 2019, only 41 lakhs individual claims were received and 46% got the title deeds. In case of community rights, there was less than 1.5 lakh claims and 51% got the title deeds (see Table 16.1).

According to Haque (2020), of the estimated community forest resources (CFRs) between 35 and 40 million ha, only 3.56 million ha was recognized. He mentions the following factors responsible for the poor implementation—no defacto security of forest tenure; forest department functioning as veto by denying claims at the screening stage itself; irregular meetings of the screening committees; reluctance of forest officials to recognize community rights; delay in conversion of forest villages to revenue villages; inability of many traditional forest dwellers to produce documentary evidence of living in forest for 75 years; lack of institutional mechanism for procurement of minor forest produce at minimum support prices.

Sundar (2019) says if the potential area of 34.6 million ha is recognized as CFRs, the potential beneficiaries would be 200 million STs and other traditional forest dwellers in over 1,70,000 villages and they could get collective rights under FRA.

Table 16.1 Statement of claims and distribution of title deeds under forest rights act 2006 as on 30 September 2019

Item	Individuals	Community	Total
No. of claims received	41,01,689	1,48,913	42,50,602
No distributed	19,05,726	76,377	19,82,103
Distributed % of received	46.46	51.29	46.63
Acres of forest land distributed	41,65,431	88,05,364	129,70.695
Average acre	2.19	115.29	

Source Ministry of Tribal Affairs, Status report on the implementation of scheduled tribes and other traditional forest dwellers act of 2006, period ending 31 January 2020

She says that 'locals must be treated as important stakeholders and not just cast aside on the grounds that forest are a national resource......the courts, conservationists and communities can worktogether for forest preservation'. Whatever rights given, ILPs must have the freedom to make decisions on production/collection, exchange and use. Even in case of community rights, or user rights they need voluntary but binding regulations so that the problem of tragedy of commons can be avoided and sustainability of resource use is achieved.

Important problems in design of long term benefit-sharing arrangement in ABS are (i) information uncertainty about the resource usage over time and (ii) information asymmetry between the providers and the users—the buyer knows more about the potential benefits than the providers. These problems arise when the accessed biological resources and traditional knowledge are used for commercialization of the derived products and patenting. The longtime span uncertainty and information asymmetry increase the transaction costs of the negotiations. To cope up with these problems ILPs need an organization and partnerships with firms and research institutions.

16.3 NBA and ABS Forms

Under the Biological Diversity Act 2002, India has a three-tier institutional mechanism: National Biodiversity Authority (NBA) at the national level, State Biodiversity Boards (SBBs) at the state level and Biodiversity Management Committees (BMCs) at the local level. According to the Act,the purpose of BMC is 'promoting conservation, sustainable use and documentation of biodiversity including preservation of habitats, conservation of land races, folk varieties and cultivars, domesticated stocks and breeds of animals and microorganism, and chronicling of knowledge related to biodiversity' (p. 27). The People Biodiversity Register (PBR) is intended for getting baseline database; develop a legal document for intellectual property rights (IPRs) and patent issues, and also information for distribution of the benefits to potential providers of biological resources and traditional knowledge. Apart from general information about the locality, it should contain list of individual healers and practitioners, forest biodiversity, agro biodiversity, aquatic biodiversity, traditional knowledge, past ABS schemes; it should not be a mere listing of species.

The preparation of PBRs has been slow. In view of COVID-19, the National Green Tribunal has extended the deadline for compliances with respect to creation of BMCs and preparation of PBRs to 30, June 2021. Some localities have entrusted preparation of PBRs to consultants and questions are being raised about the extent of peoples' participation in the preparation process. One possible way to add the missing information is to make the reports available in gram sabhas/local libraries/schools and ask the public to provide additional information needed. NBA reports that, as of January 4, 2021, only 228.543 PBRs have been completed. There are doubts about the quality of information in PBRs for use in ABS regime. One major reason for the funds collected by NBA and SBBs not reaching the providers of biological resources and traditional knowledge is the difficulty in identifying the providers. This problem arises because the same biological resource/traditional knowledge may be found in more than one locality. We suggest the following procedure to resolve the problem. A committee of experts in each district may be asked to identify which biological resource and associated traditional knowledge is shared by more than one locality, identify the localities and note them in the district level register. A similar effort may be made to document biological resources and TK common to more than one district.

NBA issued guidelines on access to biological (BRs) and associated traditional knowledge (TK) and benefit sharing regulations on 21 November 2014. It suggests different forms for different purposes (see Table 16.2). Form I is to be used for access to BRs and/or TK for research or bio-survey and bio-utilization for research. An up-front payment by the applicant is required in case of BRs having high value. For commercial utilization, Form A has to be completed by the applicant and it suggests a fee of 1–3%of the purchase price of the BR; in case the buyer is a manufacturer the fee is between 3 and 5% of the purchase price.

In case of IPRs, form III has to be completed. If the applicant commercializes, he has to share 0.2–1.0% of the annual gross ex-factory sale minus the gross tax; if the applicant assigns or licenses to a third party for commercialization, the applicant has to pay NBA 3–5% of the fee received and 2–5% of the royalty amount received. The guidelines mention various monetary and non-monetary benefit-sharing options.

Laladhas et al. (2020) document successful livelihood models through BMCs. These models are: cultivating and conserving tuberous plants by Shaji from Waynad; Sangam Women Farmers group of Telangana in conserving grains and millets; Raipassa in Tripura's ABS agreement with regard to sale of broom grass; Pithora-band BMC (MP) marketing of organic wheat and rice varieties; Eraviperooor BMC'S efforts in construction of rain water harvesting pits, biogas plants, composting pits; Modi village (Telangana) initiatives to conserve breeds of cattle, goats, sheep and poultry and cultivation of pigeon pea, community seed bank; Sikkaripari (Tripura) development of broom industry; Peoples artificial reefs for fishermen in Kerala; Irula cooperative Venom Bank, Tamilnadu for conservation of snakes and extraction of snake venom (the society with 366 members sold venom worth Rs. 3.5 crore per annum); Malajkhand BMC (MP) ABS agreement for medicinal herbs with Bio resources Co; BIO-India Biologicals, Hyderabad benefit sharing agreement for export of neem leaves to Japan; Indian Institute of Oil Research, Hyderabad's development of eco-friendly insect and plant disease management and results to two companies; and export of bovine cattle embryos of Ongole bull to Brazil with upfront access fee of Rs. 1.2 crore. These examples show that community management in the form of self-help group or society with research/institutional support can enhance livelihood opportunities for ILPs.

Individuals with property rights have incentive for sustainable use of BRs. Individual owners can also form groups at Gram Sabha level to coordinate production, harvesting and marketing decisions. This will make it easy for them to seek scientific advice from near-by research institutes/firms on the use of the BRs and TK in their area and explore opportunities for commercialization/patenting of their BRs. They can also negotiate with forest department and other government departments

Table 16.2 **ABS** applications status as on 19 March 2021

	Purpose	Who should apply	Approvals as on 19 March 2021	
			No	%
I	Access of BRs occurring in or obtained from India/and associated TK	Non-Indians, NRI, foreign entity or Indian entity having non-Indian participation in share capital or management	307	13.5
A	If the applicant is a trader/manufacturer/company	Trader/manufacturer/company		
II	Transfer the results of research	Any Indian/non-Indian or entity of any Indian, NRI, foreign entity having non-Indian participation in share capital or management	28	1.2
III	Applying for IPRs for invention based on any research or information in a BR obtained from India	Any Indian/or non-Indian entity	1776	78.2
IV	Transfer of BR/knowledge already accessed to a third party	After approval in Form to Indian/or non-Indian or entities	30	1.3
B	Conducting of non-commercial research or research for emergency purpose outside India by Indian researchers/govt institutions	Indian researchers/govt institutions	131	5.8
C	Deposition of microorganisms innon-Indian repositories for claim of novel	Indian scientists/researchers		
	Total		2272	100

Source National Biodiversity Authority of India

for obtaining funds from programmes such as MGNREGA and CAMPA for regeneration of degraded forest lands. Case studies of ABS schemes in South America and other developing countries reveal that the providers' partnership with local research institutions and firms have reduced transaction costs of bargaining with the users and expedited the processes of patenting and commercialization and yielded positive benefits to the providers of BRs and TK.

16.4 Community Forest Resource Management

In India, of the estimated forest area of about 70 million ha, at present, only 5% come under community forests. This is because of the delays in recognition of community forest rights of ILPS, inability of some of the forest dwellers to prove their living for

75 years or more in the forest area and the weak capacities of the grass root democratic organizations and government departments in providing time-bound solutions for guaranteeing the rights of ILPs. Even with regard to MFPs, some states have monopoly over items such as Tendu and Sai seeds. At present there is no incentive for the community forest dwellers to use their labour time or undertake new investment for augmentation of BRs in their communities.

Section 5 of the Forest Rights Act states that the Gram Sabha or village level institution shall.

(a) carry out all executive functions as decided by the Gram Sabha
(b) be accountable and answerable to the Gram Sabha for its action and decisions
(c) prepare a draft conservation and management plan for the community forest resource and place before the Gram Sabha for final approval
(d) ensure sustainable harvesting of Minor Forest Products (MFPs)within the villageboundary; implement the CFR conservation and management plan

The draft working committee of the MTAs envisages the following functions for the Community Forest Resource Management Committee (CFRMC):

(a) carry out all executive functions as decided by the Gram Sabha
(b) accountable and answerable to the Gram Sabha for its action and decisions
(c) prepare a draft conservation and management plan for the community forest resource and place before the Gram Sabha for final approval
(d) ensure sustainable harvesting of MFPswithin the village boundary; implement the CFR conservation and management plan
(e) shall place the CFRMC report before the Gram Sabha

Even when one restricts attention to the collection of MFPs during a year/season, we need a sustainable allocation rule for the members to avoid the problem of 'the tragedy of the commons'. Ostrom (1990) poses the central question of her study as 'how a group of principals who are in an interdependent situation can organize and govern themselves to obtain continuing joint benefits when all face temptations to free ride, shirk or otherwise act opportunistically' (p. 141). We need a community-based providers' organization to allocate the fixed supply of MFPs in a season among its members. This may be in the form of an individual quota or entitlement with penalties for the members exceeding the quota/entitlement. The Gram Sabha or a provider group must be entrusted with the responsibility for allocation, monitoring and enforcement. Equity issue may arise in the provision of minimum entitlement to a member. The Gram Sabha/producer group must take this responsibility.

Kumar (2006) notes that lessons from the community-based conservation approach has been that that biodiversity conservation cannot be done in isolation and that approaches must involve effective partnerships at various levels. According to him, coalitions composed of conservationists, local people, the forest department and social activists are likely to achieve far more than the current fragmented approach.

In a recent paper, Sahu (2020) documents the factors responsible for the success of CFRM efforts in Vidharbha region of Maharastra. He says that farmers in this region have begun to exercise their rights in harvesting and selling bamboo and tendu leaf, increasing forest protection, reducing destructive practices, managing fires, regenerating degraded patches and identifying patches that require strict conservation. He explains that key factors like collective organization of Gram Sabhas, the role of NGOs, and facilitations by Maharastra Tribal Development Department and the state implementation agencies contributed to the success.

The Vidharbha experience in CFRM suggests that tribal department, forest department and other government departments working together with Gram Sabhas and CFRMCs can achieve conservation and increase livelihood opportunities of ILPs. The ABS regime opens the door for achieving sustainable use of biodiversity, new fruitful collaborations with Indian firms and research institutions and also the institutions of the user countries, and thereby more benefits accruing to the provider communities. However, this requires capacity building of the providers, Gram Sabhas and CFRMCs, institutional partnerships with nearby research institutes and biotechnology firms, and above all incentive—based policies for fruitful cooperation among the participants. Recent literature on management of common pool resources with closed membership by Ostrom and her collegues show that successful outcomes are feasible with right institutions and incentive structures.

16.4.1 Community Forest Management Institution for ABS

What needs to be done to enable community forest dependent ILPs to benefit from the ABS and enhance their livelihoods?

First, ensure that the FRA is implemented in a time-bound manner guaranteeing community forest rights to ILPs.

Second, Gram Sabhas and CFRMCs are constituted as viable democratic institutions. The state governments must ensure that the Tribal Affairs Department, the Forest Department and other welfare departments cooperate in establishing democratic, decentralized grassroot institutions for effective management of CFRM and welfare of ILPs.

Third, identify ayurvedic and pharmaceutical companies in each community interested in producing products based on BRs and traditional knowledge locally available.

Fourth, conduct training programmes for Gram Sabha members and CFRMC members on the objectives of CBD, features of ABS in India, case studies of successful ABS programmes in India and other developing countries, and issues in (a) patenting biological products, (b) reducing transaction costs of the negotiations and (c) incentive-based cost allocation methods[2] and (d)the desired types of monetary and non-monetary benefits to the providers. The training costs may be borne by central and state governments.

[2] Lloyd Shapley won the Nobel Prize in economics in 2012 for suggesting a fair way of allocating gains from a cooperative game to its members. The Shapley value ensures each player gain as much or more as they have gain from acting independently.

Fifth, as for management of CFRMCs, allocation of collection rights of MFPs be made on the basis of transparent mechanisms such as auctions, transferable quotas and licenses for harvesting. The CFRMC is responsible for monitoring MFPs collection and must have power to penalize the violators who overharvest.
Sixth, in the management of CFRs engage local ILPs for wildlife conservation, ecotourism and other non-extractive consumption. ILPs have been living in their communities for generations and their experience in wild life conservation and ecological knowledge be fully utilized.
Seventh, explore funding sources such as MGNREGA, CAMPA, and NBA for forest regeneration, sustainable harvesting of bio-resources and infrastructural development of the community.
Eighth, seek advice of biotechnology research institutions and biotechnology firms in near-by areas for preparing and implementing ABS agreements. This will strengthen the complementarity between 'diachronic' traditional knowledge and 'synchronic' scientific knowledge and signing of ABS agreements easier (Table 16.3).
Ninth, develop partnerships among Gram Sabhas, CFRMC, biotechnology research institutes and biotechnology firms for successful execution of ABS schemes leading to commercialization of bio-products and patenting. This partnership can reduce uncertainty in the transactions and lower the negotiation and monitoring costs (Table 16.3).
Tenth, when commercialization of certain bio-products or joint production require new investments/more forest land, negotiate with concerned government departments for funds/lands for long-term supply of bio resources, which may be justified for enhanced livelihood of ILPs.
Eleventh, use ABS CFR receipts for development of forest community infrastructure, sustainable harvesting, value addition and education and training of ILPs.

16.5 Concluding Remarks

The Biodiversity Act 2002 confers sovereign rights over BRs and TK in India for conservation and sustainable use of biodiversity. The Forest Rights Act 2006 would correct the 'historical injustice' done to the ILPs and make them stakeholders in conservation and sustainable use of biodiversity. The ABS regime provides an opportunity to providers of BRs and possessors of TK in India to explore new markets for their products and TK and thereby enhances their livelihoods. This is feasible with government support in the form of guaranteeing individual and community rights of ILPs and facilitating them in opportunities for regeneration/conservation/joint forest management. When ILPs organize themselves as viable grass root organizations having partnership with biotechnology research institutions and firms for preparation of ABS agreements and contractual agreements relating to commercialization, patenting and joint production, they can improve their livelihoods.

Table 16.3 Desirable benefit options to providers of biological resources and TK

Benefit options	Providers	Remarks
One-time payment	Identified Individuals/community based organizations	When transaction and monitoring costs are high
Milestone payment	” ”	When transaction stages known; no uncertainty
Share of royalties/license fees	“ “ Long term association between providers and users beneficial	Institutional support from firms & research institutions
Joint Ventures/Joint ownership of relevant IPRs	” ”	” ”
Technology transfer/sharing R&D results	Individuals/communities partnership with firms/research institutions	Consider transfer of appropriate technologies needed in India
Location of production/R&D units in India	Institutional partnership with firms/research institutes	Ensure local benefits i.e. jobs, markets
Sharing of scientific information on conservation and sustainable use of BRs	Research support from Indian institutions	Complementarity of diachronic and synchronic knowledge for sustainable development of BRs
Providing institutional capacity building, infrastructure development	When provider identification is difficult, develop community-based institutions	Assist conservation and sustainable use of regulating, supporting and cultural ecosystem services
Funding national, state and local biodiversity funds	“ “	Earmark part of funds for education and training of community-based organization members

Source Benefit options are from National Biodiversity Authority (2014). Guidelines on access to BRs and associated TK and benefit sharing regulations. New Delhi

References

Kumar C (2006) Whither ‘community-based conservation’. Econ Polit Weekly 41(52)

Council of Scientific and Industrial Research (2021) Traditional Knowledge Digital Library. https://www.csir.res.in

Gadgil M, Berkesand F, Folke C (1993) Indigenous knowledge for biodiversity conservation. Ambio 22(2/3). Biodiversity: ecology, economics policy. Springer, pp 151–156

Government of India, Ministry of Environment and Forests (2002) The biological diversity act, New Delhi

Government of India, Ministry of Tribal Affairs (2006) Scheduled tribes and other traditional forest dwellers (recognition of forest rights) act 2006, New Delhi

Government of India. Planning Commission (2011) Report of the Sub-Group II on NTFP and their sustainable management in the12th 5-year plan. https://www.niti.gov.in/planningcommission.govt.in/docs/aboutu/committeeswrkgrp12/enf/wg/subntfp.pdf

Haque T (2020) Securing forest rights and livelihoods of tribals. National Institute of Rural Development and Panchayat Raj, Hyderabad. https://www.nirdpr.org.in/nird_docs/srsc/310820n.pdf. Accessed 28 March 2021

Laladhas KP, John R, Saraswathy V, Bharucha E, Nair AS, Oommen OV (2020) Successful livelihood models through BMCs in India. In: Oommen V, Laladhas KP, Bharucha E (eds)

Nagoya Protocol on Access to Genetic Resources and the Fair and Equitable Sharing of Benefits arising from their Utilization to the Convention on Biodiversity (2011). https://www.cbd.int/abs/.doc/nagoya-protocol-en.pdf. Accessed 2 Apr 2021

National Biodiversity Authority (2014) Guidelines on access to biological resources and associated traditional knowledge and benefit sharing regulations, Chennai

Oommen OV, Laladhas KP, Bharucha E (2020) Biodiversity and livelihood lessons from community research in India. Bentham Books

Ostrom E (1990) An institutional approach to the study of self-organization and self-governance in CPR Situations, in governing the commons: the evolution of institutions for collective action. Cambridge University Press. Reprinted in U Sankar (ed) Environmental economics (2001) Readers in economics. Oxford University Press, New Delhi

Sahu G (2020) Implementation of community forest rights, experiences in the Vidharbha Region of Maharastra. Econ Political Weekly 55(18)

Sundar N (2019) A Mahagathbandan in the forests is the need of the hour. The Hindu 23 April

United Nations Convention on Biodiversity (1992) https://www.cbd.int/convention/text. Accessed 15 April 2021

United Nations (2007) United Nations Declaration on Rights of Indigenous Peoples). https://www.un.org/development/desa/indigenouspeoples/declaration-on-the-rights-of%20indigenous-peoples.html

Chapter 17
Access and Benefit Sharing: Scope of Indian Medicinal Plants

Anirban Roy and Prakash Pradhan

Abstract India with varied topography, edaphic and climatic conditions has led to flourish mosaic habitats sustaining rich biodiversity linked with diverse people escape, especially the floral wealth of the country having associated traditional knowledge and cultural heritage. Since time immemorial, besides daily life support, the floral wealth of the country has been used in traditional medicine for local health care. But over time, these medicinal plants had been explored and exploited by many foreign entities, especially during the pre-independence era leading to the rapid depletion of these resources, promoting biopiracy along with erosion of associated traditional knowledge. To counter global loss of biodiversity and to reaffirm sovereign rights of the states over their biological resources, United Nation's Convention on Biological Diversity (CBD) was ratified in 1992. India being a signatory of the CBD, has enacted the Biological Diversity Act, 2002 (BD Act) and Biological Diversity Rules, 2004, implementing through a three tier system (National Biodiversity Authority, State Biodiversity Boards and Biodiversity Management Committees at grass root level. Complying to BD Act, for access to biological resources for commercial utilization, before undertaking such activities applicants (with international participation) are to apply to NBA (matter of Section 3(2)) and Indian applicants (matter of Section 7) are to apply to SBBs in requisite format and with desired fees. Though the Sections 3–7 (both inclusive) came into force from 1st day of July 2004, one of the major impetus to Access and Benefit Sharing (ABS) scenario in India came from Nagoya Protocol, 2010 and formulation of ABS guidelines, 2014 in pursuance of the said protocol. Out of more than 20,000 species of vascular plants, so far documented from India, the Indian classical and the folk health care traditions are dependent upon the raw materials derived from estimated 6,500 species, out of which around 1178 species of medicinal plants are reported to be traded, and 242 of them are used in annual quantities of more than 100 MT. Records show that Ayurveda, Unani, Homoeopathy and Siddha systems of medicine utilizes around 800, 305, 425 and 250 species of medicinal plants in India, respectively. India's domestic herbal industry is represented by 8610 licensed herbal units, thousands of cottage level

A. Roy (✉) · P. Pradhan
West Bengal Biodiversity Board, Prani Sampad Bhavan, 5th Floor, LB-2, Sector-III, Salt Lake City, Kolkata 700106, India
e-mail: dr.anirbanroy@yahoo.co.in

O. V. Oommen et al. (eds.), *Biodiversity Conservation Through Access and Benefit Sharing (ABS)*, https://doi.org/10.1007/978-3-031-16186-5_17

unregulated herbal set-up and millions of folk healers. The complex trade web hints towards prominent wholesale markets and mandis across 14 cities from 12 states that channelizes herbal raw materials from various supply sources to the end users. The consolidated commercial demand of herbal raw materials for the year 2014–15 has been estimated at 5,12,000 MT. Even though with wide spread awareness on ABS, it is unfortunate that most commercial entities have not come forward to be a part of ABS mechanism; major reason being misperception of ABS as additional monetary burden, while Section 21 (2) of BD Act and the ABS guidelines has clear outlines for non-monetary provisions also. Further, the provisions of Section 7 of the BD Act, in some cases is not attracted by bio-resource users and there is also a misconception that all medicinal plants come under the ABS provisions, though there are exemption on 174 biological resources belonging to 108 species of medicinal plants which are ascribed under normally traded commodities as notified by Govt. of India vide Section 40 of the BD Act. The number of BD Act compliant commercial entities pertaining to Section 3(2) who have signed e.g. ABS with NBA (2006–2007 to 2020–2021) is only 311 which suggests low participation of the industrial fraternity to the ABS provisions of the BD Act. The unresponsiveness of many commercial users of bioresources (including medicinal plants; pertaining to Section 7) to the notices of SBBs/BMCs have led to a number of ABS related court cases. Concerted effort in this regard is necessary for proper sensitization of BMCs, associated line departments, bioresource conservers/cultivators, benefit claimers and commercial users alike for involvement in ABS process, so that a win-win-win situation may be created for both the bioresource, benefit claimers and the industry. The diversified medicinal plant wealth in Indian land mass, both actual and potential, has immense scope in bioprospecting especially in pharmaceutical and nutraceutical industries which would hopefully lead to enter into ABS chain towards their conservation.

Keywords Biopiracy · Medicinal plants · Traditional knowledge · Cultural heritage · Traditional medicine

17.1 Introduction

The diversity of plants depends on the climate, soil and biotic interaction. The more variation in physical conditions of the surrounding environment of plant population causes the more development of variable characters-morphological, anatomical, physiological, genetical etc. leading to the flash of diversity of plants, both species and infra-species level. The wide array of plant diversity (species and infra species) has various secondary metabolites which are essential for their physiological activities and withstanding other stressful conditions, may be of biotic defense or abiotic stress (climatic and/or edaphic). The plants with such variations through evolutionary processes in the form of different chemical compounds have the immense potentiality of curing several diseases and had been used by humans since the prehistoric period. The use of such plants dates back, perhaps to the origin of modern man and at that period it was instinctive, likely to be that of the other animals. Later it was mostly experience based and flowed to the generations through the then communication skills. But due to lack of evidence of the use of medicinal plants or in any form of records of prehistoric time, so far known, the exact period of the use of medicinal plants and the methods of treatment practiced is still unknown. The oldest evidence of preparation of drugs from medicinal plants in written form has been discovered from Nagpur on a Sumerian clay slab which is nearly 5000 years old. This includes 12 formulations of drug preparation referring to over 250 species of plants, some of which are poppy, henbane and mandrake. In India, the earliest record of the use of plants as medicine is the Rigveda which is estimated in the period to be in between 3500 and 1800 BC. Though the references to the plants in the Rigveda are very short, still in the Hymns, the identity of several plants was determined with reasonable certainty like Semal, Palash, Pithvan and Pipal. But in the Atharva Veda a more detailed account of medicinal plants is reflected.

India with varied topography, edaphic and climatic conditions in ten biogeographic zones (Rodger and Panwar 1988) has led to flourish mosaic habitats, sustaining rich biodiversity especially floral wealth of the country and thus placed as one of the seventeen megadiverse countries of the world. Besides, Indian landmass also covers four biodiversity HotSpot regions (Himalaya, Indo-Burma, Western Ghat- Sri Lanka and Sundaland) out of the global thirty-five, particularly for the high endemicity of vascular plants. Moreover, India is one of the twelve Centers of origin of crop plants (Indo-Burma Centre) (Vavilov 2009) indicating the nurturing ground of crops wild relatives.

There is a culture and tradition of every country or many countries within a broad landscape of using natural resources especially the plants with their different parts for curing and preventing various diseases. The plant resources used for such purposes are generally available in that particular landscape and climatic conditions. The use of these plant resources as medicine is the outcome of a high degree of perception through power of observation, generalization and analysis combined with the convalescence of patients and resources applied. The successful medicines with necessary formulations have been flowing through practicing over the generations. Since the ancient time to modern dates, this medicinal science has been nurtured and upgraded with necessary scientific input. India being one of the oldest civilizations of the world having rich medicinal plant heritage is a treasure of medicinal sciences, both wisdom and resources. Bass (1889) in the 'outlines of the history of medicine and medical profession' considered Indian medicine as the oldest of the world with a continuous development of its own. Neuburger (1925) in his History of Medicine (translated by Ernest Playfair) commented "the medicine of Indians, if it does not equal the best achievements of the race, at least nearly approaches them, and owing to the wealth of knowledge, depth of speculation and systematic construction, takes an outstanding position in the history of oriental medicine."

Since time immemorial, besides daily life support, the floral wealth of the country has been used in traditional medicine for local health care. But over time, these medicinal plants had been explored and exploited through unsustainable harvesting by many commercial users including foreign entities, especially during the pre-independence era leading to the rapid depletion of these resources and promoting biopiracy. Moreover, traditional knowledge related to medicinal plants had also been used commercially, in many cases, without providing its protection for benefit share to knowledge holders.

To counter global loss of biodiversity and to reaffirm sovereign rights of the states over their biological resources, United Nations Convention on Biological Diversity (CBD) was ratified in 1992. India being a signatory of the CBD, has enacted the Biological Diversity Act, 2002 (BD Act) and Biological Diversity Rules, 2004, implementing through three tier system viz. National Biodiversity Authority (NBA), State Biodiversity Boards (SBBs) and Biodiversity Management Committees (BMCs) at grass root level. Complying to BD Act, for access to biological resources for commercial utilization, before undertaking such activities applicants (with international participation) are to apply to NBA (matter of Section 3(2)) and Indian applicants (matter of Section. 7) are to apply to SBBs in requisite format and with requisite fees.

Though the Sections. 3–7 (both inclusive) came into force from 1st day of July 2004, one of the major impetus to Access and Benefit Sharing (ABS) scenario in India came from Nagoya Protocol, 2010 and formulation of ABS guidelines, 2014 in pursuance of the said protocol.

17.2 Documentation and Distribution of Medicinal Plants in India

India has a rich heritage of knowledge about medicinal plants and their uses. The documentation of medicinal plants in India has been done both in utilitarian and distributional aspects in India. Out of more than 20,000 species of vascular plants, so far documented from India, the Indian classical and the folk health care traditions are dependent upon the raw materials derived from estimated 6,500 species, out of which around 1178 species of medicinal plants are reported to be traded, and 242 of them are used in annual quantities of more than 100MT (NMPB 2021a). The Indian Medicinal Plants Database developed by National Medicinal Plant Board (NMPB) and Foundation for Revitalisation of Local Health Traditions (FRLHT) has reported Andhra Pradesh to have 1559 species of medicinal plants, similarly Karnataka is reported to be having 1838, while Kerala has 2052,, Maharashtra—2630, Sikkim—1681, Tamil Nadu 1840 and West Bengal has 2800 (IMPD 2021a, b, c, d, e, f, g). It is reported that the habitat-wise, 36% of the medicinal bio-resources are coming from trees, 28% from herbs, 21% from shrubs, 7% from herbaceous climbers, 8% from lianas (Sarkar 2012). Records show that Ayurveda, Siddha and Unani systems of medicine have more than 90% formulations which are plant based and Ayurveda, Folk, Homeopathy, Modern, Siddha, Tibetan, Unani system of medicine utilizes around 2351, 5137, 506, 204, 1785, 350 and 979 species of medicinal plants in

India, respectively (Kala et al. 2006). The Unani system of Medicine uses 109 plant species with non-Indian origin (Khare 2007).

17.2.1 Utilitarian Documentation

In modern times, many valuable utilitarian knowledge have been compiled into compendia and books by notable organizations. Foremost, Ministry of Health, Govt. of India has compiled (1) Ayurvedic Pharmacopoeia of India (Part I, Vol I-IV) and (2) National Formulary of Unani Medicine (Part I, 1st Ed.); Central Drug Research Institute has compiled Compendium of Indian Medicinal Plants; Central Institute of Medicinal and Aromatic Plants has prepared Dictionary of Indian Medicinal Plants; Chaukhambha Bharati Academy, Varanasi has compiled Dravya-guna Vigyaana; The Indian Medical Practitioners' Cooperative Pharmacy and Stores Ltd. has prepared Formulary of Siddha Medicines; Regional Research Laboratory, Jammu and Indian Drug Manufacturers' Association, Mumbai has compiled Indian Herbal Pharmacopoeia, Vol. I and II; Rashtriya Ayurveda Vidyapeeth (National Academy of Ayurveda) has compiled (1) Medicinal Plants used in Ayurveda, (2) Plants of Sharangadhara Samhita and (3) Plants of Bhava Prakash; Central Council for Research in Ayurveda and Siddha (CCRAS) has prepared documents on 1) Phytochemical Investigations of Certain Medicinal Plants used in Ayurveda, and 2) Standard Nomenclature of Ayurvedic Medicinal Plants; Ayurvedic and Tibbi Academy, Uttar Pradesh has compiled Unani Dravyagunaadarsha; While NISCAIR has compiled (1) The Wealth of India, (2) The Treatise on Indian Medicinal Plants, (3) Glossary of Indian Medicinal Plants with Active Principles and hosts the journal Medicinal & Aromatic Plant Abstracts. Most of the commercial users of bio resources refer to these sources/Pharmacopoeias for knowledge about ingredients for bulk drug manufacturing and dosage consensus of the crude drug (Khare 2007).

17.2.2 Distributional Documentation

17.2.2.1 Biogeographic Distribution of Medicinal Plants in India

The bio-geographic zone (Rodgers and Panwar 1988) wise distribution of medicinal plants in India are depicted in Fig. 17.1. The Deccan Peninsula has the highest number of medicinal plant species (3000), followed by Western Ghats (2000), North-East India (2000), Himalayan (1700), Semi-Arid (1000), Gangetic Plain (1000), Islands (1000), Trans Himalayan (700), Desert (500), Coasts (500) (Sarkar 2012).

Characteristic medicinal plants of the Trans Himalayan bio-geographic zone include *Arnebia euchroma*, *Ephedra gerardiana*, *Hippophae rhamnoides*

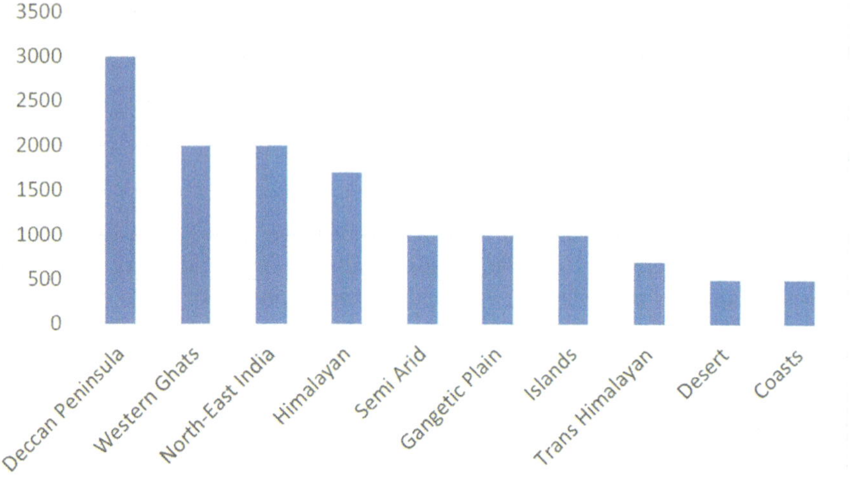

Fig. 17.1 Distribution of Medicinal plants in different Indian Bio-geographic zones

(=*Elaeagnus rhamnoides*) etc. Similarly, Himalayan biogeographic zone is characterized by *Aconitum heterophyllum*, *Ferula jaeschkeana*, *Nardostachys grandiflora* (=*Nardostachys jatamansi*), *Panax pseudoginseng*, *Rhododendron anthopogon*, *Saussurea costus*, *Taxus wallichiana* etc. Desert biogeographic zone is characterized by *Citrullus colocynthis*, *Convolvulus microphyllus*, *Cressa cretica*, *Tecomella undulata* etc. Semi-Arid biogeographic zone is characterized by *Balanites aegyptiaca*, *Caesalpinia bonduc*, *Commiphora wightii*, *Tribulus rajasthanensis* etc. Western Ghats biogeographic zone is characterized by *Garcinia indica*, *Myristica malabarica*, *Utleria salicifolia*, *Vateria indica* etc. The Deccan Peninsula biogeographic zone is characterized by *Decalepis hamiltonii*, *Pterocarpus santalinus*, *Shorea tumbuggaia*, *Terminalia pallida* etc. Gangetic plains biogeographic zone is characterized by *Holarrhena pubescens*, *Mallotus philippensis*, *Peganum harmala*, *Pluchea lanceolata* etc. Coastal biogeographic zone is characterized by *Acanthus ilicifolius*, *Avicennia marina*, *Rhizophora mucronata*, *Sonneratia caseolaris* etc. North East India biogeographic zone is characterized by *Abroma augusta*, *Aquilaria malaccensis*, *Hydnocarpus kurzii*, *Smilax glabra* etc. The Island's biogeographic zone is characterized by *Adenanthera pavonina*, *Aesandra butyracea* (=*Diploknema butyracea*), *Barringtonia asiatica*, *Calophyllum inophyllum* etc. (Fig. 17.1).

17.2.2.2 Ecological Niche Modeling/Species Distribution Modeling of Threatened Medicinal Plants in India

Ecological Niche Modeling (ENM)/Species Distribution Modeling (SDM) is a moderately new technique to understand distribution of species in terms of its environmental correlates. This technique mostly utilizes species occurrences (and pseudo-absences) along with various explanatory variables such Normalized Difference Vegetation Index (NDVI), Potential evapotranspiration, Solar Radiation, Wind Speed, Altitude, Slope, Aspect etc. and more importantly bioclimatic variables which expresses biologically eloquent spatially variable parameters in terms of annual means, seasonality and extremes/limiting factors of Temperature and Precipitation (Pradhan 2015, 2016, 2019; Pradhan and Setyawan 2021). In India, a number of ENM/SDM based studies on threatened medicinal plants have been conducted e.g. *Pterocarpus santalinus* (Babar et al. 2012), *Justicia adhatoda* (Yang et al. 2013), *Monotropa uniflora* (Pradhan 2015), *Aristolochia indica* (Sarma et al. 2018), *Ephedra gerardiana, Lilium polyphyllum, Crepidium acuminatum, Pittosporum eriocarpum* and *Skimmia anquetilia* (Sourabh et al. 2018), *Brucea mollis* (Mipun et al. 2019), *Paris polyphylla* (Lepcha et al. 2019), *Illicium griffithii* (Bapu and Nimasow 2021), *Lilium polyphyllum* (Dhyani et al. 2021), *Oroxylum indicum* (Kumar et al. 2021), *Picrorhiza kurroa* (Chandra et al. 2021). Further, Botanical Survey of India and National Botanical Research Institute are conducting dedicated SDM studies on threatened plants. Among various goals of SDM, identification of potential areas for prioritizing conservation of target species is one of the prime objectives. In the present context, the use of the said objective may add significantly to the conservation of threatened medicinal plants in India.

17.2.2.3 State Wise Threat Status Assessment of Prioritized Medicinal Plants

During 1995–2014, 14 Conservation Assessment & Management Prioritization (CAMP) workshops covering 17 states of India were conducted by the Conservation Breeding Specialist Group (CBSG), India, in collaboration with FRLHT and sponsored by National Medicinal Plants Board, Government of India for rapid assessment of threat status of prioritized medicinal plant species of conservation concern in different states/regions of India. The participating state and the year when the CAMP workshop(s) were conducted include—Karnataka, Kerala and Tamil Nadu (1995, 1996, 1997 and 1999), Jammu and Kashmir (1998, 2003), Andhra Pradesh and Maharashtra (2001), Arunachal Pradesh, Assam and Chhattisgarh (2003), Madhya Pradesh (2003, 2006), Himachal Pradesh (2003, 2010), Meghalaya and Uttarakhand (2003), Sikkim (2003, 2014), Odisha, Rajasthan and West Bengal (2007).

State-wise numbers of threatened medicinal plants of India based on CAMP exercise represented from state with highest number of species to state with lowest number of species under different threat categories (a) Vulnerable, (b) Endangered, (c) Critically endangered respectively (Fig. 17.2, 17.3 and 17.4) include—Andhra Pradesh (11, 23, 5); Arunachal Pradesh (17, 12, 6); Assam (6, 8, 1); Chhattisgarh (30, 5, 1); Himachal Pradesh (17, 19, 11); Jammu & Kashmir (28, 22, 9); Karnataka (46, 21, 6); Kerala (41, 25, 9); Madhya Pradesh (31, 6, 3); Maharashtra (10, 14, 2); Meghalaya (13, 7, 3); Odisha (25, 12, 2); Rajasthan (19, 12, 6); Sikkim (26, 5, 0); Tamil Nadu (33, 23, 13); Uttarakhand (23, 18, 7); West Bengal (15, 19, 6) (Gowthami et al. 2021).

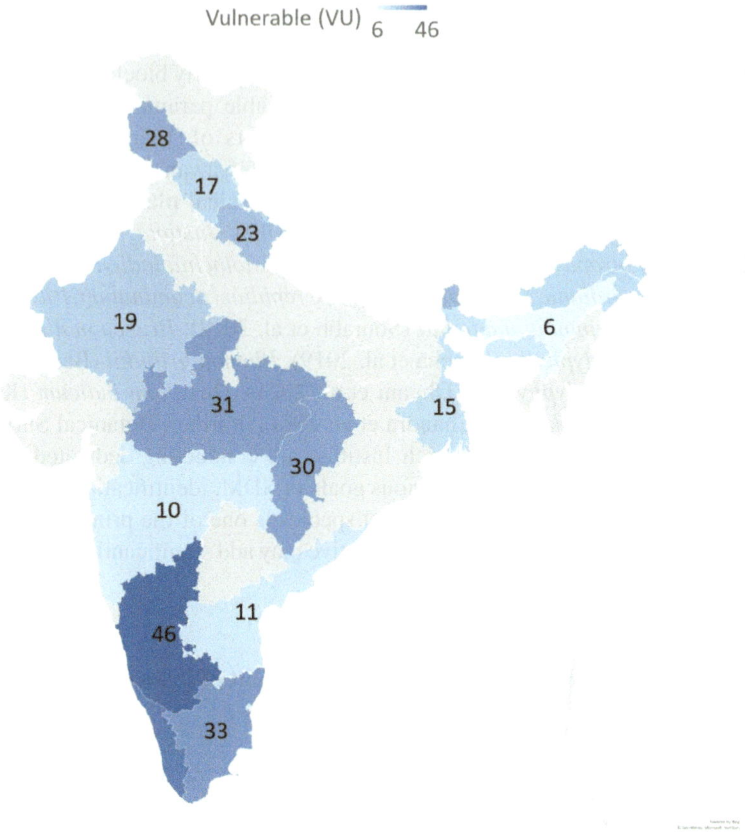

Fig. 17.2 Distribution of locally vulnerable medicinal plants in India as identified by CAMP workshop(s)

From the above enumeration the state wise locally critically endangered species are as follows: **Andhra Pradesh**: *Cycas beddomei*, *Embelia ribes*, *Litsea glutinosa*, *Pterocarpus santalinus*, *Rauvolfia serpentina*; **Arunachal Pradesh**: *Amentotaxus assamica*, *Aquilaria malaccensis*, *Gymnocladus assamicus*, *Malaxis muscifera*, *Rauvolfia serpentina*, *Smilax glabra*; **Chhatishgarh:** *Rauvolfia serpentina*; **Assam**: *Smilax glabra*; **Himachal Pradesh**: *Aconitum deinorrhizum* (=*Aconitum heterophylloides*), *A. heterophyllum*, *Atropa acuminata*, *Dactylorhiza hatagirea*, *Gentiana kurroo*, *Habenaria edgeworthii* (=*Platanthera edgeworthii*), *Jurinea dolomiaea*, *Lilium polyphyllum*, *Malaxis muscifera*, *Picrorhiza kurroa*, *Swertia chirayita*; **Jammu & Kashmir**: *Aconitum chasmanthum*, *A. heterophyllum*, *Arnebia benthamii*, *Betula utilis*, *Dactylorhiza hatagirea*, *Gentiana kurroo*, *Lilium polyphyllum*, *Saussurea costus*, *Swertia chirayita*;

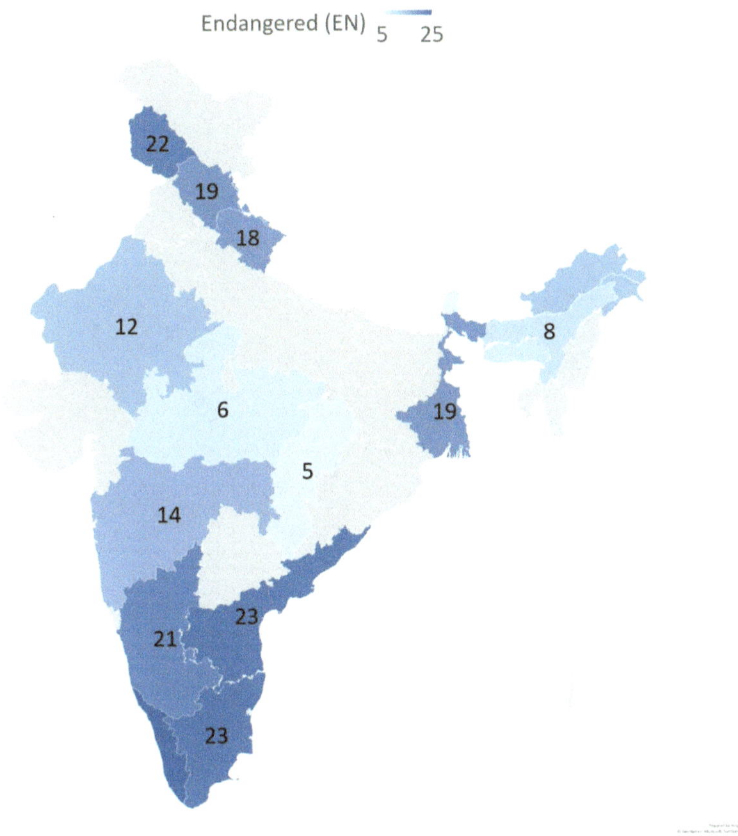

Fig. 17.3 Distribution of locally endangered medicinal plants in India as identified by CAMP workshop(s)

Karnataka: *Coscinium fenestratum, Cycas circinalis, Eulopia cullenii, Paphiopedilum druryi, Pueraria tuberosa, Valeriana leschenaultii*; **Kerala**: *Coscinium fenestratum, Eulophia cullenii (=Eulophia flava), Heliotropium keralense, Janakia arayalpathra (=Decalepis arayalpathra), Paphiopedilum druryi, Piper barberi, Utleria salicifolia, Valeriana leschenaultii, Vateria macrocarpa*; **Madhya Pradesh**: *Alectra parasitica* subsp. *chitrakutensis, Commiphora wightii, Psilotum nudum*; **Maharashtra**: *Holostemma ada-kodien, Rauvolfia serpentina*; **Meghalaya**: *Illicium griffithii, Smilax glabra, Taxus wallichiana*; **Odisha**: *Saraca asoca, Symplocos racemosa*; **Rajasthan**: *Chlorophytum borivilianum, Cochlospermum religiosum, Commiphora wightii, Eulophia ochreata, Pterocarpus marsupium, Tribulus rajasthanensis*; **Tamil Nadu**: *Adhatoda beddomei (=Justicia beddomei), Coscinium fenestratum, Cycas circinalis, Eulophia cullenii (=Eulophia flava), Helminthostachys zeylanica, Janakia arayalpathra (=Decalepis arayalpathra), Paphiopedilum druryi, Piper barberi, Pterocarpus santalinus,*

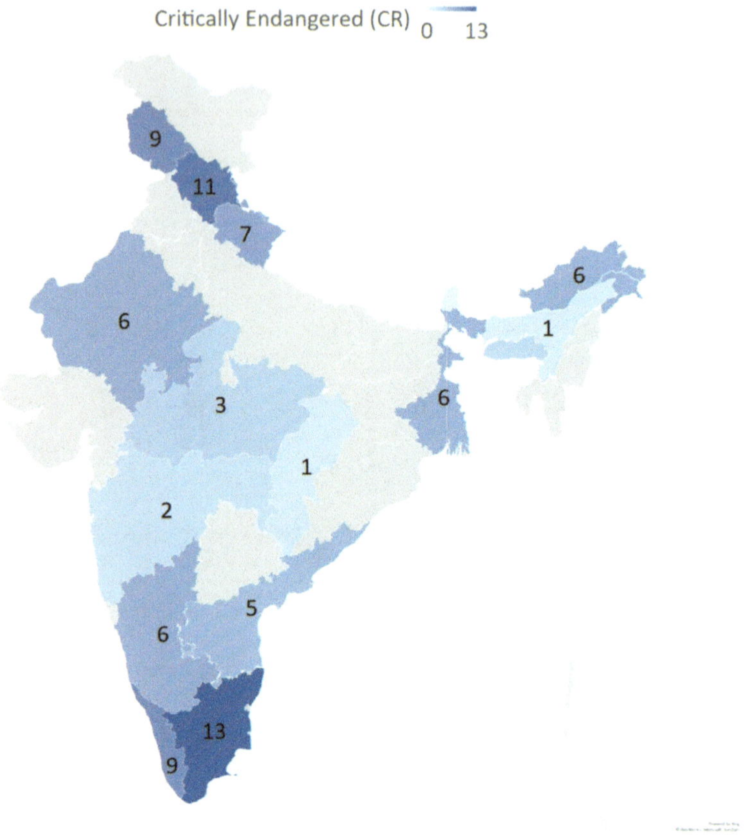

Fig. 17.4 Distribution of locally critically endangered medicinal plants in India as identified by CAMP workshop(s)

Shorea tumbuggaia, *Utleria salicifolia*, *Valeriana leschenaultii*, *Vateria macrocarpa*; **Uttarakhand**: *Aconitum heterophyllum*, *Arnebia benthamii*, *Dactylorhiza hatagirea*, *Gentiana kurroo*, *Lilium polyphyllum*, *Nardostachys grandiflora* (=*Nardostachys jatamansi*), *Picrorhiza kurroa*; **West Bengal**: *Panax pseudoginseng*, *Persea glaucescens*, *Picrorhiza kurroa*, *Podophyllum hexandrum*, *Swertia chirayita*, *Taxus wallichiana* (Figs. 17.2, 17.3 and 17.4).

17.3 Grass Root level Documentation of Medicinal Plants and Associated Traditional Knowledge in India: People's Biodiversity Register as a Tool

In the three-tier framework for implementation of the BD Act, at the grass root focal points, Biodiversity Management Committees (BMCs) are closest to the local bio-resource associated knowledge base (Shah 2012). The main function of the BMC is to prepare the People's Biodiversity Register (PBR) in consultation with the local people. Besides containing comprehensive information on availability and knowledge of local biological resources, their medicinal or any other use, PBRs are good resource for details on local traditional health care (human and livestock) practitioners and local knowledge holders such as vaids and hakims using biological resources (Annexure 2 of NBA 2013). NBA's PBR format also specifically focuses on medicinal plants (herbs, shrubs, trees etc.) [Format 12], wild plants of medicinal importance [Format 22] and markets/fairs for domesticated animals, medicinal plants and other products [Format 17].

In case the area is rich in medicinal plants, National Biodiversity Authority (NBA) (2013) emphasizes special focus to be given to document the resources and its current use pattern. Based on the results of the documentation, the BMC will be in a position to determine the resource potential, prospects of cultivation, sustainable use and trading (NBA 2013).

However, with ample capacity building of BMCs there is inadequacy of collected and compiled data in PBRs in several cases mainly due to absence of minimum standards of compiled data, while also due to inadequate reach outs and local group discussions by volunteers collecting the data. The compiled database on locally collected/cultivated and used medicinal plants is crucial armament in the hands of BMC for local sustenance of medicinal plant resources. The details on local players on commercial utilization of bioresources opens direct avenue for grassroot level ABS implementation, while as per Section 41(3) of the BD Act empowers BMCs to levy charges by way of collection of fees from any person for accessing or collecting biological resources for commercial purposes from the areas falling within its territorial jurisdiction. However, it is very rare that BMCs have come forward by charging such levy as well as negotiating ABS with commercial users on their own potential. On the positive side, many SBBs have been funding BMCs for post PBR conservation efforts including conservation, propagation and cultivation of medicinal plants (Shah 2012).

Preparation of PBR in India is nearly complete at the level of all the local bodies and there is ongoing process of preparation of E-PBR framework in Biodiversity Management Information System (BIOMIS) platform which is a web enabled repository of biodiversity information for State Biodiversity Boards. The main aim of BIOMIS is to ensure widespread citizen participation for documentation of biodiversity information of the state and generation of Electronic—People's Biodiversity Register and provide open access biodiversity related data of the state. Documentation of Indian medicinal plants will get rapid momentum after full functioning of such a biodiversity information system, which is necessary for the prioritizing conservation and management of these valuable bio resources.

17.4 ABS Scope of Indian Medicinal Plants

India's domestic herbal industry is represented by 8610 licensed herbal units, thousands of cottage level unregulated herbal establishments and millions of folk healers. The complex trade-web hints towards prominent wholesale markets and mandis across more than 14 cities from 12 states that channelizes herbal raw materials from various supply sources to the end users. The consolidated commercial demand of herbal raw materials for the year 2014–15 has been estimated at 5,12,000 MT out of which around 22% were sourced through cultivation. The export demand of medicinal plants was estimated to be about 1,34,500 MT with an export value of ₹ 3211 Crores during 2014–15 (NMPB 2021a).

The majority of the 'commercial utilization of bioresources' attracting ABS provisions of the BD Act, 2002 has been conceptually ascribed to the commercial use of medicinal bioresources in the manufacturing chain. In addition, the majority of 'medicinal bio-resources' are coming from medicinal plants and very few from fauna side eg. honey from *Apis* spp., cantharidine from Blister Beetles etc.

The principal motto of ABS provisions is the conservation and sustenance of the commercially utilized bio resources/associated traditional knowledge and to support livelihood of the local conservers of the bio resources and/or local people holding such unique traditional knowledge regarding use of the bio resources.

ABS scope of Indian medicinal plants may be identified by analysing cross-sectional linkages regarding their trade volume, threat status, and promotional initiatives by government, and more importantly, applicability of ABS provisions of the BD Act, 2002.

The list of Normally Traded as Commodities (NTCs) notified by Govt. of India vide Section 40 of the BD Act has made it easier to identify medicinal bio resources where ABS provisions of BD Act will not be applicable. Primarily, those species which are not listed under NTC are well coming under the ABS provisions. Interestingly, species listed in NTC may attract ABS provisions **(i)** if the source is other than that is listed in NTC notification (e.g. *Azadirachta indica* [Neem] is NTC when leaves are collected from cultivated sources, but it may attract ABS provisions if the leaves are collected from wild), **(ii)** if the plant part is other than that is listed in NTC notification (e.g. *Azadirachta indica* [Neem] leaves collected cultivated sources and fruits collected from cultivated as well as wild sources are treated as NTC, however, bark of Neem is not listed as NTC, hence it is a potential bio resource for ABS).

Based upon the National level study on Medicinal Plants trade undertaken by FRLHT during 2008 and supported by NMPB, 85 such species could be identified which are of conservation concern in various states, yet the trade volume of which have been recorded in quantity exceeding 100 MT/year (dry wt.) in India (Table 17.1).

Table 17.1 State wise medicinal plants of conservation concern but with trade volume > 100 MT/yr (dry wt.); along with Global IUCN Red List status (as on July 2021) and NTC status as per Biological Diversity Act, 2002. **State Names**: ANP = Andhra Pradesh, ARP = Arunachal Pradesh, ASM = Assam, CHT = Chhatishgarh, HP = Himachal Pradesh, J&K = Jammu and Kashmir, KRT = Karnataka, KRL = Kerala, MP = Madhya Pradesh, MAH = Maharashtra, MEG = Meghalaya, ODI = Odisha, RAJ = Rajasthan, SKM = Sikkim, TN = Tamil Nadu, UTK = Uttarakhand, WB = West Bengal; **Threat Status**: CR = Critically Endangered; EN = Endangered; VU = Vulnerable; NT = Near Threatened; LC = Least Concern, DD = Data Deficient, **Habit**: H = Herb, S = Shrub, C = Climber, T = Tree

Sl. no	Species	State & threat status	Habit	Global IUCN red list status and NTC status
1	*Abelmoschus moschatus* Medik	WB-NT	S	
2	*Abrus precatorius* L	CHT-NT, MP-NT	C	
3	*Aconitum ferox* Wall. ex Ser	ARP-EN, SKM-EN (2003), VU (2014), WB-EN	H	
4	*Aconitum heterophyllum* Wall. ex Royle	HP-CR, J&K-CR, SKM-EN, UTK-CR	H	EN
5	*Acorus calamus* L	ANP-EN, CHT-EN, KRL-EN, MP-EN, TN-VU	H	LC; Cultivated Rhizomes as NTC
6	*Aegle marmelos* (L.) Corrêa	ANP-VU, KRT-VU, MAH-VU, TN-VU	T	NT
7	*Alpinia calcarata* (Haw.) Roscoe	WB-EN	H	Cultivated Rhizomes as NTC
8	*Andrographis paniculata* (Burm.f.) Nees	CHT-VU, MP-VU	H	
9	*Aquilaria sinensis* (Lour.) Spreng. (=*Aquilaria malaccensis* Benth.)	ARP-CR, ASM-EN, MEG-EN	T	VU
10	*Asparagusracemosus* Willd	CHT-NT, MP-NT, WB-EN	C	

(continued)

Table 17.1 (continued)

Sl. no	Species	State & threat status	Habit	Global IUCN red list status and NTC status
11	*Baliospermum solanifolium* (Burm.) Suresh (= *Baliospermum montanum* (Willd.) Müll.Arg.)	CHT-NT, KRT-VU, KRL-VU, MP-NT	S	LC
12	*Berberis aristata* DC	SKM-DD (2014), WB-VU	S	LC
13	*Bergenia ciliata* (Haw.) Sternb	MEG-VU, SKM-VU, ARP-VU	H	LC
14	*Boswellia serrata* Roxb. ex Colebr	CHT-VU, MP-VU, RAJ-EN	H	
15	*Buchanania cochinchinensis* (Lour.) M.R.Almeida (=*Buchanania lanzan* Spreng.)	CHT-NT, MP-NT, RAJ-VU	T	
16	*Butea monosperma* var. *lutea* (Witt.) Maheshwari	ANP-EN, RAJ-VU	T	DD
17	*Celastrus paniculatus* Willd	ANP-VU, CHT-VU, KRT-NT, KRL-VU, MP-VU, ODI-VU, RAJ-VU, TN-NT, WB-EN	C	
18	*Chlorophytum arundinaceum* Baker	MAH-EN	H	
19	*Chlorophytum borivilianum* Santapau & R.R. Fern	MAH-EN, RAJ-CR	H	CR; Cultivated root tuber as NTC
20	*Chlorophytum tuberosum* (Roxb.) Baker	CHT-VU, MP-VU	H	LC
21	*Cinnamomum sulphuratum* Nees	KRT-VU, KRL-VU, TN-VU	T	VU

(continued)

However, there are some high trade volume species which are listed NTC, viz. *Acorus calamus* (cultivated rhizomes); *Alpinia calcarata* (cultivated rhizomes); *Chlorophytum borivilianum* (root tubers from cultivated sources); rhizomes of *Curcuma angustifolia* and *C. zedoaria* from cultivated sources; *Gloriosa superba*

Table 17.1 (continued)

Sl. no	Species	State & threat status	Habit	Global IUCN red list status and NTC status
22	*Cinnamomum tamala* (Buch.-Ham.) T. Nees & Eberm	ARP-VU, HP-VU, J&K-EN, MEG-VU, UTK-VU	T	LC; Cultivated bark and leaf as NTC
23	*Citrullus colocynthis* (L.) Schrad	CHT-VU, MP-VU	C	Cultivated fruit and root as NTC
24	*Commiphora wightii* (Arn.) Bhandari	MP-CR, RAJ-CR	H	CR
25	*Coscinium fenestratum* (Goetgh.) Colebr	KRT-CR, KRL-CR, TN-CR	C	DD
26	*Curcuma angustifolia* Roxb	MP-VU	H	Cultivated rhizomes as NTC
27	*Curcuma zedoaria* (Christm.) Roscoe	MP-VU	S	DD; Cultivated rhizomes as NTC
28	*Decalepis hamiltonii* Wight & Arn	ANP-EN, KRT-EN, TN-EN, KRL-EN	C	EN
29	*Embelia ribes* Burm.f	ANP-CR, ARP-NT, KRT-VU, KRL-NT, ODI-VU, TN-VU	C	
30	*Embelia tsjeriam-cottam* (Roem. & Schult.) A.DC	CHT-NT, HP-VU, J&K-VU, KRT-VU, KRL-VU, MP-VU, MAH-VU, ODI-VU, TN-VU, UTK-VU	S	

(continued)

(seeds and roots from cultivated sources); *Phyllanthus emblica* (fruits from both cultivated and wild sources); *Piper longum* (cultivated fruit, root and stem); *Citrullus colocynthis* (cultivated fruit and root); *Cinnamomum tamala* (bark and leaf from cultivated

Table 17.1 (continued)

Sl. no	Species	State & threat status	Habit	Global IUCN red list status and NTC status
31	*Ephedra gerardiana* Wall. ex Stapf	HP-EN, SKM-VU (2014), UTK-EN	S	VU
32	*Fumaria indica* (Hausskn.) Pugsley	MAH-EN	H	
33	*Garcinia indica* (Thouars) Choisy	KRT-VU, KRL-VU, MAH-NT	T	VU
34	*Gardenia resinifera* Roth	MAH-NT	T	
35	*Gloriosa superba* L	ANP-VU, CHT-VU, HP-VU, J&K-VU, KRT-VU, KRL-VU, MP-VU, MAH-VU, ODI-EN, RAJ-VU, UTK-VU, WB-VU	C	LC; cultivated root and seed as NTC
36	*Gymnema sylvestre* (Retz.) R.Br. ex Sm	ANP-VU, CHT-VU, MP-VU, MAH-NT, RAJ-EN, WB-VU	C	
37	*Holostemma ada-kodien* Schult	ANP-NT, KRT-VU, KRL-EN, MAH-CR, RAJ-VU, TN-NT	C	
38	*Ipomoea mauritiana* Jacq	WB-NT	C	

(continued)

sources); *Plectranthus barbatus* (cultivated leaf and root) and *Trichosanthes cucumerina* (cultivated leaf, stem, seed, fruit and tuber) are listed as NTC, even though they were listed under different threat categories in various states.

Interestingly, from the list high trade volume species, *Chlorophytum borivilianum* (commonly Musali) is listed under Critically Endangered category of IUCN with decreasing global population trend (Ved et al. 2015), yet its root tuber is listed under NTC (from cultivated sources) (Table 17.1).

Table 17.1 (continued)

Sl. no	Species	State & threat status	Habit	Global IUCN red list status and NTC status
39	*Jurinea dolomiaea* Boiss	HP-EN, J&K-EN, UTK-EN	H	
40	*Jurinea macrocephala* DC	SKM-VU	H	
41	*Litsea glutinosa* (Lour.) C.B.Rob	ANP-CR, CHT-VU, HP-VU, J&K-EN, MP-VU, ODI-VU, UTK-NT	T	LC
42	*Merremia turpethum* (L.) Shah & Bhat	KRT-VU, KRL-EN	C	
43	*Mesua ferrea* L	ODI-EN, WB-EN	T	
44	*Mucuna pruriens* (L.) DC	CHT-NT, MP-NT, WB-EN	C	
45	*Nardostachys jatamansi* (D. Don) DC	ARP-EN, HP-EN, SKM-EN (2003), VU (2014), UTK-CR	H	CR
46	*Nilgirianthus ciliatus* (Nees) Bremek	KRL-EN, KRT-EN, TN-EN	S	VU
47	*Operculina turpethum* (L.) Silva Manso	CHT-NT, KRT-VU, KRL-EN, MP-NT, MAH-EN, ODI-VU, TN-NT	C	

(continued)

17.5 Medicinal Plants, the Cultivation of Which is Supported by Government Subsidies

Cultivation of some medicinal plants are being encouraged from the Government side. National Ayush Mission (NAM) under the scheme of National Medicinal Plants Board (NMPB), supports medicinal plant cultivators involved in cultivation of NAM

Table 17.1 (continued)

Sl. no	Species	State & threat status	Habit	Global IUCN red list status and NTC status
48	*Oroxylum indicum* (L.) Kurz	ANP-VU, ARP-VU, ASM-VU, CHT-VU, KRT-VU, KRL-EN, MP-VU, MAH-EN, MEG-VU, ODI-EN, RAJ-EN, SKM-VU (2003), NT (2014)	T	
49	*Phyllanthus emblica* L	CHT-VU, MP-VU	T	LC; Fruits as NTC from mixed source
50	*Picrorhiza kurrooa* Royle	ARP-EN, HP-EN, J&K-EN, SKM-VU, UTK-CR, WB-CR	H	
51	*Piper longum* L	CHT-VU, KRL-NT, ODI-EN, TN-EN	H	Cultivated fruit, root and stem as NTC
52	*Piper peepuloides* Roxb	ARP-VU	C	
53	*Plectranthus barbatus* Andrews	ANP-EN	H	Cultivated leaf and root as NTC
54	*Pluchea lanceolata* (DC.) C.B. Clarke	MP-NT	S	
55	*Plumbago zeylanica* L	CHT-VU, MP-VU, RAJ-VU	S	

(continued)

prioritized plants with subsidies ranging from 30 to 75%. Table 17.2 shows the said list where their NTC and threat status are screened to identify medicinal plants which are coming under the purview of ABS (non-NTC) and which are either cultivated or have potential of cultivation.

Table 17.1 (continued)

Sl. no	Species	State & threat status	Habit	Global IUCN red list status and NTC status
56	*Pseudarthria viscida* (L.) Wight & Arn	KRT-VU, KRL-VU, TN-NT	H	
57	*Pterocarpus marsupium* Roxb	CHT-VU, MP-VU, MAH-VU, ODI-EN, RAJ-CR, WB-EN	T	NT
58	*Pterocarpus santalinus* L.f	ANP-CR, TN-CR	T	EN
59	*Rauvolfia serpentina* (L.) Benth. ex Kurz	ANP-CR, ARP-CR, ASM-VU, CHT-CR, HP-CR, J&K-VU, KRT-EN, KRL-EN, MAH-CR, MEG-VU, ODI-EN, TN-EN, UTK-VU, WB-EN	H	
60	*Rheum australe* D. Don (=*Rheum emodii* Wall. ex Meisn.)	HP-EN, J&K-EN, UTK-EN	H	
61	*Rheum moorcroftianum* Royle	HP-EN, J&K-VU, UTK-NT	H	
62	*Rheum webbianum* Royle	HP-VU, J&K-VU, UTK-VU	H	

(continued)

The non-NTC medicinal plants in the category of 75% subsidy support by NMPB, include *Aconitum ferox*, *A. balfourii* (=*A. lethale*), *A. heterophyllum*, *Aquilaria agallocha*, *Berberis aristata* (=*B. umbellata*), *Commiphora wightii*, *Nardostachys jatamansi* (=*N. grandiflora*), *Oroxylum indicum*, *Picrorhiza kurroa*, *Podophyllum hexandrum* (=*Sinopodophyllum hexandrum*), *Pterocarpus santalinus*, *Santalum album*, *Saussurea costus* (=*Saussurea lappa*) and *Swertia chirayita*. Some above medicinal plants are listed in the negative list of exports issued by DGFT (Directorate General of Foreign Trade) list, appendix II of CITES (Convention on International Trade

Table 17.1 (continued)

Sl. no	Species	State & threat status	Habit	Global IUCN red list status and NTC status
63	*Rhododendron anthopogon* D. Don	ARP-VU, HP-VU, J&K-VU, SKM-EN (2003), VU (2014), UTK-NT	S	
64	*Rubia cordifolia* L	ANP-VU, CHT-VU, MP-VU, MAH-VU	H	
65	*Santalum album* L	ANP-EN, KRT-VU, KRL-EN, MAH-EN, TN-EN	T	VU
66	*Saraca asoca* (Roxb.) Willd	ANP-EN, KRT-EN, MAH-EN, ODI-CR	T	VU
67	*Saussurea costus* (Falc.) Lipsch	J&K-CR	H	CR
68	*Schrebera swietenioides* Roxb	KRT-VU, ODI-VU, RAJ-VU	T	
69	*Shorea robusta* Gaertn	ANP-NT	T	LC
70	*Smilax glabra* Roxb	ARP-CR, ASM-CR, MEG-CR	C	
71	*Firmiana simplex* (L.) W. Wight (=*Sterculia urens* Roxb.)	ANP-VU, CHT-VU, MP-VU, RAJ-EN	T	LC
72	*Stereospermum chelonoides* (L.f.) DC	MP-NT	T	

(continued)

in Endangered Species of Wild Fauna and Flora). Some are designated as threatened plants by Jammu & Kashmir, Himachal Pradesh, Uttarakhand, West Bengal, while *Saussurea costus* (=*Saussurea lappa*) is also enlisted under Indian Wildlife (Protection) Act, 1972; and *Pterocarpus santalinus* is enlisted as Endangered and *Commiphora wightii*, *Nardostachys jatamansi* and *Saussurea costus* are enlisted as Critically Endangered by IUCN with decreasing trend of global population.

The non-NTC medicinal plants in the category of 50% subsidy support by NMPB, include *Pterocarpus marsupium*, *Rauvolfia serpentina*, *Aegle marmelos*,

Table 17.1 (continued)

Sl. no	Species	State & threat status	Habit	Global IUCN red list status and NTC status
73	*Stereospermum tetragonum* DC. (=*Stereospermum colais* (Buch.-Ham. ex Dillwyn) Mabb.)	ODI-EN, RAJ-EN, WB-VU	T	
74	*Strychnos nux-vomica* L	CHT-VU	T	
75	*Strychnos potatorum* L.f	ODI-VU	T	
76	*Swertia chirayita* (Roxb.) Buch.-Ham. ex C.B. Clarke	ARP-VU, HP-CR, J&K-CR, MEG-VU, SKM-VU (2003), EN (2014), UTK-EN, WB-CR	H	
77	*Symplocos racemosa* Roxb	KRT-VU, MAH-VU, ODI-CR, TN-NT, UTK-VU	T	
78	*Taxus wallichiana* Zucc	ARP-EN, HP-EN, J&K-EN, MEG-CR, SKM-EN, UTK-EN, WB-CR	T	EN
79	*Terminalia arjuna* (Roxb. ex DC.) Wight & Arn	CHT-NT, KRT-NT, MP-NT, MAH-NT, RAJ-VU	T	
80	*Terminalia chebula* Retz	CHT-VU, MP-VU	T	LC
81	*Tinospora sinensis* (Lour.) Merr	KRT-VU, KRL-NT, MAH-NT	C	

(continued)

Albizia lebbeck, *Alstonia scholaris*, *Crateva nurvala*, *Desmodium gangeticum*, *Gmelina arborea*, *Hippophae rhamnoides* (=*Elaeagnus rhamnoides*), *Mesua ferrea*, *Plumbago zeylanica*, *Pueraria tuberosa*, *Premna integrifolia* (=*P. serratifolia*), *Rheum emodii*, *Saraca asoca*, *Smilax china*, *Stereospermum suaveolens* (=*S. chelonoides*), *Tecomella undulata*, *Uraria picta*, *Valeriana wallichii* (=*V. jatamansi*)

Table 17.1 (continued)

Sl. no	Species	State & threat status	Habit	Global IUCN red list status and NTC status
82	*Trichosanthes cucumerina* L	ANP-NT	C	Cultivated Leaf, Stem, Seed, Fruit and Tuber as NTC
83	*Valeriana hardwickii* Wall	ARP-VU, MEG-VU, SKM-VU	H	
84	*Valeriana jatamansi* Jones	ARP-VU, HP-VU, J&K-VU, MEG-VU, SKM-VU, UTK-VU	H	
85	*Vateria indica* L	KRT-VU, KRL-VU, TN-VU	T	VU

and *Zanthoxylum alatum* (=*Z. armatum*). Out of those, *Pterocarpus marsupium* and *Rauvolfia serpentina* are listed in the negative list of exports issued by DGFT list and appendix II of CITES.

The list of non-NTC medicinal plants where 30% subsidy support is provided by NMPB for cultivation are *Abrus precatorius*, *Alpinia calcarata*, *Andrographis paniculata*, *Asparagus racemosus*, *Bergenia ciliata*, *Celastrus paniculatus*, *Cinnamomum camphora*, *Clerodendrum phlomidis*, *Clitoria ternatea*, *Convolvulus microphyllus*, *Cryptolepis buchananii* (=*C. dubia*), *Dioscorea bulbifera*, *Embelia ribes*, *Gymnema sylvestre*, *Hemidesmus indicus*, *Holarrhena antidysenterica*, (=*Holarrhena pubescens*), *Ipomoea turpethum* (=*Operculina turpethum*), *Mucuna prurita* (=*M. pruriens*), *Cullen corylifolium* (=*Psoralea corylifolia*), *Rubia cordifolia*, *Solanum anguivi*, *Terminalia arjuna*, *T. bellirica* and *T. chebula*.

Besides, there are several medicinal plants which are non-NTC and in market demand thus having high ABS potential. Bio resources from *Acacia indica*, *Azadirachta indica* (Seed, bark), *Baliospermum montanum*, *Berberis asiatica*, *Clerodendrum indicum*, *Datura metel*, *Fumaria parviflora*, *Glycosmis pentaphylla*, *Hygrophila auriculata* (=*Asteracantha longifolia*), *Leonotis nepetifolia*, *Nymphaea stellata*, *Phyllanthus fraternus*, *Phyllanthus niruri*, *Prunus cerasoides*, *Salmalia malabarica*, *Shorea robusta* (Seed), *Solanum indicum*, *Symplocos racemosa*, *Zizyphus mauritiana* etc. are used by various companies for their medicinal products (Table 17.2).

Further there are unique cases like that of the use of *Saussurea costus* (=*Saussurea lappa*) needs to be addressed. The plant is listed under DGFT, Schedule VI of Indian Wildlife (Protection) Act, 1972 and notified as threatened in Jammu &

Table 17.2 Medicinal plants attracting ABS provisions of BD Act, 2002 (not enlisted under NTC), their DGFT, CITES and IUCN enlistment status and various slabs of subsidies provided by National Medicinal Plants Board, Government of India for their cultivation (NMPB 2021a, b, c, d)

Sl. no	Scientific name	Regional/national/global threat status and DGFT, CITES enlistment	Subsidy provided by NMPB (%)
1	*Aconitum ferox* Wall. ex Ser	DGFT, Threatened in West Bengal	75
2	*Aconitum balfourii* Stapf (=*Aconitum lethale* Griff.)	DGFT, notified threatened in Uttarakhand, threatened in West Bengal	75
3	*Aconitum heterophyllum* Wall. ex Royle	DGFT, notified threatened in Himachal Pradesh, Jammu & Kashmir, Uttarakhand, IUCN (EN)	75
4	*Aquilaria agallocha* Roxb	CITES Appendix II	75
5	*Berberis aristata* Sims (=*Berberis umbellata* Wall. ex G. Don)	IUCN (LC)	75
6	*Commiphora wightii* (Arn.) Bhandari	IUCN (CR)	75
7	*Nardostachys jatamansi* (D. Don) DC. (=*Nardostachys grandiflora* DC.)	Notified threatened in Himachal Pradesh, Uttarakhand, IUCN (CR)	75
8	*Oroxylum indicum* (L.) Kurz	NA	75
9	*Picrorhiza kurroa* Royle ex Benth	DGFT, notified threatened in Jammu & Kashmir	75
10	*Podophyllum hexandrum* Royle (=*Sinopodophyllum hexandrum* (Royle) T.S. Ying)	DGFT, notified threatened in Jammu & Kashmir, threatened West Bengal	75
11	*Pterocarpus santalinus* L.f	DGFT, CITES Appendix II, IUCN (EN)	75
12	*Santalum album* L	IUCN (VU)	75
13	*Saussurea costus* (Falc.) Lipsch. (=*Saussurea lappa* (Decne.) Sch. Bip.)	DGFT, notified threatened in Jammu & Kashmir, Schedule VI of Indian Wildlife (Protection) Act, 1972, IUCN (CR)	75
14	*Swertia chirayita* Buch.-Ham. ex Wall	DGFT, Threatened in West Bengal	75
15	*Pterocarpus marsupium* Roxb	DGFT, CITES Appendix II, IUCN (NT)	50
16	*Rauvolfia serpentina* (L.) Benth. ex Kurz	DGFT, CITES Appendix II	50
17	*Aegle marmelos* (L.) Corrêa	IUCN (NT)	50
18	*Albizia lebbeck* (L.) Benth	IUCN (LC)	50

(continued)

Table 17.2 (continued)

Sl. no	Scientific name	Regional/national/global threat status and DGFT, CITES enlistment	Subsidy provided by NMPB (%)
19	*Alstonia scholaris* (L.) R. Br	IUCN (LC)	50
20	*Crateva nurvala* Buch.-Ham	–	50
21	*Desmodium gangeticum* (L.) DC	–	50
22	*Gmelina arborea* Roxb	IUCN (LC)	50
23	*Hippophae rhamnoides* L. (=*Elaeagnus rhamnoides* (L.) A. Nelson)	IUCN(LC)	50
24	*Mesua ferrea* L	–	50
25	*Plumbago zeylanica* L	–	50
26	*Pueraria tuberosa* (Willd.) DC	–	50
27	*Premna integrifolia* L. (=*Premna serratifolia* L.)	IUCN (LC)	50
28	*Rheum emodii* Wall.ex Meisn.	–	50
29	*Saraca asoca* (Roxb.) Willd	IUCN (VU)	50
30	*Smilax china* L	–	50
31	*Stereospermum suaveolens* (Roxb.) DC. (=*Stereospermum chelonoides* (L.f.) DC.)	–	50
32	*Tecomella undulata* (Sm.) Seem	–	50
33	*Uraria picta* (Jacq.) Desv	IUCN (LC)	50
34	*Valeriana wallichii* DC. (=*Valeriana jatamansi* Jones)	–	50
35	*Zanthoxylum alatum* Roxb. (=*Zanthoxylum armatum* DC.)	IUCN (LC)	50
36	*Abrus precatorius* L	–	30
37	*Andrographis paniculata* (Burm.f.) Nees	–	30
38	*Asparagus racemosus* Willd	–	30
39	*Bergenia ciliata* (Haw.) Sternb	IUCN (LC)	30
40	*Celastrus paniculatus* Willd	–	30
41	*Cinnamomum camphora* (L.) J. Presl	–	30
42	*Clerodendrum phlomidis* L.f	IUCN (LC)	30
43	*Clitoria ternatea* L. (Blue & White varieties)	–	30
44	*Convolvulus microphyllus* Sieber ex Spreng	–	30

(continued)

Table 17.2 (continued)

Sl. no	Scientific name	Regional/national/global threat status and DGFT, CITES enlistment	Subsidy provided by NMPB (%)
45	*Cryptolepis buchananii* Roem. & Schult. (=*Cryptolepis dubia* (Burm.f.) M.R. Almeida)	–	30
46	*Dioscorea bulbifera* L	–	30
47	*Embelia ribes* Burm.f	–	30
48	*Gymnema sylvestre* (Retz.) R.Br. ex Sm	–	30
49	*Hemidesmus indicus* (L.) R. Br. ex Schult	–	30
50	*Holarrhena antidysenterica* (Roth) Wall. ex A.DC. (=*Holarrhena pubescens* Wall. ex G. Don)	IUCN (LC)	30
51	*Ipomoea turpethum* (L.) R. Br. (=*Operculina turpethum* (L.) Silva Manso)	–	30
52	*Mucuna prurita* (L.) Hook. (=*Mucuna pruriens* (L.) DC.)	–	30
53	*Psoralea corylifolia* L. (=*Cullen corylifolium* (L.) Medik.)	IUCN (LC)	30
54	*Rubia cordifolia* L	–	30
55	*Solanum anguivi* Lam	IUCN (LC)	30
56	*Terminalia arjuna* (Roxb. ex DC.) Wight & Arn	–	30
57	*Terminalia bellirica* (Gaertn.) Roxb	–	30
58	*Terminalia chebula* Retz	IUCN (LC)	30

Kashmir but is among the bio resources (Table 17.1) whose trade volume is more than 100 MT/yr (dry weight). Rule 14 of BD Rules, 2004 provides for restriction on activities related to access for any endemic and rare species (NBA 2021d), while the Ayurvedic Pharmacopoeia of India (API) mentions the use of the root in cough, bronchitis, dyspnoea; erysipelas and gout (Khare, 2007). As because this medicinal bio resource is listed in API, Drug Licencing authorities allow its use by various drug manufacturing entities. Therefore, there is a need for national streamlined consensus on access and use of such notified threatened species, and if access is by such national consensus permissible, then there is need for national enlistment of their cultivators so that their use may be promoted through such enlisted cultivated sources.

17.6 Streamlining ABS in Terms of Medicinal Plants

The ABS guideline 2014 highlights many monetary and non-monetary ABS sharing outlines. However, commercial users of medicinal plants are mostly known to opt for two main prevalent benefit sharing modes which include (i) based upon gross ex-factory sale, (ii) based upon purchase price, though bioresource auction-based models have also been adopted. Although every commercial user during their visit, conversation, and communication with the SBB are made aware about the ABS and the regulatory monetary and non-monetary provisions under ABS guidelines 2014, ex-factory sale price-based. Turnover based ABS amount calculation is opted by default by most of the commercial users of medicinal plant. This mode of ABS calculation is easy for both the parties as ABS guidelines 2014 has well defined criteria for this mode of ABS calculation, besides balance sheets and Profit & Loss account offer readily derivable sale figures.

Calculation of ABS based upon purchase price is another option but ABS guidelines doesn't offer proper contextual outline for this mode of ABS calculation. Following this mode, the commercial users have to segregate medicinal plants species wise and compile the purchase volume corresponding to the financial year, which is more impossible than hectic for handling by a chartered accountant. Sweat aside, this method offers maximum monetary savings to the commercial users, as purchase price of species from the Normally Traded Commodities (NTCs) could be deducted from the final purchase price thus minimizing final ABS amount.

The auction of high value medicinal bioresources such as Red sanders (*Pterocarpus santalinus*) and the caterpillar fungus (*Ophiocordyceps sinensis*) are also being carried out. Applicants pertaining to Section 3(2) intending to access Red sanders are to apply to NBA with requisite fees and successful bidder are required to pay 5.0% of the auctioned amount (excluding taxes and other Government dues) to the National Biodiversity Fund, only after which they can lift the auctioned Red sanders. Alpine-Subalpine areas of Eastern Himalayan state of Sikkim are abode of caterpillar fungus (*Ophiocordyceps sinensis*) and the harvest is auctioned by Territorial Circle (Headquarter), Forest, Environment and Wildlife Management Department (FEWMD), Sikkim at offset price of 5 Lakh/Kg. The bidders are required to register with requisite fee and successful ones are to pay (i) Rs. 10/gram as royalty to FEWMD, and (ii) 3% of the total auctioned value to the Sikkim SBB as ABS. Such auctioning offers higher dividend to the stakeholders and as a result play a key role in better sustenance of the bioresource, while the successful bidders have access to transit challans for easy transport of bioresource to their destinations (FEMWD 2016).

Even though with wide spread awareness on ABS, it is unfortunate that most commercial entities have not come forward to be a part of ABS mechanism; major reason being misperception of ABS as additional monetary burden, while Section 21 (2) of BD Act and the ABS guidelines has clear outlines for non-monetary provisions also. Further, the provisions of Section 7 of the BD Act, in some cases is not attracted by bio-resource users and there is also a misconception that all medicinal plants come

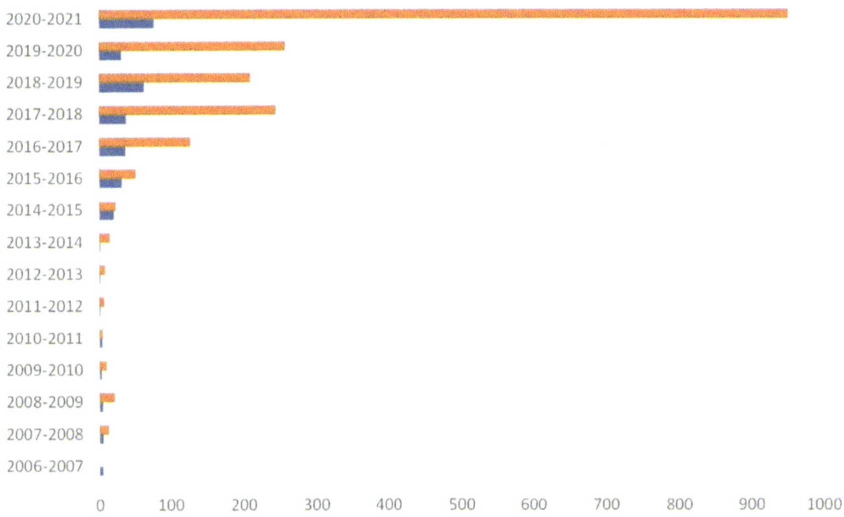

Fig. 17.5 Financial year wise details of Form I and Form III approvals by NBA. Orange coloured bars = Form III: approval for obtaining IPR; Blue bars = Form I: Access to Bio-resources for Research/Commercial Purpose

under the ABS provisions, though there are exemption on 108 species of medicinal plants and 174 biological resources derived from it, which are ascribed under NTCs as notified by Govt. of India vide S.O. 1352(E)., dated 07.04.2016, as per the provisions of Section 40 of the BD Act (Fig. 17.5).

The number of BD Act compliant commercial entities pertaining to Section 3(2) who have signed e.g. ABS with NBA (2006–2007 to 2020–2021) is only 314 which suggests low participation of the industrial fraternity to the ABS provisions of the BD Act. The unresponsiveness of many commercial users of bioresources (including medicinal plants; pertaining to Section 7) to the notices of SBBs/BMCs have led to a number of ABS related court cases (NBA 2021a). On the positive note, the number of Form I applications received by NBA (2021b) and subsequent approvals granted are increasing annually ($R^2 = 0.438$, $P = 0.005$), especially from the period of 2014–2015 (Fig. 17.5).

Majority of medicinal plant markets are driven by middle men and traders. Middle men further may have multiple layers of transit reaching upto the collector. However, traders are overall the key players in channelization of medicinal plant bioresources to the commercial entities. The Form I and Form A prescribed by BD Act has specific context for the source origin of the bioresources, but most of the time the commercial entities mention name of the trader as source, may be because their understanding of the traders as the point source or inability (secrecy) of the traders to reveal the point sources to the commercial entities.

SBBs have been encouraging commercial users to procure medicinal plants directly from BMCs, and the major issues plaguing such endeavour are reported to be (i) transportation of raw herbs from BMC area to the commercial users, (ii) the volume of raw materials required by the commercial users may not be possible for a BMC to cater, and (iii) BMCs generally being non-commercial entity (no trade license, no GST registration), the commercial users of the bioresources would prefer supplies with proper GST bill which would minimize their tax payable on output, while being able to claim input tax credit.

17.7 Bioprospecting

The diversified medicinal plant wealth in Indian land mass, both actual and potential has immense scope in bio prospecting especially in pharmaceutical and nutraceutical industries. Intellectual Property Right based on bio resources, including medicinal plants are routed through NBA and subsequently concerned SBBs and BMCs for scrutiny of disclosure of origin, Traditional Knowledge involving biological material and the approval/opposition/revocation of the patent for lack of compliance including ABS provisions (UEBT 2019). During 2006–2007 to 2020–21 period, NBA has received total of 1935 Form III applications (Fig. 17.1) and subsequent approvals granted for IPR are also increasing annually ($R^2 = 0.466$, $P = 0.005$) (NBA 2021b), especially from the period of 2014–2015 (Fig. 17.5).

The potential of Indian medicinal plants may also be highlighted from the fact that 4000 patent references in the US Patent and Trademark Office alone were related to seven medicinal plants of Indian origin. Further, 50% of the drugs recorded in the British Pharmacopoeia have their reported origins in medicinal plants from the Western Himalayas, a region so rich in genetic resources of medicinal plants which supports bio-resources for 80% of *Ayurvedic*, 46% of *Unani* and 33% of allopathic medicinal formulations (Bavikatte and Tvedt 2015).

Another aspect of use of medicinal plants is that it can be directly used as resource or its derivatives may act as base for development of synthetic molecules or analogues, and it is necessary to adequately address the issue as to applicability of ABS to the synthetic molecules or analogues (Suneetha et al. 2009).

India being an early entrant to countries developing bio-prospecting regulations through robust legislation like BD Act, much is still to be achieved in terms of managing the 'unregulated' collection, sourcing, trading of medicinal plants (Bavikatte and Tvedt 2015). Further, the reliability of business upon common trade names may not always be authentic viz. **Kapoor Kachari/Shati** may mean *Hedychium spicatum* as well as *Kaempferia galanga*, name **Chirata/Bhuunimba** is used in common practice for *Swertia chirayita* and *Andrographis paniculata* (Kalmeg), the latter being more common adulterant of the former; both *Pluchea lanceolata* (Compositae) and *Alpinia galanga* (Zingiberaceae) are considered equivalent to **Rasna** (Khare 2007).

17.8 Medicinal Plants-Based ABS in India: Notable Success Stories

The classic example of medicinal plant-based ABS mechanism may be referred to the development of a poly-herb anti-fatigue drug, Jeevani. During botanical exploration guided by Kani tribes, the scientists of Tropical Botanic Garden and Research Institute (TBGRI) noticed the use of leaves of *Trichopus zeylanicus* ssp. *travencorius* (Arogyapacha) for use against fatigue. Arogyapacha, which the Kani regard as sacred, upon request the knowledge was transferred to the TBGRI scientists on goodwill and an oral commitment to share any returns accrued from commercial use of the plant. On a good note, this knowledge transfer led to the development of formulation of anti-fatigue drug Jeevani, whose 50% of the receipts of sale goes to the Trust Fund in the name of Kani community (Pushpangadan et al. 1998; Suneetha et al. 2009).

Another example relates to how a domestic company can involve local communities in the development of sustainable products, with an emphasis on the sustainable use of genetic resources and equity in transactions. Gram Mooligai Co. Ltd. (GMCL), a Madurai-originated NGO manages Self-Help Groups of women who are actively involved in the collection, cultivation and processing of medicinal plants for herbal medicines. GMCL is active primarily in the drought-prone areas of Tamil Nadu state in the south and four central Indian states (Madhya Pradesh, Maharashtra, Chhattisgarh & Odisha). It promotes medicinal plants and agriculture based sustainable livelihoods for poverty alleviation in the drylands and forest areas, for the women, the landless and the small farmers, including the backward/scheduled castes/tribes. In the working model, GMCL procures plants and plant products directly from these Self-help Groups, at remunerative rates ensuring sustainable harvesting practices among the communities, and then the company sells the herbs and shares 70% of the returns with the communities. In addition, GMCL is also involved in the production of simple medicinal formulations based on Traditional Knowledge. GMCL has shared ABS with Madhya Pradesh's State Biodiversity Board and Government of India, Ministry of Environment, Forest and Climate Change in association with UNDP India has conferred GMCL with Biodiversity Award 2016 in the category of Successful Mechanisms/Models for Access and Benefit Sharing (Torri 2009; MoEF&CC 2018; NLSABS 2021).

17.9 Way Ahead

Much before the ABS guidelines 2014 came into force, Joint Forest Management Committees (JFMCs) were provided with monetary support for their sustainable collection of forest derived bioresources such as non-timber forest produces (NTFPs), by Forest Development Corporations. State Medicinal Plant Boards (SMPB) have been facilitating cluster farming-based cultivation of medicinal plants and helping in channelizing the harvest directly to commercial entities, especially manufacturers, while the Horticulture Department has various schemes for cultivation of medicinal plants. Comprehensive Area Development Corporation (CADC) of the state (under Panchayet and Rural Development Department) also has ample scope in cultivation of medicinal plants in convergence with various schemes involving the self-help groups, having proper market linkage towards livelihood promotion, conservation of medicinal plant resources and promoting ABS potentiality.

The Medicinal Plant Conservation Areas (MPCAs), the areas of special interest outside the Protected Area Network which are under the jurisdiction of the Forest Department are pristine landscapes for medicinal plants specific for the concerned topography, climate and soil conditions. As per the Biodiversity Heritage Site (BHS) guidelines issued by NBA, the MPCAs are ideal candidates for declaration as BHS by the concerned state government. The genetic resources of medicinal plants of MPCA may be tapped sustainably through JFMC-BMC coalition for in situ conservation, ex situ cultivation, local livelihood promotion and potentiating ABS through establishing linkage of cultivator-commercial user channel.

On the regulatory framework, commercial entities are required to apply to State Biodiversity Board prior to access of biological resources and commercial utilization, but there are other key players in this administrative setup, including (i) SMPB and Drug Control offices have direct regulatory control over concerned commercial entities in terms of regulating Drug Licences, (ii) State Pollution Control Board have direct control over issuance of Consent to Establish and Consent to Operate, while (iii) local bodies, in most cases whose executives are also the members of the concerned local body's BMC, are the issuing authorities of trade licences to the commercial users of bio-resources. Though these administrative setups have their own mandates to fulfil, if they could come together to support the State Biodiversity Board for the cause of facilitating ABS then the rate of compliance of BD Act by commercial users of bioresources would rise drastically.

As the Section 7 implementation and the onus of consideration of NTCs are at the helm of Biodiversity Boards/Councils of State and UTs, any concerned expert member scrutinizing ABS application would rather consider bio-resources enlisted under NTCs not to be part of BD Act, 2002, than anticipate their use as a resource in a process or for the development of a product, when in such case the item will not be '*exempted as a normally traded commodity*' and shall be treated as a biological resource, the access and obtainment of which is regulated by the ABS provisions of BD Act (NBA 2021c). Straightforward guidelines from NBA on NTCs with case scenarios would help all the stakeholders, especially regulators and applicants in reorienting their stance in ABS matters.

Concerted effort in this regard is necessary for proper sensitization of Cultivators, Collectors/JFMCs, BMCs, SBBs, SMPBs, Drug Controll, Forest Department, Horticulture Department, Panchayet and Rural Development Department, Commercial users (Traders and manufacturers) for involvement in ABS process, so that a win–win situation may be created for both the bioresource-benefit claimants and the industry.

Acknowledgements Authors are indebted to West Bengal Biodiversity Board for providing necessary facilities to carry out the present work.

Conflict of Interest The views and opinions expressed in this article are those of the author in his personal capacity. None of the observations, opinions, or conclusions expressed herein are to be attributed to or construed as the views or positions of any organization, association, or company. No conflict of interest.

References

Baas JH (1889) Outlines of the history of medicine and medical profession. Translated, revised and enlarged in conjunction with author by Handerson, M.A. J.H. Vail & Co., New York

Babar S, Amarnath G, Reddy CS, Jentsch A, Sudhakar S (2012) Species distribution models: ecological explanation and prediction of an endemic and endangered plant species (Pterocarpus santalinus L.f.). Curr Sci 102(8):1157–1165

Bapu TD, Nimasow G (2021) An assessment of the population status of the threatened medicinal plant Illicium griffithii Hook.f. & Thomson in West Kameng District of Arunachal Pradesh, India 13(1):17504–17512. https://doi.org/10.11609/jott.6082.13.1.17504-17512

Bavikatte KS, Tvedt MW (2015) Beyond the thumb rule approach: regulatory innovations for bioprospecting in India. Law Environ Dev J 11(1):1–20. http://www.lead-journal.org/15001.pdf. Accessed 12 May 2021

Chandra N, Singh G, Lingwal S, Bisht M, Tiwari LM (2021) Population assessment and habitat distribution modelling of the threatened medicinal plant Picrorhiza kurroa Royle ex Benth. in the Kumaun Himalaya, India. J Threatened Taxa 13(7):18868–18877. https://doi.org/10.11609/jott.5603.13.7.18868-18877

Dhyani A, Kadaverugu R, Nautiyal BP, Nautiyal MC (2021) Predicting the potential distribution of a critically endangered medicinal plant *Lilium polyphyllum* in Indian Western Himalayan Region. Reg Environ Change 21:30. https://doi.org/10.1007/s10113-021-01763-5

FEMWD (2016) Public Auction Notice, Forest, Environment and Wildlife Management Department, Sikkim. http://sbbsikkim.nic.in/pdf/news/auction.pdf. Accessed 28 May 2021

Gowthami R, Sharma N, Pandey R, Agarwal A (2021) Status and consolidated list of threatened medicinal plants of India. Genet Resour Crop Evol 68:2235–2263. https://doi.org/10.1007/s10722-021-01199-0

IMPD (2021a) Indian Medicinal Plants Database. Developed by NMPB and FRLHT. http://www.medicinalplants.in/statewiselist/karnataka. Accessed 28 July 2021

IMPD (2021b) Indian Medicinal Plants Database. Developed by NMPB and FRLHT. http://www.medicinalplants.in/statewiselist/kerala. Accessed 28 July 2021

IMPD (2021c) Indian Medicinal Plants Database. Developed by NMPB and FRLHT. http://www.medicinalplants.in/statewiselist/tamilnadu. Accessed 28 July 2021

IMPD (2021d) Indian Medicinal Plants Database. Developed by NMPB and FRLHT. http://www.medicinalplants.in/statewiselist/andhrapradesh. Accessed 28 July 2021

IMPD (2021e) Indian Medicinal Plants Database. Developed by NMPB and FRLHT. http://www.medicinalplants.in/statewiselist/sikkim. Accessed 28 July 2021

IMPD (2021f) Indian Medicinal Plants Database. Developed by NMPB and FRLHT. http://www.medicinalplants.in/statewiselist/rajasthan. Accessed 28 July 2021

IMPD (2021g) Indian Medicinal Plants Database. Developed by NMPB and FRLHT. http://www.medicinalplants.in/statewiselist/westbengal. Accessed 28 July 2021g

Kala CP, Dhyani PP, Sajwan BS (2006) Developing the medicinal plants sector in northern India: challenges and opportunities. J Ethnobiol Ethnomed 2:32. https://doi.org/10.1186/1746-4269-2-32

Khare CP (2007) Indian medicinal plants, an illustrated dictionary. Springer, Berlin/Heidelberg

Kumar D, Rawat S, Joshi R (2021) Predicting the current and future suitable habitat distribution of the medicinal tree *Oroxylum indicum* (L.) Kurz in India. J Appl Res Med Aromatic Plants 23:100309. https://doi.org/10.1016/j.jarmap.2021.100309

Lepcha DL, Pradhan A, Chhetri DJ (2019) Population assessment and species distribution modeling of *Paris polyphylla* in Sikkim Himalaya, India. Biodiversitas 20:1299–1305. https://doi.org/10.13057/biodiv/d200508

Mipun P, Adhikari D, Bora A, Bhat NA, Kumar Y (2019) Species distribution modelling of *Brucea mollis* wall. ex kurz in Northeast India for its conservation. Plant Arch 19(2):3191–3196

MoEF&CC (2018) Implementation of Nagoya protocol on access and benefit sharing: India's Experience. Ministry of Environment, Forest & Climate Change, Government of India. http://nbaindia.org/uploaded/pdf/Implementation%20of%20Nagoya%20Protocol%20in%20India.pdf. Accessed 28 June 2021

NBA (2021a) Court–directives/orders. http://nbaindia.org/content/730/66/1/CourtOrders.html. Accessed 28 June 2021

NBA (2021b). http://nbaindia.org/content/683/61/1/approvals.html. Accessed 29 June 2021

NBA (2021c) Explanatory Notes on Certain Sections of the Biological Diversity Act, 2002. http://nbaindia.org/content/565/56/1/explanatorynote.html. Accessed 28 June 2021

NBA (2021d) Biological Diversity Rules 2004. http://nbaindia.org/content/17/20/1/rules.html. Accessed 29 June 2021

NBA (2013) Peoples Biodiversity Register, revised PBR guidelines 2013 (based on the guideline issued by National Biodiversity Authority in 2009)

Neuburger M (1925) History of medicine. Translated by Ernest Playfair. Oxford University Press, London

NLSABS (2021) https://nlsabs.com/?page_id=219. Accessed 28 June 2021

NMPB (2021a) https://www.nmpb.nic.in/content/medicinal-plants-fact-sheet. Accessed 28 June 2021

NMPB (2021b) https://www.nmpb.nic.in/sites/default/files/cultivation/Species_for_30_subsidy.pdf. Accessed 21 June 2021

NMPB (2021c) https://www.nmpb.nic.in/sites/default/files/cultivation/Species_for_50_subsidy.pdf. Accessed 21 June 2021

NMPB (2021d) https://www.nmpb.nic.in/sites/default/files/cultivation/Species_for_70_subsidy.pdf. Accessed 21 June 2021

Pradhan P (2015) Potential distribution of Monotropa uniflora L. as a surrogate for range of Monotropoideae (Ericaceae) in South Asia. Biodiversitas 16(2):109–115. https://doi.org/10.13057/biodiv/d160201

Pradhan P (2016) Strengthening MaxEnt modelling through screening of redundant explanatory bioclimatic variables with variance inflation factor analysis. Researcher 8(5):29–34. https://doi.org/10.7537/marsrsj080516.05

Pradhan P (2019) Testing equivalency of interpolation derived bioclimatic variables with actual precipitation: a step towards selecting more realistic explanatory variables for species distribution modelling. Res J Chem Environ 23:38–41

Pradhan P, Setyawan AD (2021) Filtering multi-collinear predictor variables from multi-resolution rasters of WorldClim 2.1 for Ecological Niche Modeling in Indonesian context. Asian J For 5(2):111–122. https://doi.org/10.13057/asianjfor/r050207

Pushpangadan P, Rajashekaran S, Ratheshkumar PK, Jawahar CR, Nair VV, Lakshmi N, Amma LS (1998) 'Arogyapacha' (Trichopus zeyalnicus Gaertn), The 'Ginseng' of Kani Tribes of Agasthyar Hills (Kerala) for Ever green Health and Vitality. Anc Sci Life 8(1):13–16

Rodgers WA, Panwar HS (1988) Planning a wildlife protected area network in India. 2 volumes. Project FO: IND/82/003. FAO, Dehradun, India, p 339

Sarkar MK (2012) Biodiversity governance for managing endemic and threatened medicinal plants in India—a geoinformatic approach, 2nd edn. National Biodiversity Authority, Chennai

Sarma B, Baruah PS, Tanti B (2018) Habitat distribution modeling for reintroduction and conservation of Aristolochia indica L.—a threatened medicinal plant in Assam, India. J Threatened Taxa 10(11):12531–12537. https://doi.org/10.11609/jott.3600.10.11.12531-12537

Shah R (2012) Medicinal plant conservation and ABS: Horizontal and vertical convergence. Uttarakhand Biodiversity Board, presented in XI Conference of parties, Convention on Biological Diversity, Hyderabad, India. https://www.cbd.int/abs/side-events/icnp2/undp-india-01-en.pdf. Accessed 12 May 2021

Sourabh P, Thakur J, Priti Sharma P, Uniyal PL, Pandey AK (2018) Habitat distribution modelling for reintroduction of endangered medicinal plants—Ephedra gerardiana, Lilium polyphyllum, Crepidium acuminatum, Pittosporum eriocarpum and Skimmia anquetilia in India. Int J Ecol Environ Sci 44(2):207–216

Suneetha MS, Pisupati B, Kumar S (2009) Framework for benefit sharing guidelines for India. Asian Biotechnol Dev Rev 11(2):55–88

Torri MC (2009) Beyond benefit-sharing agreements: bioprospecting for the poor? Int J Technol Manag Sustain Dev 8(2):103–127. https://doi.org/10.1386/ijtm.8.2.103/1

UEBT (2019) Union for Ethical BioTrade (UEBT): India ABS fact sheet. https://www.ethicalbiotrade.org/resource-pages/abs-in-india. Accessed 29 June 2021

Vavilov NI (2009) Origin and geography of cultivated plants. Translated by Doris Love. Cambridge University Press

Ved D, Saha D, Ravikumar K, Haridasan K (2015) Chlorophytum borivilianum. The IUCN Red List of Threatened Species 2015:e.T50126575A50131300. https://doi.org/10.2305/IUCN.UK.2015-2.RLTS.T50126575A50131300.en. Accessed 30 July 2021

Yang XQ, Kushwaha SPS, Saran S, Xu J, Roy PS (2013) Maxent modeling for predicting the potential distribution of medicinal plant, *Justicia adhatoda* L. Lesser Himal Foothills Ecol Eng 51:83–87. https://doi.org/10.1016/j.ecoleng.2012.12.004

Chapter 18
The Need for a Nagoya Protocol 'Plus'—Access and Benefit Sharing in the Context of Digital Sequence Information

Balakrishna Pisupati and Sachin Sathyarajan

Abstract The Nagoya Protocol on access to genetic resources and the fair and equitable sharing of benefits arising from their utilization (the Nagoya Protocol) reaffirms the sovereign rights of States over their natural resources and calls for fair and equitable sharing of benefits when such resources and associated traditional knowledge are used. It explicitly discourages anyone from using such resources and associated traditional knowledge without prior informed consent and mutually agreed terms for such use. With 133 countries acceding to the Protocol as of January 2022, the Protocol receives much prominence in realizing the objective of the Convention on Biological Diversity (CBD) and its Article 15. Several countries, both developing and developed, are currently working on national implementation of the Protocol by designing administrative, regulatory and legal regimes. Parallel to this, recent advances in science and technology, especially biotechnology and synthetic biology, have been evolving with new ways of using the genetic sequence information which, in a number of instances, is displacing the need to have physical access to any genetic resource. Developments in the use of digital sequence information (DSI) over the past decade make it possible to synthesize new organisms and products using DSI that are currently available publicly and are also held in private. This has now resulted in debates within the CBD and the Nagoya Protocol, as well as other fora such as the United Nations Food and Agricultural Organization (FAO) and the discussions under the United Nations Convention on Law of the Seas (UNCLOS) on issues of Areas beyond National Jurisdiction (ABNJ). The implementation as to whether the developments in use of DSI and synthetic biology would potentially make the Protocol redundant when countries will have limited control over access to genetic sequence information and their subsequent use, over-riding the need for prior informed consent and mutually agreed terms in having access to genetic resources. Added to this is the emerging dimension of the complex intellectual property regimes while dealing with

B. Pisupati—The views presented here are personal and are not those of UNEP.

B. Pisupati
Law Division, United Nations Environment Programme (UNEP), Nairobi, Kenya

S. Sathyarajan (✉)
TDU, Yelahanka, Bengaluru, India
e-mail: sachinsathyarajan@gmail.com

O. V. Oommen et al. (eds.), *Biodiversity Conservation Through Access and Benefit Sharing (ABS)*, https://doi.org/10.1007/978-3-031-16186-5_18

DSI and synthetic biology and the impacts of major global initiatives such as the Earth BioGenome Project and the Amazon Bank of Codes on the future of Nagoya Protocol and its implementation.

Keyword CBD · Nagoya protocol · DSI · FAO · UNCLOS · ABNJ · Biogenome

18.1 Introduction

The Convention on Biological Diversity (CBD) created a pivot shift in the understanding of genetic resources in international law—from the "common heritage of humanity" to national sovereignty over territorial resources. The potential economic benefits to the States were framed as an incentive for the efforts taken for conservation and sustainable use of genetic resources within their borders. To facilitate the third objective of the Convention, the "Nagoya Protocol on Access to Genetic Resources and the Fair and Equitable Sharing of Benefits Resulting from their Utilization" was adopted in 2010.

The Nagoya Protocol reaffirms the sovereign rights of States over their natural resources and calls for fair and equitable sharing of benefits when genetic resources (GR) and associated traditional knowledge (TK) is used. It requires identification of the legal provider of the genetic resources from whom users have to obtain prior informed consent and mutually agreed terms (PIC and MAT) for use of the genetic resources.

With 138 countries being part of the Protocol as of July 2022, the Protocol receives much prominence in realising the objective of the Convention on Biological Diversity (CBD) and its Article 15.[1] The Protocol requires effective implementation at the domestic level and the Member States are currently working on national implementation of the Protocol by designating national focal points, competent national authorities and other administrative regimes.

Parallel to the developments at the CBD, emerging technologies like new generation sequencing and synthetic biology applications have created possibilities for utilizing the "digital sequence information" (DSI). The developments in technology has led to "dematerialization", displacing the need to have physical access to biological resources for scientific applications. The growing databases of DSI over the past decade coupled with gene editing technologies like CRISPR have created opportunities for synthesizing new organisms and products.

Next generation sequencing (NGS) or deep sequencing describes a DNA sequencing technology which has led to proliferation of genetic information in databases.[2] There is an explosion of knowledge where we are creating digital databases that contains the collective biological intelligence of billions of years of

[1] https://treaties.un.org/pages/ViewDetails.aspx?src=IND&mtdsg_no=XXVII-8-b&chapter=27&clang=_en.

[2] Sam Behjati and Patrick S. Tarpey, "What is next generation sequencing?" available at https://www.ncbi.nlm.nih.gov/pmc/articles/PMC3841808/.

evolutionary history. NGS platforms produce massive amounts of sequencing data because millions of DNA fragments can be sequenced simultaneously. The cost of DNA synthesis is progressively decreasing and several large scale NGS projects are being a reality.

The ability to represent DNA in a digital manner and the ease of making genetic modifications with tools such as CRISPR/Cas9 has been pathbreaking. The opportunities of designing and editing biological models and components can lead to unprecedented developments in several critical sectors. It can thus create innovative solutions for some of the most pressing issues of today's times including-pharmaceuticals and healthcare, agribiotech and food security, biofuels and clean energy, and bioplastics and material sciences.

The scientific and technological developments have now resulted in debates within the CBD and the Nagoya Protocol, as well as other global fora such as the United Nations Food and Agricultural Organization (FAO) and the discussions under the United Nations Convention on Law of the Seas (UNCLOS) on issues of Areas Beyond National Jurisdiction (BBNJ).[3] The discussions at the CBD explore developments in the fields of DSI and synthetic biology and its impact overtheState control over access to genetic sequence information and their subsequent use.

The CBD Conference of Parties (COP 13) had extensive discussions on synthetic biology as a "new and emerging issue" and invited the Parties and the relevant organizations to apply the precautionary approach to the field of synthetic life.[4] "Digital sequence information" (DSI) is widely acknowledged as a placeholder term for which no consensus on a replacement or precise definition exists to date. The term "digital sequence information" (DSI) was introduced in decisions CBD XIII/16 and Nagoya Protocol NP-2/14.[5]

18.2 Major Challenges

Several countries have, so far not addressed DSI as a part of their legislative, administrative, or policy measures related to access and benefit sharing (ABS). At present there is still debate issues related to whether DSI falls outside the scope of the definition of "genetic resources" found in the CBD and the Nagoya Protocol. Many national ABS frameworks, except for few, have limited or no provisions to deal with ABS while using DSI. Lack of awareness on how to deal with DSI especially in the public domain, and the lack of technical capacities to track and trace innovation based on DSI seems to be major roadblocks.

The Ad Hoc Technical Expert Group (AHTEG) on synthetic biology and the CBD reports on technological developments on DSI on genetic resources have

[3] Thambisetty (2019).

[4] Decision X/13, para. 4 (2010).

[5] Convention on Biological Diversity (2018).

been discussing some of these issues.[6] The electronic databases containing DNA sequences have been developing since the 1980s. The genetic information available on large databases such as the International Nucleotide Sequence Database Collaboration (INSDC)[7] are widely accessed by scientists and researchers. Such information can be recreated and modified to create new technological applications or even new organisms.

There is no credible way to retrace the provenance of the genetic information that forms the basis of the discovery. The developments in synthetic biology and DSI, is coupled with ABS bypass tools that has disrupted the CBD Nagoya framework. Amidst the development of ABS bypass tools, there is a fear of a new wave of "digital piracy" in the Global South.

The scale and volume of DSI data is immense and are published in a multilateral manner. A major technical issue is that sequences are highly repetitive. The CBD Nagoya framework is a bilateral approach with a heavy reliance on national capacities to track and implement ABS. In an era of expanding bio-economy strategies in the EU, USA, Canada and Australia, the successful utilization of genetic resources can be vital for economic growth.[8]

Regeneron Pharma, a US based company developed a new drug for Ebola using the C15 strain of the Ebola virus that was sequenced and shared by the scientists from Guinea.[9] There was no ABS framework for accessing the DSI and Guinea, the provider country gained no benefit. The global inequity in accessing vaccines and therapeutics continue in the scenario of the COVID-19. However without the open and fast sharing of genomic information, development of medical countermeasures such as vaccines and therapeutics is impossible.

The territorialization and national sovereignty over genetic resources, advanced through the CBD Nagoya framework can be made obsolete through the wide availability of ABS bypass tools. The CBD-Nagoya framework was not originally designed to accommodate the "virtual" concerns that DSI creates. A majority of the DSI published online is free to access and download, with no country tag or PIC/MAT obligations. For instance, there is no dedicated PIC/MAT metadata field associated with the country of origin in the biggest online database—INSDC submission form.[10]

One of the direct consequence of the digitization, is that the effective benefit sharing arising out of DSI is far more difficult to ascertain compared to tangible genetic resources. These challenges increase over time as sequences pass through

[6] Report of the Ad Hoc Technical Expert Group on Synthetic Biology Montreal, Available at https://www.cbd.int/doc/c/aa10/9160/6c3fcedf265dbee686715016/synbio-ahteg-2017-01-03-en.pdf.

[7] International Nucleotide Sequence Database Collaboration (INSDC) is a long-standing foundational initiative that operates between DDBJ, EMBL-EBI and NCBI. INSDC covers the spectrum of data raw reads, through alignments and assemblies to functional annotation, enriched with contextual information relating to samples and experimental configurations. Available at https://www.insdc.org.

[8] Staffas et al. (2013).

[9] Bagley (2022).

[10] https://www.absbiotrade.info/fileadmin/Downloads/3.%20TOPICS/SPECIFIC%20ISSUES/DSI/National_Approaches_to_DSI/Brochure-ABS-National-Approaches-DSI-2022.pdf.

multiple improvisations and the direct link for benefit sharing erodes. The key question in such situations include whether the benefit sharing apply to an indefinite number of transactions or is there an expiration point for establishing the link.

Adding fuel to the fire, is the additional concerns with regards to the patenting regime. The digital libraries allow any entity to access genetic information for private use which can be commercialized as patented inventions. Most of synthetic biology applications are eligible for utility patent protection in many countries. With this current model, commercialization possibilities of biological inventions employing DSI may not benefit the country of origin.

18.3 "Nagoya plus" Measures

The countries that do not have a domestic framework for ABS will have to employ private contract law to ascertain the conditions of use of the DSI. There are instances of absence of specific ABS legislation governing the use and deposit of DSI. In other instances, such as the United States, the country of origin may be a non-party to the Convention. Benefit sharing obligations have to be negotiated privately in such scenarios. At present, most Parties to the CBD Nagoya Protocol do not have measures on DSI.

The next part of the paper discusses the examples of several national initiatives which are exploring "Nagoya plus" options to regulate DSI. The digital biopiracy concerns have led to some countries explicitly incorporating DSI-related language into their ABS legislative, administrative, and policy frameworks. Some countries are interpreting the language in their existing ABS measures to include DSI in conjunction with access to a genetic resource. Other countries are requiring PIC and MAT to access "DSI" independently of access to a physical genetic resource.

We can find some examples of these in the national implementation of ABS in countries such as Costa Rica and Namibia which are now enforcing access requirements for DSI in conjunction with the utilization of tangible genetic resources. However, they do not include any benefit sharing possibilities on top of the access requirement. The access permit is granted by a competent national authority, under the relevant ABS regulation.

Let's consider the instance of Costa Rica, where the national biodiversity law has designated "CONAGEBIO" as the competent national authority and it has granted serval ABS permits for basic research, bioprospecting and commercial use.[11] In the opinion of the Costa Rican Government, DSI forms part of "access to genetic resources.[12] Since DSI is not covered under the definition of genetic resources, and

---

[11] Cabrera (2020).

[12] Costa Rica's Biodiversity Law No. 7788 of 1998 deals with the ABS framework and the Government of Costa Rica has indicated to the CBD Secretariat that "DSI" is covered under the definition of access to genetic resources.

is covered under "access to genetic resources," there is no regulation for DSI in non-commercial research beyond the obligation to require PIC and MAT.[13]

CONAGEBIO can impose restrictions on access permits on any further dissemination of genetic information in public databases. This protects Costa Rica's sovereign authority over DSI, resulting from any authorized access.[14] However most research publications need mandatory depositing of sequence information in public databases and this could potentially create confusion.[15]

In the African region, Namibia has opted for a similar approach regarding access permits.[16] Namibia's ABS legislation of 2017, requires users who intend to access biological and genetic resources and their intangible components, which can include read DSI to get permits. In 2016, Namibia issued permits for genome sequencing, including restrictions on the commercialisation and publication of research outputs based on DSI.[17]

Some other countries such as Columbia, Peru, Bhutan, Kenya, Malaysia etc. have redesigned their ABS frameworks to regulate access to DSI irrespective of access to the tangible genetic resource. Members of the Andean Community like Peru and Columbia, have been working on implementing the Andean Decision 391 which

---

Refer p. 45 of the Fact-finding Study on How Domestic Measures Address Benefit-sharing Arising from Commercial and Non-commercial Use of Digital Sequence Information on Genetic Resources and Address the Use of Digital Sequence Information on Genetic Resources for Research and Development, As requested by decision 14/20 (paragraph 11 (e)) from the 14th CBD COP, 25 October 2019. Available at https://www.cbd.int/abs/DSI-peer/Study4_domestic_measures.pdf.

[13] Until now no access permit has been granted for the commercial use of DSI/genetic information per se not involving access to the physical material (genetic or biochemical compound).

[14] Refer p. 45 of the Fact-finding Study on How Domestic Measures Address Benefit-sharing Arising from Commercial and Non-commercial Use of Digital Sequence Information on Genetic Resources and Address the Use of Digital Sequence Information on Genetic Resources for Research and Development, As requested by decision 14/20 (paragraph 11 (e)) from the Fourteenth Conference of the Parties to the Convention on Biological Diversity, 25 October 2019. Available at https://www.cbd.int/abs/DSI-peer/Study4_domestic_measures.pdf.

[15] Cabrera (2020).

[16] Access to Biological and Genetic Resources and Associated Traditional Knowledge (No. 2 of 2017).

Art 8. *Access permit.*

(1) A person who intends to access biological and genetic resources in Namibia, found in both in situ or ex situ conditions or parts of such biological and genetic resources, their derivatives, products, including synthetic products, and wherever applicable, intangible components, including genetic information or gene sequences, and associated traditional knowledge, must apply for an access permit in the prescribed form and manner prior to carrying out any. Available at https://namiblii.org/akn/na/act/2017/2.

[17] Fact-finding Study on How Domestic Measures Address Benefit-sharing Arising from Commercial and Non-commercial Use of Digital Sequence Information on Genetic Resources and Address the Use of Digital Sequence Information on Genetic Resources for Research and Development, As requested by decision 14/20 (paragraph 11 (e)) from the 14th CBD COP,25 October 2019. Available at https://www.cbd.int/abs/DSI-peer/Study4_domestic_measures.pdf.

defines genetic resources as including "intangible components" which is interpreted to cover DSI.[18] This interpretation allows DSI to be subject to ABS requirements.

Some other countries have no access requirements for DSI, but benefit sharing is required from its utilization. Countries like Brazil and India have opted for benefit sharing for the utilization of DSI. This regime prioritises the results of the use of DSI and not the process of accessing DSI. Monetary sharing obligations kick in the moment there is a commercial product that utilises DSI is reported in the provider country.

Megadiverse countries like Brazil and India have included benefit sharing provisions for utilization of DSI. Brazil, have included in their domestic legislation provisions exerting their rights over DSI even if held outside their territorial borders. The Brazilian law of 2015[19] expands the interpretation of "access to genetic resources" to include the use of information from genetic sequences published in databases.[20] The Brazilian system requires no PIC for access or use of DSI but requires monetary benefit sharing when the final product is commercialised. Thus the Brazilian system regulates the results and not the processes.

18.4 Are We Heading for a Nagoya Protocol "Plus"?

Brazil has adopted the "Nagoya plus" approach defining the term "genetic heritage" that includes information on "genetic resources". It has also expanded to include genetic resources accessed from within the Brazilian territory and also extending to the continental shelf or exclusive economic zone. 'Access' has been expanded to all research or "technological development" conducted on a sample of genetic heritage or associated traditional knowledge. Brazil requires users of DSI to register their use through the "SisGen" online system prior to applying for exclusive patent rights.

---

[18] Decision 391, Common Regime on Access to Genetic Resources http://www.sice.oas.org/trade/junac/decisiones/dec391e.asp.

Art 1 Access: the obtaining and use of genetic resources conserved in situ and ex situ, of their by-products and, if applicable, of their intangible components, for purposes of research, biological prospecting, conservation, industrial application and commercial use, among other things.

[19] Law (13.123/15) Provisional Act, No 2,186-16 2001 (Brazil), Title II, Art 7, para ILaw No. 13.123 of May 20, 2015 (Access and Benefits Sharing of Genetic Resources and Associated Traditional Knowledge). Available at http://www.wipo.int/wipolex/en/details.jsp?id=15741.

[20] PNAS, "Earth BioGenome Project: Sequencing life for the future of life".

India requires benefit-sharing for DSI, with each individual case being assessed by the National Biodiversity Authority.[21] The term "associated knowledge" within the national legislation, Biodiversity Act can be interpreted to include DSI.[22] Benefit-sharing obligations in India are specified as a percentage of the user's commercial gains.[23] Similarly, South Africa goes beyond genetic resources to include all indigenous biological resources.[24] It includes any derivative or genetic material of plants, animals or organisms.[25]

Some Nagoya Protocol "Plus" mechanisms are adopted by net users of genetic resources as well, such as the member states of the European Union (EU) and Japan.[26] The EU Guidance Document, which is non-binding in nature, explains that users must

---

[21] P. 26 of the Fact-finding Study on How Domestic Measures Address Benefit-sharing Arising from Commercial and Non-commercial Use of Digital Sequence Information on Genetic Resources and Address the Use of Digital Sequence Information on Genetic Resources for Research and Development, As requested by decision 14/20 (paragraph 11 (e)) from the Fourteenth Conference of the Parties to the Convention on Biological Diversity, 25 October 2019. Available at https://www.cbd.int/abs/DSI-peer/Study4_domestic_measures.pdf.

[22] Section 2 (c) of the BD Act defines "biological resources" as plants, animals and micro-organisations or parts thereof, their genetic material and by-products (excluding value-added products) with actual or potential use or value but not including human genetic material.

[23] Indo-German Biodiversity Programme/Access and Benefit Sharing Partnership project (2019) Access and Benefit Sharing Monitoring Tool. Available at http://indo-germanbiodiversity.com/pdf/publication/publication07-06-2019-1559912567.pdf.

[24] National Environmental Management: Biodiversity Act, 2004 (Act No. 10 of 2004) (S. Afr.) [herein after NEMBA]; Bioprospecting, Access and Benefit-Sharing Regulations, 2008.

Indigenous biological resources comprise "any living or dead organism of an indigenous species, any genetic material or derivatives of such organisms, or any chemical compounds and products obtained through use of biotechnology that have been altered with genetic material or chemical compounds found in indigenous species".

[25] https://www.dffe.gov.za/projectsprogrammes/bioprospectingaccess_benefitsharing_babs_clearinghouse#:~:text=South%20African%20indigenous%20biological%20resources,such%20plants%2C%20animals%20or%20organisms.

[26] Under the EU Regulation 511/2014.

[u]sers shall exercise due diligence to ascertain that genetic resources and traditional knowledge associated with genetic resources which they utilise have been accessed in accordance with applicable access and benefit-sharing legislation or regulatory requirements, and that benefits are fairly and equitably shared upon mutually agreed terms, in accordance with any applicable legislation or regulatory requirements.

respect any conditions in MAT that deal with DSI[27] while EU clearly stated that open access to DSI is a significant non-monetary benefit.[28]

Given the developments and debates, both within the Nagoya Protocol and its implementation, as well as those beyond the Protocol, it is clear that DSI related issues need to be discussed as an additional and critical dimension of future implementation and compliance to the objectives of CBD and the Protocol.

It is time for countries that are in the final stages of developing their national ABS frameworks to consider whether they would prefer to include DSI related components for access and benefit sharing as the Nagoya Protocol 'Plus' approach since any immediate and future decisions from the Protocol on DSI could potentially affect national implementation significantly.

If there is one issue that will sure emerge to challenge the Nagoya Protocol and its efficiency to deal with implementing the third objective of the CBD, it will be the decisions that will be made by the Parties to Protocol and the CBD on how to consider the DSI and synthetic biology related dimensions. This 'Plus' approach need to discussed as soon as possible, especially when countries are working towards a new global biodiversity framework beyond 2020 and when the Parties decided to enhance more synergies between the Convention and its Protocols.

In order to do so, discussions within the Nagoya Protocol should focus on the following:

1. Clear and evidence based considerations on impacts of DSI and synthetic biology on the Nagoya Protocol and implications for national implementation,
2. Linking discussions between the Nagoya and Cartagena Protocols on issues related to DSI and synthetic biology that are to an extent being discussed separately. The forth-coming meeting of the third Subsidiary Body on Implementation (SBI 3) in 2022 should create an opportunity to have a joint expert group that will identify options for the synergies,
3. Develop elements for the Nagoya 'Plus' approach under the Protocol focusing on specifics such as bringing DSI related issues under the purview of the Protocol,

---

[27] EU Guidance document on the scope of application and core obligations of Regulation (EU) No. 511/2014 of the European Parliament and of the Council on the compliance measures for users from the Nagoya Protocol on Access to Genetic Resources and the Fair and Equitable Sharing of Benefits Arising from their Utilisation in the Union (2016/C 313/01).

[T]he use or publication of such data might be covered by conditions set in the mutually agreed terms, which should be respected. In particular, those who accessed the genetic resources and obtain sequence data from them should respect the conditions of the agreement entered into, and inform subsequent actors about any rights and obligations attached to the data obtained and related to any further uses of it. Available at https://eur-lex.europa.eu/legal-content/EN/TXT/PDF/?uri=CELEX:52016XC0827(01)&from=EN.

[28] P. 19 of the Fact-finding Study on How Domestic Measures Address Benefit-sharing Arising from Commercial and Non-commercial Use of Digital Sequence Information on Genetic Resources and Address the Use of Digital Sequence Information on Genetic Resources for Research and Development, As requested by decision 14/20 (paragraph 11 (e)) from the Fourteenth Conference of the Parties to the Convention on Biological Diversity, 25 October 2019. Available at https://www.cbd.int/abs/DSI-peer/Study4_domestic_measures.pdf.

re-designing the research and publication criteria for results involving genetic sequence information,

4. Draw a set of options for dealing with intellectual property issues that combine the focus on issues of ABS and emerging technological developments,
5. Have a committed discussion that clarifies the future of the Nagoya Protocol that will be responsive and supportive of the objectives of the Protocol that calls for fair and equitable benefit sharing that is ethical, moral and legally correct, and
6. Map the capacity needs of Parties to the Nagoya Protocol to deal with the recommendations emanating from SBI 3 and CBD COP 15 meetings.

References

Bagley M (2022) "Just" sharing: the virtues of digital sequence information benefit-sharing for the common good. Harv Int Law J

Cabrera J (2020) Costa Rica ABS legislation and practice after the Nagoya Protocol, April 2020

Convention on Biological Diversity (2018) Fact-finding and scoping study on digital sequence information on genetic resources in the context of the convention on biological diversity and the Nagoya Protocol. https://www.cbd.int/meetings/COP-15

Staffas L, Gustavsson M, McCormick K (2013) Strategies and policies for the bio-economy and bio-based economy: an analysis of official national approaches

Thambisetty S (2019) Biodiversity beyond national jurisdiction: (intellectual) property heuristics

The manufacturer's authorised representative in the EU is Springer Nature Customer Service Centre GmbH, Europaplatz 3, 69115 Heidelberg, Germany. If you have any concerns regarding our products, please contact ProductSafety@springernature.com

Printed and bound by CPI Group (UK) Ltd, Croydon, CR0 4YY
28/09/2026
02224056-0005